# DATA MINING AND KNOWLEDGE DISCOVERY VIA LOGIC-BASED METHODS

# Springer Optimization and Its Applications

VOLUME 43



*Aims and Scope*
Optimization has been expanding in all directions at an astonishing rate during the last few decades. New algorithmic and theoretical techniques have been developed, the diffusion into other disciplines has proceeded at a rapid pace, and our knowledge of all aspects of the field has grown even more profound. At the same time, one of the most striking trends in optimization is the constantly increasing emphasis on the interdisciplinary nature of the field. Optimization has been a basic tool in all areas of applied mathematics, engineering, medicine, economics and other sciences.

The series *Springer Optimization and Its Applications* publishes undergraduate and graduate textbooks, monographs and state-of-the-art expository works that focus on algorithms for solving optimization problems and also study applications involving such problems. Some of the topics covered include nonlinear optimization (convex and nonconvex), network flow problems, stochastic optimization, optimal control, discrete optimization, multiobjective programming, description of software packages, approximation techniques and heuristic approaches.

For other titles published in this series, go to
http://www.springer.com/series/7393

# DATA MINING AND KNOWLEDGE DISCOVERY VIA LOGIC-BASED METHODS

Theory, Algorithms, and Applications

By

EVANGELOS TRIANTAPHYLLOU
Louisiana State University
Baton Rouge, Louisiana, USA

Springer

Evangelos Triantaphyllou
Louisiana State University
Department of Computer Science
298 Coates Hall
Baton Rouge, LA 70803
USA
trianta@lsu.edu

ISBN 978-1-4614-2613-4 ISBN 978-1-4419-1630-3 (eBook)
DOI 10.1007/978-1-4419-1630-3
Springer New York Dordrecht Heidelberg London

Mathematics Subject Classification (2010): 62-07, 68T05, 90-02



Printed on acid-free paper

Springer is part of Springer Science+Business Media (www.springer.com)

This book is dedicated to a number of individuals and groups of people for different reasons. It is dedicated to my mother Helen and the only sibling I have, my brother Andreas. It is dedicated to my late father John (Ioannis) and late grandfather Evangelos Psaltopoulos.

The unconditional support and inspiration of my wife, Juri, will always be recognized and, from the bottom of my heart, this book is dedicated to her. It would have never been prepared without Juri's continuous encouragement, patience, and unique inspiration. It is also dedicated to Ragus and Ollopa ("Ikasinilab") for their unconditional love and support. Ollopa was helping with this project all the way until to the very last days of his wonderful life. He will always live in our memories. It is also dedicated to my beloved family from Takarazuka. This book is also dedicated to our very new inspiration of our lives.

As is the case with all my previous books and also with any future ones, this book is dedicated to all those (and they are many) who were trying very hard to convince me, among other things, that I would never be able to graduate from elementary school or pass the entrance exams for high school.

# Foreword

The importance of having efficient and effective methods for data mining and knowledge discovery (DM&KD), to which the present book is devoted, grows every day and numerous such methods have been developed in recent decades. There exists a great variety of different settings for the main problem studied by data mining and knowledge discovery, and it seems that a very popular one is formulated in terms of binary attributes. In this setting, states of nature of the application area under consideration are described by Boolean vectors defined on some attributes. That is, by data points defined in the Boolean space of the attributes. It is postulated that there exists a partition of this space into two classes, which should be inferred as patterns on the attributes when only several data points are known, the so-called positive and negative training examples.

The main problem in DM&KD is defined as finding rules for recognizing (classifying) new data points of unknown class, i.e., deciding which of them are positive and which are negative. In other words, to infer the binary value of one more attribute, called the goal or class attribute. To solve this problem, some methods have been suggested which construct a Boolean function separating the two given sets of positive and negative training data points. This function can then be used as a decision function, or a classifier, for dividing the Boolean space into two classes, and so uniquely deciding for every data point the class to which it belongs. This function can be considered as the knowledge extracted from the two sets of training data points.

It was suggested in some early works to use as classifiers threshold functions defined on the set of attributes. Unfortunately, only a small part of Boolean functions can be represented in such a form. This is why the normal form, disjunctive or conjunctive (DNF or CNF), was used in subsequent developments to represent arbitrary Boolean decision functions. It was also assumed that the simpler the function is (that is, the shorter its DNF or CNF representation is), the better classifier it is. That assumption was often justified when solving different real-life problems. This book suggests a new development of this approach based on mathematical logic and, especially, on using Boolean functions for representing knowledge defined on many binary attributes.

Next, let us have a brief excursion into the history of this problem, by visiting some old and new contributions. The first known formal methods for expressing logical reasoning are due to Aristotle (384 BC–322 BC) who lived in ancient Greece, the native land of the author. It is known as his famous *syllogistics*, the first deductive system for producing new affirmations from some known ones. This can be acknowledged as being the first system of logical recognition. A long time later, in the 17th century, the notion of binary mathematics based on a two-value system was proposed by Gottfried Leibniz, as well as a combinatorial approach for solving some related problems. Later on, in the middle of the 19th century, George Boole wrote his seminal books *The mathematical analysis of logic: being an essay towards a calculus for deductive reasoning* and *An Investigation of the Laws of Thought on Which are Founded the Mathematical Theories of Logic and Probabilities*. These contributions served as the foundations of modern Boolean algebra and spawned many branches, including the theory of proofs, logical inference and especially the theory of Boolean functions. They are widely used today in computer science, especially in the area of the design of logic circuits and artificial intelligence (AI) in general.

The first real-life applications of these theories took place in the first thirty years of the 20th century. This is when Shannon, Nakashima and Shestakov independently proposed to apply Boolean algebra to the description, analysis and synthesis of relay devices which were widely used at that time in communication, transportation and industrial systems. The progress in this direction was greatly accelerated in the next fifty years due to the dawn of modern computers. This happened for two reasons. First, in order to design more sophisticated circuits for the new generation of computers, new efficient methods were needed. Second, the computers themselves could be used for the implementation of such methods, which would make it possible to realize very difficult and labor-consuming algorithms for the design and optimization of multicomponent logic circuits. Later, it became apparent that methods developed for the previous purposes were also useful for an important problem in artificial intelligence, namely, data mining and knowledge discovery, as well as for pattern recognition.

Such methods are discussed in the present book, which also contains a wide review of numerous computational results obtained by the author and other researches in this area, together with descriptions of important application areas for their use. These problems are combinatorially hard to solve, which means that their exact (optimal) solutions are inevitably connected with the requirement to check many different intermediate constructions, the number of which depends exponentially on the size of the input data. This is why good combinatorial methods are needed for their solution. Fortunately, in many cases efficient algorithms could be developed for finding some approximate solutions, which are acceptable from the practical point of view. This makes it possible to sufficiently reduce the number of intermediate solutions and hence to restrict the running time.

A classical example of the above situation is the problem of minimizing a Boolean function in disjunctive (or conjunctive) normal form. In this monograph, this task is pursued in the context of searching for a Boolean function which separates two given subsets of the Boolean space of attributes (as represented by collections of positive and negative examples). At the same time, such a Boolean function is desired to be as simple as possible. This means that incompletely defined Boolean functions are considered. The author, Professor Evangelos Triantaphyllou, suggests a set of efficient algorithms for inferring Boolean functions from training examples, including a fast heuristic greedy algorithm (called OCAT), its combination with tree searching techniques (also known as branch-and-bound search), an incremental learning algorithm, and so on. These methods are efficient and can enable one to find good solutions in cases with many attributes and data points. Such cases are typical in many real-life situations where such problems arise. The special problem of guided learning is also investigated. The question now is which new training examples (data points) to consider, one at a time, for training such that a small number of new examples would lead to the inference of the appropriate Boolean function quickly.

Special attention is also devoted to monotone Boolean functions. This is done because such functions may provide adequate description in many practical situations. The author studied existing approaches for the search of monotone functions, and suggests a new way for inferring such functions from training examples. A key issue in this particular investigation is to consider the number of such functions for a given dimension of the input data (i.e., the number of binary attributes).

Methods of DM&KD have numerous important applications in many different domains in real life. It is enough to mention some of them, as described in this book. These are the problems of verifying software and hardware of electronic devices, locating failures in logic circuits, processing of large amounts of data which represent numerous transactions in supermarkets in order to optimize the arrangement of goods, and so on. One additional field for the application of DM&KD could also be mentioned, namely, the design of two-level (AND-OR) logic circuits implementing Boolean functions, defined on a small number of combinations of values of input variables.

One of the most important problems today is that of breast cancer diagnosis. This is a critical problem because diagnosing breast cancer early may save the lives of many women. In this book it is shown how training data sets can be formed from descriptions of malignant and benign cases, how input data can be described and analyzed in an objective and consistent manner and how the diagnostic problem can be formulated as a nested system of two smaller diagnostic problems. All these are done in the context of Boolean functions.

The author correctly observes that the problem of DM&KD is far from being fully investigated and more research within the framework of Boolean functions is needed. Moreover, he offers some possible extensions for future research in this area. This is done systematically at the end of each chapter.

The descriptions of the various methods and algorithms are accompanied with extensive experimental results confirming their efficiency. Computational results are generated as follows. First a set of test cases is generated regarding the approach to be tested. Next the proposed methods are applied on these test problems and the test results are analyzed graphically and statistically. In this way, more insights on the problem at hand can be gained and some areas for possible future research can be identified.

The book is very well written in a way for anyone to understand with a minimum background in mathematics and computer science concepts. However, this is not done at the expense of the mathematical rigor of the algorithmic developments. I believe that this book should be recommended both to students who wish to learn about the foundations of logic-based approaches as they apply to data mining and knowledge discovery along with their many applications, and also to researchers who wish to develop new means for solving more problems effectively in this area.

Professor Arkadij Zakrevskij

Minsk, Belarus
Corresponding Member of the National Academy of
Sciences of Belarus

Summer of 2009

# Preface

There is already a plethora of books on data mining. So, what is new with this book? The answer is in its unique perspective in studying a series of interconnected key data mining and knowledge discovery problems both in depth and also in connection with other related topics and doing so in a way that stimulates the quest for more advancements in the future. This book is related to another book titled *Data Mining and Knowledge Discovery Approaches Based on Rule Induction Techniques* (published by Springer in the summer of 2006), which was co-edited by the author. The chapters of the edited book were written by 40 authors and co-authors from 20 countries and, in general, they are related to rule induction methods.

Although there are many approaches to data mining and knowledge discovery (DM&KD), the focus of this monograph is on the development and use of some novel mathematical logic methods as they have been pioneered by the author of this book and his research associates in the last 20 years. The author started the research that led to this publication in the early 1980s, when he was a graduate student at the Pennsylvania State University.

During this experience he has witnessed the amazing explosion in the development of effective and efficient computing and mass storage media. At the same time, a vast number of ubiquitous devices are selecting data on almost any aspect of modern life. The above developments create an unprecedented challenge to extract useful information from such vast amounts of data. Just a few years ago people were talking about megabytes to express the size of a huge database. Today people talk about gigabytes or even terabytes. It is not a coincidence that the terms *mega*, *giga*, and *tera* (not to be confused with *terra* or earth in Latin) mean in Greek "large," "giant," and "monster," respectively.

The above situation has created many opportunities but many new and tough challenges too. The emerging field of data mining and knowledge discovery is the most immediate result of this extraordinary explosion on information and availability of cost-effective computing power. The ultimate goal of this new field is to offer methods for analyzing large amounts of data and extracting useful new knowledge embedded in such data. As K. C. Cole wrote in her seminal book *The Universe and the Teacup: The Mathematics of Truth and Beauty*, "… nature bestows her blessings buried in mountains of garbage." An anonymous author expressed a closely related concept by stating poetically that "today we are giants of information but dwarfs of new knowledge."

On the other hand, the principles that are behind many data mining methods are not new to modern science. The danger related with the excess of information and with its interpretation already alarmed the medieval philosopher William of Occam (also known as Okham) and motivated him to state his famous "razor": *entia non sunt multiplicanda praeter necessitatem* (entities must not be multiplied (i.e., become more complex) beyond necessity). Even older is the story in the Bible of the Tower of Babel in which people were overwhelmed by new and ultraspecialized knowledge and eventually lost control of the most ambitious project of that time.

People dealt with data mining problems when they first tried to use past experience in order to predict or interpret new phenomena. Such challenges always existed when people tried to predict the weather, crop production, market conditions, and the behavior of key political figures, just to name a few examples. In this sense, the field of data mining and knowledge discovery is as old as humankind.

Traditional statistical approaches cannot cope successfully with the heterogeneity of the data fields and also with the massive amounts of data available today for analysis. Since there are many different goals in analyzing data and also different types of data, there are also different data mining and knowledge discovery methods, specifically designed to deal with data that are crisp, fuzzy, deterministic, stochastic, discrete, continuous, categorical, or any combination of the above. Sometimes the goal is to just use historic data to predict the behavior of a natural or artificial system. In other cases the goal is to extract easily understandable knowledge that can assist us to better understand the behavior of different types of systems, such as a mechanical apparatus, a complex electronic device, a weather system or an illness.

Thus, there is a need to have methods which can extract new knowledge in a way that is easily verifiable and also easily understandable by a very wide array of domain experts who may not have the computational and mathematical expertise to fully understand how a data mining approach extracts new knowledge. However, they may easily comprehend newly extracted knowledge, if such knowledge can be expressed in an intuitive manner.

The methods described in this book offer just this opportunity. This book presents methods that deal with key data mining and knowledge discovery issues in an intuitive manner and in a natural sequence. These methods are based on mathematical logic. Such methods derive new knowledge in a way that can be easily understood and interpreted by a wide array of domain experts and end users. Thus, the focus is on discussing methods which are based on Boolean functions; which can then easily be transformed into rules when they express new knowledge. The most typical form of such rules is a decision rule of the form: IF ⟨*some condition(s) is (are) true*⟩ THEN ⟨*another condition should also be true*⟩.

Thus, this book provides a unique perspective into the essence of some fundamental data mining and knowledge discovery issues. It discusses the theoretical foundations of the capabilities of the methods described in this book. It also presents a wide collection of illustrative examples, many of which come from real-life applications. A truly unique characteristic of this book is that almost all theoretical developments are accompanied by an extensive empirical analysis which often involves the solution of a very large number of simulated test problems. The results of these empirical analyses are tabulated, graphically depicted, and analyzed in depth. In this way, the theoretical and empirical analyses presented in this book are complementary to each other, so the reader can gain both a comprehensive and deep theoretical and practical insight of the covered subjects.

Another unique characteristic of this book is that at the end of each chapter there is a description of some possible research problems for future research. It also presents an extensive and updated bibliography and references of all the covered subjects. These are very valuable characteristics for people who wish to get involved with new research in this field.

Therefore, the book *Data Mining and Knowledge Discovery via Logic-Based Methods: Theory, Algorithms, and Applications* can provide a valuable insight for people who are interested in obtaining a deep understanding of some of the most frequently encountered data mining and knowledge discovery challenges. This book can be used as a textbook for senior undergraduate or graduate courses in data mining in engineering, computer science, and business schools; it can also provide a panoramic and systematic exposure of related methods and problems to researchers. Finally, it can become a valuable guide for practitioners who wish to take a more effective and critical approach to the solution of real-life data mining and knowledge discovery problems.

The philosophy followed on the development of the subjects covered in this book was first to present and define the subject of interest in that chapter and do so in a way that motivates the reader. Next, the following three key aspects were considered for each subject: (i) a discussion of the related theory, (ii) a presentation of the required algorithms, and (iii) a discussion of applications. This was done in a way such that progress in any one of these three aspects would motivate progress in the other two aspects. For instance, theoretical advances make it possible to discover and implement new algorithms. Next, these algorithms can be used to address certain applications that could not be addressed before. Similarly, the need to handle certain real-life applications provides the motivation to develop new theories which in turn may result in new algorithms and so on. That is, these three key aspects are parts of a continuous closed loop in which any one of these three aspects feeds the other two.

Thus, this book deals with the pertinent theories, algorithms, and applications as a closed loop. This is reflected on the organization of each chapter but also on the organization of the entire book, which is comprised of two sections. The sections are titled “Part I: Algorithmic Issues” and “Part II: Application Issues.” The first section focuses more on the development of some new and fundamental algorithms along with the related theory while the second section focuses on some select applications and case studies along with the associated algorithms and theoretical aspects. This is also shown in the Contents.

The arrangement of the chapters follows a natural exposition of the main subjects in rule induction for DM&KD theory and practice. Part I (“Algorithmic Issues”) starts with the first chapter, which discusses the intuitive appeal of the main data mining and knowledge discovery problems discussed throughout this monograph. It pays extra attention to the reasons that lead to formulate some of these problems as optimization problems since one always needs to keep control on the size (i.e., for size minimization) of the extracted new rules or when one tries to gain a deeper understanding of the system of interest by issuing a small number of new queries (i.e., for query minimization).

The second and third chapters present some sophisticated branch-and-bound algorithms for extracting a pattern (in the form of a compact Boolean function) from collections of observations grouped into two disjoint classes. The fourth chapter presents some fast heuristics for the same problem.

The fifth chapter studies the problem of guided learning. That is, now the analyst has the option to decide the composition of the observation to send to an expert or "oracle" for the determination of its class membership. Apparently, the goal now is to gain a good understanding of the system of interest by issuing a small number of inquiries of the previous type.

A related problem is studied in the sixth chapter. Now it is assumed that the analyst has two sets of examples (observations) and a Boolean function that is inferred from these examples. Furthermore, it is assumed that the analyst has a new example that invalidates this Boolean function. Thus, the problem is how to modify the Boolean function such that it satisfies all the requirements of the available examples plus the new example. This is known as the incremental learning problem.

Chapter 7 presents an intriguing duality relationship which exists between Boolean functions expressed in CNF (conjunctive normal form) and DNF (disjunctive normal form), which are inferred from examples. This dual relationship could be used in solving large-scale inference problems, in addition to other algorithmic advantages.

The chapter that follows describes a graph theoretic approach for decomposing large-scale data mining problems. This approach is based on the construction of a special graph, called the *rejectability graph*, from two collections of data. Then certain characteristics of this graph, such as its minimum clique cover, can lead to some intuitive and very powerful decomposition strategies.

Part II ("Application Issues") begins with Chapter 9. This chapter presents an intriguing problem related to any model (and not only those based on logic methods) inferred from grouped observations. This is the problem of the reliability of the model and it is associated with both the number of the training data (sampled observations grouped into two disjoint classes) and also the nature of these data. It is argued that many model inference methods today may derive models that cannot guarantee the reliability of their predictions/classifications. This chapter prepares the basic arguments for studying a potentially very critical type of Boolean functions known as *monotone* Boolean functions.

The problems of inferring a monotone Boolean function from inquiries to an expert ("oracle"), along with some key mathematical properties and some application issues are discussed in Chapters 10 and 11. Although this type of Boolean functions has been known in the literature for some time, it was the author of this book along with some of his key research associates who made some intriguing contributions to this part of the literature in recent years. Furthermore, Chapter 11 describes some key problems in assessing the effectiveness of data mining and knowledge discovery models (and not only for those which are based on logic). These issues are referred to as the "three major illusions" in evaluating the accuracy of such models. There it is shown that many models which are considered as highly successful, in reality may even be totally useless when one studies their accuracy in depth.

Chapter 12 presents how some of the previous methods for inferring a Boolean function from observations can be used (after some modifications) to extract what is known in the literature as association rules. Traditional methods suffer the problem of extracting an overwhelming number of association rules and they are doing so in exponential time. The new methods discussed in this chapter are based on some fast (of polynomial time) heuristics that can derive a compact set of association rules.

Chapter 13 presents some new methods for analyzing and categorizing text documents. Since the Web has made possible the availability of immense textual (and not only) information easily accessible to anyone with access to it, such methods are expected to attract even more interest in the immediate future.

Chapters 14, 15, and 16 discuss some real-life case studies. Chapter 14 discusses the analysis of some real-life EMG (electromyography) signals for predicting muscle fatigue. The same chapter also presents a comparative study which indicates that the proposed logic-based methods are superior to some of the traditional methods used for this kind of analysis.

Chapter 15 presents some real-life data gathered from the analysis of cases suspected of breast cancer. Next these data are transformed into equivalent binary data and then some diagnostic rules (in the form of compact Boolean functions) are extracted by using the methods discussed in earlier chapters. These rules are next presented in the form of IF-THEN logical expressions (diagnostic rules).

Chapter 16 presents a combination of some of the proposed logic methods with fuzzy logic. This is done in order to objectively capture fuzzy data that may play a key role in many data mining and knowledge discovery applications. The proposed new method is demonstrated in characterizing breast lesions in digital mammography as lobular or microlobular. Such information is highly significant in analyzing medical data for breast cancer diagnosis.

The last chapter presents some concluding remarks. Furthermore, it presents twelve different areas that are most likely to experience high interest for future research efforts in the field of data mining and knowledge discovery.

All the above chapters make clear that methods based on mathematical logic already play an important role in data mining and knowledge discovery. Furthermore, such methods are almost guaranteed to play an even more important role in the near future as such problems increase both in complexity and in size.

*Evangelos Triantaphyllou*
Baton Rouge, LA
April 2010

# Acknowledgments

Dr. Evangelos Triantaphyllou is always deeply indebted to many people who have helped him tremendously during his career and beyond. He always recognizes with immense gratitude the very special role his math teacher played in his life; Mr. Lefteris Tsiliakos, along with Mr. Tsiliakos' wonderful family (including his extended family). He also recognizes the critical assistance and valuable encouragement of his undergraduate Advisor at the National Technical University of Athens, Greece; Professor Luis Wassenhoven.

His most special thanks go to his first M.S. Advisor and Mentor, Professor Stuart H. Mann, currently the Dean of the W.F. Harrah College of Hotel Administration at the University of Nevada in Las Vegas. He would also like to thank his other M.S. Advisor, Distinguished Professor Panos M. Pardalos currently at the University of Florida, and his Ph.D. Advisor, Professor Allen L. Soyster, former Chairman of the Industrial Engineering Department at Penn State University and former Dean of Engineering at the Northeastern University for his inspirational advising and assistance during his doctoral studies at Penn State.

Special thanks also go to his great neighbors and friends; Janet, Bert, and Laddie Toms for their multiple support during the development of this book and beyond. Especially for allowing him to work on this book in their amazing Liki Tiki study facility. Many special thanks are also given to Ms. Elizabeth Loew, a Senior Editor at Springer, for her encouragement and great patience.

Most of the research accomplishments on data mining and optimization by Dr. Triantaphyllou would not have been made possible without the critical support by Dr. Donald Wagner at the Office of Naval Research (ONR), U.S. Department of the Navy. Dr. Wagner's contribution to this success is greatly appreciated.

Many thanks go to his colleagues at LSU. Especially to Dr. Kevin Carman, Dean of the College of Basic Sciences at LSU; Dr. S.S. Iyengar, Distinguished Professor and Chairman of the Computer Science Department at LSU; Dr. T. Warren Liao, his good neighbor, friend, and distinguished colleague at LSU; and last but not least to his student Forrest Osterman for his many and thoughtful comments on an early version of this book.

He is also very thankful to Professor Arkadij Zakrevskij, corresponding member of the National Academy of Sciences of Belarus, for writing the great foreword for this book and his encouragement, kindness, and great patience. A special mention here goes to Dr. Xenia Naidenova, a Senior Researcher from the Military Medical Academy at Saint Petersburg in Russia, for her continuous encouragement and friendship through the years.

Dr. Triantaphyllou would also like to acknowledge his most sincere and immense gratitude to his graduate and undergraduate students, who have always provided him with unlimited inspiration, motivation, great pride, and endless joy.

*Evangelos Triantaphyllou*
Baton Rouge, LA
January 2010

# Contents

# List of Figures

# List of Tables

# Part I

# Algorithmic Issues

# Chapter 1

# Introduction

## 1.1 What Is Data Mining and Knowledge Discovery?

Data mining and knowledge discovery is a family of computational methods that aim at collecting and analyzing data related to the function of a system of interest for the purpose of gaining a better understanding of it. This system of interest might be artificial or natural. According to the Merriam-Webster online dictionary the term *system* is derived from the Greek terms *syn* (plus, with, along with, together, at the same time) and *istanai* (to cause to stand) and it means a complex entity which is comprised of other more elementary entities which in turn may be comprised of other even more elementary entities and so on. All these entities are somehow interconnected with each other and form a unified whole (the *system*). Thus, all these entities are related to each other and their collective operation is of interest to the analyst, hence the need to employ data mining and knowledge discovery (DM&KD) methods. Some illustrative examples of various systems are given in the next section.

The data (or observations) may describe different aspects of the operation of the system of interest. Usually, the overall state of the system, also known as a state of nature, corresponds to one of a number of different classes. It is not always clear what the data should be or how to define the different states of nature of the system or the classes under consideration. It all depends on the specific application and the goal of the analysis. This task may require lots of skill and experience to properly define them. This is part of the *art* aspect of the "*art and science*" approach to problem-solving in general.

It could also be possible to have more than two classes with a continuous transition between different classes. However, in the following we will assume that there are only two classes and these classes are mutually exclusive and exhaustive. That is, the system has to be in only one of these two classes at any given time. Sometimes, it is possible to have a third class called *undecidable* or *unclassifiable* (not to be confused with unclassified observations) in order to capture undecidable instances of the system. In general, cases with more than two classes can be modeled as a sequence of two-class problems. For instance, a case with four classes may be modeled as a sequence of at most three two-class problems.

E. Triantaphyllou, *Data Mining and Knowledge Discovery via Logic-Based Methods*,
Springer Optimization and Its Applications 43, DOI 10.1007/978-1-4419-1630-3_1,

It should be noted here that a widely used definition for knowledge discovery is given in the book by [Fayyad, *et al.*, 1996] (on page 6): "Knowledge discovery in databases is the non-trivial process of identifying valid, novel, potentially useful, and ultimately understandable patterns in data." More definitions can be found in many other books. However, most of them seem to agree on the issues discussed in the previous paragraphs.

The majority of the treatments in this book are centered on classification, that is, the assignment of new data to one of some predetermined classes. This may be done by first inferring a model from the data and then using this model and the new data point for this assignment. DM&KD may also aim at clustering of data which have not been assigned to predetermined classes. Another group of methods focus on prediction or forecasting. Prediction (which oftentimes is used the same way as classification) usually involves the determination of some probability measure related to belonging to a given class. For instance, we may talk about predicting the outcome of an election or forecasting the weather. A related term is that of *diagnosis*. This term is related to the understanding of the cause of a malfunction or a medical condition. Other goals of DM&KD may be to find explanations of decisions pertinent to computerized systems, extracting similarity relationships, learning of new concepts (conceptual learning), learning of new ontologies, and so on.

Traditionally, such problems have been studied via statistical methods. However, statistical methods are accurate if the data follow certain strict assumptions and if the data are mostly numerical, well defined and plentiful. With the abundance of data collection methods and the highly unstructured databases of modern society (for instance, as in the World Wide Web), there is an urgent need for the development of new methods. This is what a new cadre of DM&KD methods is called to answer.

What all the above problems have in common is the use of analytical tools on collections of data to somehow better understand a phenomenon or system and eventually benefit from this understanding and the data. The focus of the majority of the algorithms in this book is on inferring patterns in the form of Boolean functions for the purpose of classification and also diagnosing of various conditions.

The next section presents some illustrative examples of the above issues from a diverse spectrum of domains. The third section of this chapter describes the main steps of the entire data mining and knowledge discovery process. The fourth section highlights the basics of four critical research problems which concentrate lots of interest today in this area of data analysis. Finally, this chapter ends with a brief section describing some concluding remarks.

## 1.2 Some Potential Application Areas for Data Mining and Knowledge Discovery

It is impossible to list all possible application areas of data mining and knowledge discovery. Such applications can be found anywhere there is a system of interest, which can be in one of a number of states and data can be collected about this system. In the following sections we highlight some broad potential areas for illustrative purposes only, as a complete list is impossible to compile. These potential areas are grouped into different categories in a way that reflects the scientific disciplines that primarily study them.

### 1.2.1 Applications in Engineering

An example of a system in a traditional engineering setting might be a *mechanical or electronic* device. For instance, the engine of a car is such a system. An engine could be viewed as being comprised of a number of secondary systems (e.g., the combustion chamber, the pistons, the ignition device, etc.). Then an analyst may wish to collect data that describe the fuel composition, the fuel consumption, heat conditions at different parts, any vibration data, the composition of the exhaust gases, pollutant composition and built up levels inside different parts of the engine, the engine's performance measurements, and so on. As different classes one may wish to view the operation of this system as successful (i.e., it does not require intervention for repair) or problematic (if its operation must be interrupted for a repair to take place).

As another example, a system in this category might be the *hardware* of a personal computer (PC). Then data may describe the operational characteristics of its components (screen, motherboard, hard disk, keyboard, mouse, CPU, etc.). As with the previous example, the classes may be defined based on the successful and the problematic operation of the PC system.

A system can also be a *software package*, say, for a word processor. Then data may describe its functionality under different printers, when creating figures, using different fonts, various editing capabilities, operation when other applications are active at the same time, size of the files under development, etc. Again, the classes can be defined based on the successful or problematic operation of this word processor.

### 1.2.2 Applications in Medical Sciences

Data mining and knowledge discovery have a direct application in the medical diagnosis of many medical ailments. A typical example is *breast cancer diagnosis*. Data may describe the geometric characteristics present in a mammogram (i.e., an X-ray image of a breast). Other data may describe the family history, results of blood tests, personal traits, etc. The classes now might be the benign or malignant nature of the findings. Similar applications can be found in any other medical ailment or condition.

A recent interest of such technologies can also be found in the *design of new drugs*. Sometimes developing a single drug may cost hundreds of millions or even billions of dollars. In such a setting the data may describe the different components of the drug, the characteristics of the patient (presence of other medical conditions besides the targeted one and physiological characteristics of the patient), the dosage information, and so on. Then the classes could be the effective impact of the drug or not.

Another increasingly popular application area is in the discovery of conditions and characteristics that may be associated with the development of various medical ailments later in life such as *heart disease, diabetes, Alzheimer's disease, various cancer types*, etc. Data can be the family history, clinical data of the human subjects, lifestyle characteristics, environmental factors, and so on. The classes might correspond to the development of a given medical ailment or not.

### 1.2.3 Applications in the Basic Sciences

Perhaps one of the oldest applications of DM&KD is that of the *prediction of weather phenomena*, such as rain, snow, high winds, tornadoes, etc. From the early days people were observing weather conditions in an effort to predict the weather of the next few days. Data can be the cloud conditions, wind direction, air pressure, temperature readings, humidity level, and so on. Then the classes could be defined based on the presence or not of some key weather phenomena such as rain, high or low temperatures, formation of tornadoes, and high winds.

An application of interest to many coastal areas is the *prediction of coastal erosion* so appropriate measures can be taken more effectively. Data can be the rain levels, effects of rivers and lakes, geological characteristics of the land areas, oceanic conditions, the local weather, any technical measures taken by people in dealing with the problem, and so on. The classes could be the high or low level of coastal erosion.

A rather profitable application area is in the discovery of new *gas, oil, and mineral deposits*. Drilling and/or mining explorations may be excessively costly; hence, having effective prediction methods is of paramount practical importance. As data one may consider the geological characteristics of the candidate area and seismic data. The classes could be defined by the presence or not of profitable deposits.

A critical issue during many military operations is the accurate *identification of targets*. A related issue is the classification of targets as friendly or enemy. Data can be derived by analyzing images of the target and surrounding area, signal analysis, battle planning data, and so on. Then a target is either a friendly or a hostile one and these can be the two classes in this setting.

A similar application is in the *screening of travelers* when they board mass transportation media. This issue has gained special interest after the 9/11 events in the U.S. Data can be the biometric characteristics of the traveler, X-ray images of the luggage, behavior at the checking points, etc. Then travelers may be classified according to different risk levels.

### 1.2.4 Applications in Business

The world of modern business uses many means, and that includes DM&KD techniques, that can better identify *marketing opportunities of new products*. In this setting data can be the lifestyle characteristics of different consumer groups and their level of acceptance of the new product. Other data can be the marketing methods used to promote the new product, as well as the various characteristics of the product itself. As classes one may define the high or low acceptance of a given product by a particular consumer group.

A related topic is the *design of a new product*. One may wish to use elements of past successful designs in order to combine them into a new and successful product. Thus, data can relate to the design characteristics of previous products, the groups that accepted the previous products, and so on. As above, the two classes correspond to the high or low acceptance of the new product.

The huge plethora of *investment opportunities* and, at the same time, the overwhelming presence of financial information sources (especially on the Web), make DM&KD in finance an appealing application area. This was especially the case during the euphoric period of the late 1990s. Data can be any information on past performance, company and sector/industry reports, and general market conditions. The classes could be defined by the high or low level of return of a typical investment vehicle during a given period of time.

### 1.2.5 Applications in the Political and Social Sciences

A crucial issue with any *political campaign* is the study of its effectiveness on various groups of potential voters. Data can be the socioeconomic characteristics of a given group of voters. Other data may come from the issues presented in the political campaign and the means (TV ads, newspapers, radio, the Web) of presenting these issues. The classes could be defined by the change in the opinions of the voters regarding the issues discussed and the candidates who promote them.

Another application of DM&KD may come from efforts to *control crime* in urban areas. The pertinent data may describe the socioeconomic composition of a particular urban area, the existing means to control crime, the type and frequency of crime incidents, and so on. The classes could be defined by the level (type and frequency) of crime in an area or the effectiveness of different crime reduction strategies.

## 1.3 The Data Mining and Knowledge Discovery Process

As mentioned in the previous section, a very critical step in any DM&KD analysis is the proper definition of the goals of the analysis and the collection of the data. The entire process can be conceptualized as divided into a sequence of steps as depicted below in Figure 1.1.

### 1.3.1 Problem Definition

The first and single most important step of the DM&KD process deals with the problem definition. What is the system or phenomenon of interest? What are the purpose and the goals of the analysis? How could we describe the different states of nature and classes of observations? What data may be relevant to this analysis? How can the data be collected? These are some key questions that need to be addressed before any

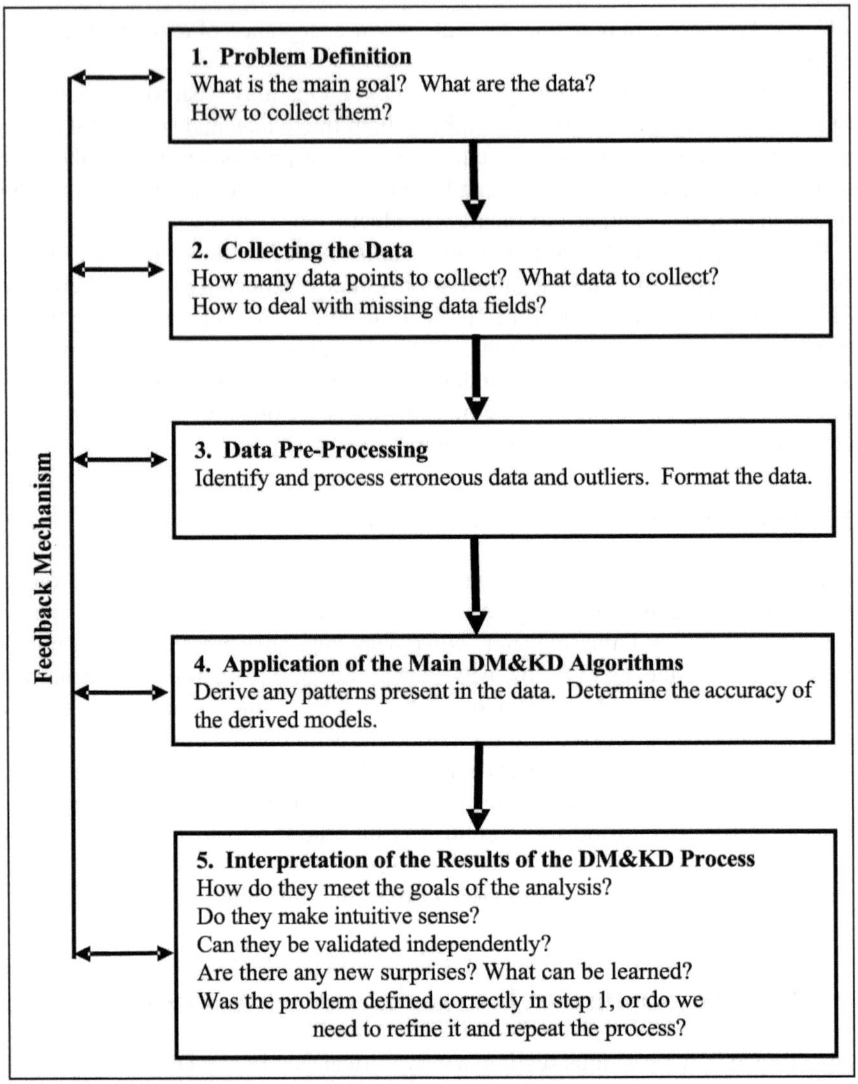


**Figure 1.1.** The Key Steps of the Data Mining and Knowledge Discovery Process.

other step is taken. If this step is not dealt with correctly, then the entire process (and problem-solving approach in general) is doomed to failure. A very common mistake is to solve *correctly* the *wrong* problem. This is what R. L. Ackoff called the type III error in his famous book [1987] *The Art of Problem Solving*.

It is always a prudent practice not to think "monolithically." That is, one always needs to keep an open mind, be flexible, and be willing to revise any previous beliefs as more information and experience in dealing with a problem become available. That is why all the boxes in Figure 1.1 are connected with each other by means of a feedback mechanism. For instance, if in a later step the analyst realizes that the system under consideration needs additional data fields in order to describe classes more accurately, then such data need to be collected.

### 1.3.2 Collecting the Data

Regarding the required data for the DM&KD analysis, such data do not need to be collected only by means of questionnaires. Data may come from past cases each of which took lots of resources to be analyzed. As mentioned earlier, in an oil well drilling situation, data may refer to the geotechnical characteristics of the drilling site. The classes might be defined according to the amount of oil that can be pumped out of the site. Or simply whether the oil well is profitable or not. Then the acquisition of data from a single site might involve lots of time, effort, and ultimately financial resources. In a different situation, data about market preferences may be collected by issuing a questionnaire and thus be very cost-effective on an individual basis.

The analyst may not know how many data points are sufficient for a given application. The general practice is to collect as many data points as possible. Even more important, the analyst may not even know what data to collect. A data point may be viewed as a data record comprised of various fields. Thus, which fields are appropriate? Again, the general approach is to consider as many fields per data point as possible provided that they appear to be somewhat relevant. However, this could be a tricky task.

For instance, consider the case of the data in Figure 1.2. These data are defined in terms of a single attribute, namely, $A_1$. There are two classes; one is represented by the solid dots and the other is represented by the gray dots. Certainly, there is a pattern in this figure of how the solid and gray dots are related to each other and this pattern could be described in terms of their values in terms of the $A_1$ attribute. However, that pattern is a rather complicated one.

Next, suppose that a second attribute, say $A_2$, is considered and when this is done, then the situation appears as in Figure 1.3 for exactly the same points. One may observe that when the data in Figure 1.3 are projected on the $A_1$ axis, then the situation depicted in Figure 1.2 emerges. Now exactly the very same points indicate a different pattern which is much easier to interpret. The new pattern indicates that if a point has an $A_2$ value higher than a given level (say some threshold value $h$), then it is a solid point. It is a gray point if its $A_2$ value is less than that threshold value $h$.

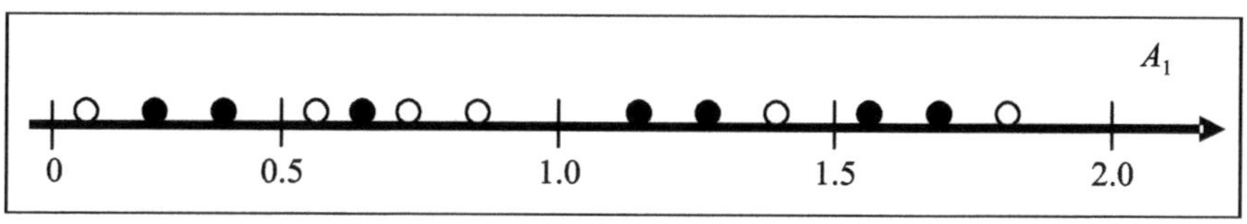


**Figure 1.2.** Data Defined in Terms of a Single Attribute.

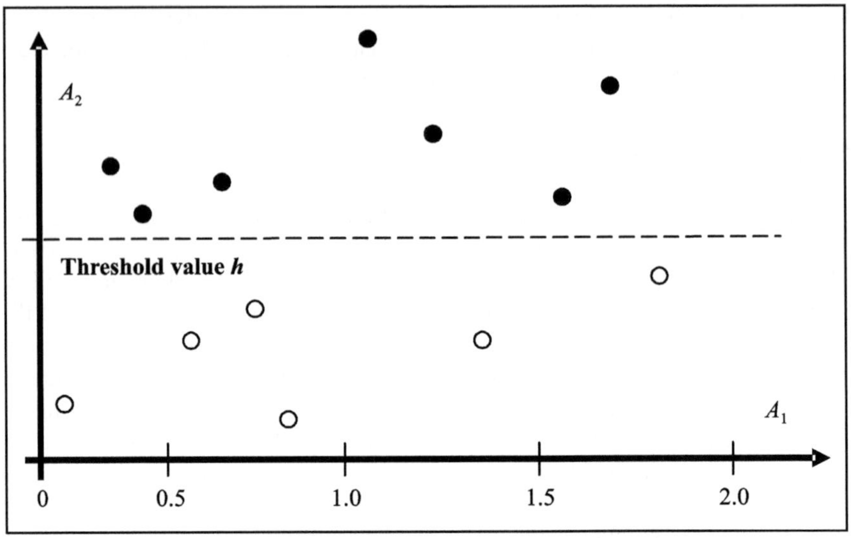


**Figure 1.3.** Data Defined in Terms of Two Attributes.

By examining Figure 1.3, one may argue that attribute $A_1$ does not offer much discriminatory value while attribute $A_2$ is the most important one. That is, the data can still be described effectively in terms of a single attribute (i.e., $A_2$ and not $A_1$) but that realization requires the examination of the data set in terms of more than one attribute.

### 1.3.3 Data Preprocessing

It is often the case for some data to include errors. Such errors could be values in some fields which are clearly out of range (say an air temperature of 125°F in the shade for some geographic location) or the combination of other values makes it clear that something is wrong. For instance, a location in Canada has an air temperature of 105°F during the month of December. Then, at least one of the three field values (Canada, December, 105°F) is erroneous. How should one deal with such cases? One approach is to try to "guess" the correct values, but that could involve introducing biases into the data. Another approach could be to discard any record which is suspected to contain erroneous data. Such an approach, however, may not be practical if the size of the data set is small and discarding records may make it too small to have any information value. A third approach might be to ignore the fields with the errors and try to analyze the information contained in the rest of the fields of records with corrupted data.

Of particular interest might be the case of having *outliers*, that is, data points which, somehow, are out of range. In other words, they are out of what one may consider as "normal." Such cases may be the result of errors (for instance, due to malfunctioning data collection devices or sensors). Another cause, however, may be that these values are indeed valid, but their mere presence indicates that something out of the ordinary takes place. A classical case is records of financial transactions. In this case, outliers may indicate fraudulent transactions which could obviously be of keen interest to the analyst. Now, a goal of the data mining approach might be how to identify such outliers as they may represent rare, but still very valuable from the analysis point of view, phenomena.

Once the data have been collected and are preprocessed, they need to be formatted in a way that would make them suitable as input to the appropriate data mining algorithm(s). This depends on the software requirements of the algorithms to be used.

### 1.3.4 Application of the Main Data Mining and Knowledge Discovery Algorithms

The main task of the data mining process is to analyze the data by employing the appropriate algorithm(s) and extract any patterns implied by the data. There are many algorithms that could be employed at this stage. This is one of the causes of the great confusion in the practical use of such methods. An increasingly popular approach is to use methods which can extract patterns in the form of classification/prediction rules. That is, logical statements of the form

IF (*some conditions are met*),
THEN (*the class of the new observation is* "*CL*"),

where "*CL*" is the name of a particular class. Next, such rules can be easily validated and implemented by domain experts who may or may not be computer or mathematically literate.

In case the patterns are used for prediction or classification of new data points of unknown class, their effectiveness needs to be checked against some test data. Of key importance is their accuracy rate. For instance, one may use weather data to develop a model that could forecast the weather. For simplicity, suppose that only two classification classes are considered: rain and no rain. Oftentimes accuracy is defined as the rate at which the proposed model accurately predicts the weather.

However, one may wish to consider *two* individual accuracy rates as follows; first how accurately the system predicts rain and second how accurately the system predicts no rain. The reason for considering this separation is because the impact of making mistakes under the previous two situations could be significantly different. This is more apparent if one considers, say, a medical diagnosis system derived from a data mining analysis of historic data.

Suppose that for such a system the two classes are "presence of a particular disease" and "no presence of the particular disease." The impact of making the mistake that this disease is *not* present while in reality it is present (i.e., when we have a *false-negative* case) could be dramatically higher when it is compared with the implications of the mistake when the system suggests that the disease *is* present while in reality it is not (i.e., when we have a *false-positive* case). For instance, for the case of a life-threatening disease (such as an aggressive type of cancer) the above situation may be detrimental to the survival of the patient.

Besides the previous two accuracy rates, there is a third rate too. This is the rate when the system responds with a *do not know* type of answer. This can happen if the system decides that a new case (data point) is quite different than any of the cases used for training when the current model was inferred. Then, conceptually, a system may have very high accuracy rates in terms of the previous two scenarios of false-positive and false-negative rates, but also a very high rate in terms of the third rate with the *do not know* responses. That is, this system would refuse too often to classify/diagnose new data points and instead declare them as undecidable cases (i.e., assign the *do not know* label). Such a system would be impractical as it too often makes useless recommendations. Thus, one has to find a way to balance the above three accuracy rates in a way which is consistent with the purpose of the data mining analysis and the type of problems to be solved and also the amount and nature of the training data. This is still an ongoing research area and systems based on Boolean functions seem to be the most promising ones to offer such control in the way they make predictions. Some related discussion on this very important subject is provided in Chapter 9 and also in Section 11.6 of Chapter 11.

### 1.3.5 Interpretation of the Results of the Data Mining and Knowledge Discovery Process

This is the step of the "moment of truth." At this point the data have been collected and analyzed, a model/models have been inferred and their performance on some test data seems to be encouraging. When the inferred model(s) can be viewed as a set of IF-THEN type of rules, then it is easy to interpret any new knowledge in a way that can be understood by domain experts who may or may not be computer or mathematically literate. Methods that rely on mathematical logic (i.e., on Boolean functions) offer an intuitive manner getting patterns that can be translated into IF-THEN rules and thus it is easier to understand their decision-making process.

If possible, the newly inferred knowledge needs to be verified and validated by the DM&KD analyst and also by the domain experts. It is always possible to detect errors that were invisible up to this point. Furthermore, the new knowledge may offer insights into issues that were too complicated before. This could be the beginning of new investigations that could not be initiated before. This is when the "eureka" ("I found it" in Greek) moment oftentimes takes place. However, as with any other step of the data mining process, one always needs to be vigilant and not hesitate to revise parts of the understanding as more and more facts and insights become available. This is why each step in Figure 1.1 is connected with each other by means of the *feedback mechanism*.

## 1.4 Four Key Research Challenges in Data Mining and Knowledge Discovery

All the application areas discussed in Section 1.2 have some key elements in common. The analyst can acquire lots of data that can be highly *heterogeneous* in nature and then use these data to make sense out of them. Traditional statistical methods can easily handle mostly *homogeneous* data. Furthermore, models derived by statistical methods may be limited in offering any new knowledge which could be easily interpreted by domain experts (i.e., people who do not necessarily have computational or mathematical/statistical backgrounds). Also, statistical models are valid subject to the satisfactory validation of certain assumptions on the nature of the data (such as following certain distributions). Such assumptions may not always be possible to validate. This is especially true when the data set is of small size. On the other hand, DM&KD methods may not rely on such assumptions.

Deriving readily interpretable models is important for the domain experts when they use the result of a DM&KD analysis and also for the very validation and fine-tuning of the derived DM&KD models. The following sections describe, in simple terms, some key computational challenges that are of key importance to the research community today and are highly likely to be so in the future. Some of these challenges led to the development of the DM&KD methods described in this book.

### 1.4.1 Collecting Observations about the Behavior of the System

A key problem with any analysis of data is what information to consider when collecting observations in order to study a system of interest. There is not an easy way to answer this question. The observations should describe the behavior of the system under consideration in a way such that when they are analyzed by DM&KD means, the extracted patterns will be effective in meeting the goals of the analysis.

Each data point is assumed to be a vector of some dimension $n$ which describes the behavior of the system in terms of *n attributes*. It could also be the case that these attributes (or just some of them) are organized in a tree-like hierarchy. After the analysis, it is possible that some of the attributes are to be dropped out as being insignificant or just irrelevant. Furthermore, collecting information about different attributes may involve entirely different costs. For instance, in a medical setting the cost of collecting information about a patient's blood pressure is far less than the cost of performing a biopsy of a lesion from, say, the liver or the brain of the patient.

The analyst may wish to first collect information about easily obtainable attributes. If the inferred model is not accurate enough and/or easily interpretable, then the analyst may wish to consider more attributes and augment the analysis. Another key problem is how to identify *noise* in the data and clean them. The danger here is that what appears to be noise in reality might be legitimate outliers and thus an excellent and rare opportunity to find evidence of some rare but very important aspects of the system under consideration might be ignored. In other words, such noise might be disguised nuggets of valuable new knowledge. As with outliers in a traditional statistical study, one has to be very careful in removing or keeping such data.

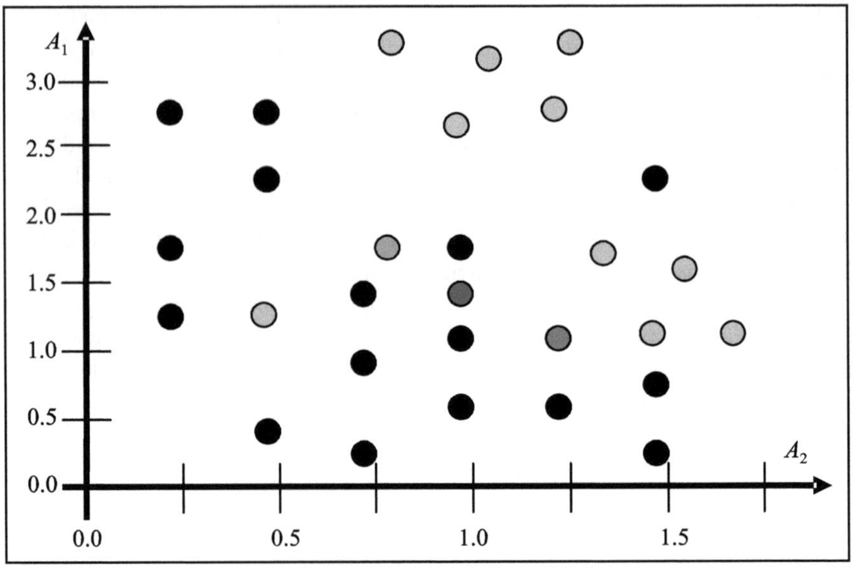


**Figure 1.4.** A Random Sample of Observations Classified in Two Classes.

### 1.4.2 Identifying Patterns from Collections of Data

In order to help fix ideas, consider the observations depicted in Figure 1.4. These observations are defined in terms of the two attributes $A_1$ and $A_2$. Each observation is represented by either a gray circle or a solid dark circle.

The main question that any DM&KD analysis tries to answer is what can one learn about these data? Such knowledge may next be used to accurately predict the class membership (in this case is it a "gray" or "solid dark" observation?) of new and unclassified observations. Such knowledge may also lead to a better understanding of the inner workings of the system under consideration, the design of the experiments to refine the current understanding of the system, and so on.

There are many ways one can define the concept of *knowledge* given a set of observations grouped into different classes. It seems that most experts agree that given observations of data grouped into different categories (classes), knowledge is any pattern implied by these data which has the potential to answer the previous main question. This understanding of knowledge makes the quest of acquiring new knowledge an ill-defined problem as there might be more than one pattern implied by the data. Thus, what is the best pattern? The direction adopted in the following developments is that the best pattern among a set of candidate patterns is the simplest one but still sufficient to answer the previous main question. This kind of philosophy is not new. As the medieval philosopher William of Occam (also known as Okham) stated in his famous "razor": *Entia non sunt multiplicanda prater necessitatem* (plurality should not be assumed without necessity).

Many DM&KD approaches interpret the above need in a way that tries to minimize the complexity of the derived pattern. Such patterns can be derived in terms of a decision tree, a set of separating planes, statistical models defined on some parameters, classification rules, etc. Then the need is to derive a decision tree with a minimum number of branches and/or nodes; in the case of separating planes, a minimum number of such planes; in the case of a statistical model, a model with the minimum number of parameters; or the minimum number of classification rules and so on.

However, obtaining a minimum number of the previous pattern entities may be computationally a very difficult, if not impossible, task. Thus, a more practical approach oftentimes is to develop fast heuristics that derive a very small number of such pattern entities. In the majority of the DM&KD methods to be discussed in the following chapters the above general objective is interpreted by deriving a minimum or near-minimum number of classification rules. The above are also consistent with the well-known *set covering problem.*

Next, suppose that instead of the rather complex data set depicted in Figure 1.4 now we have the rather simpler data set depicted in Figure 1.5. What pattern in the form of classification rules can be implied by these data?

The answer to this question may not be unique. Figure 1.6 presents a *possible* answer to this question. This answer is represented by the single square block that encloses all the solid dark points without including any of the gray points. One may consider a number of blocks that, collectively, achieve the same goal. Similarly, one may consider a number of different single-box solutions that do meet the same goal.

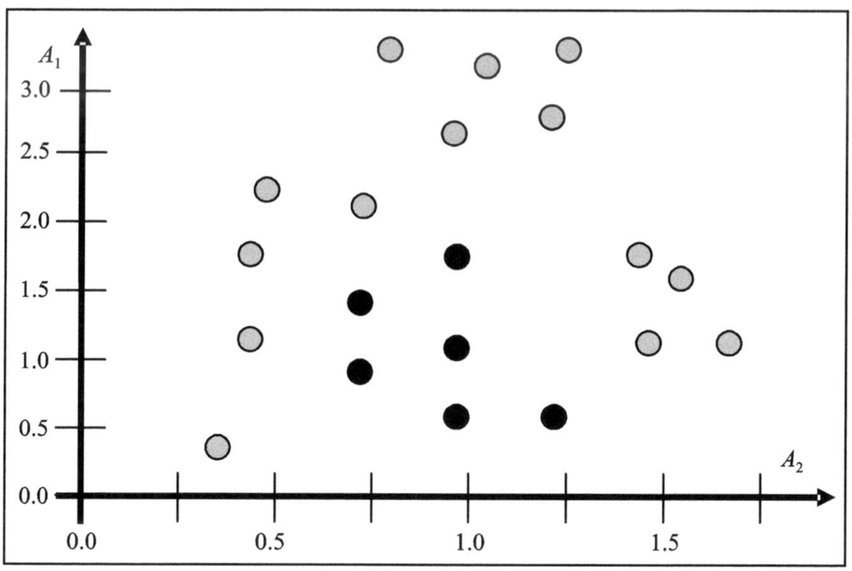


**Figure 1.5.** A Simple Sample of Observations Classified in Two Categories.

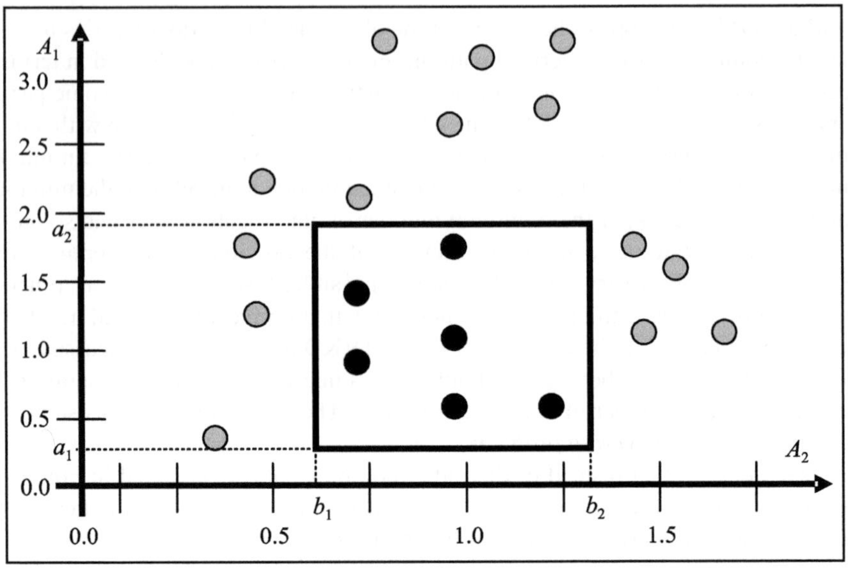


**Figure 1.6.** A Single Classification Rule as Implied by the Data.

Such a box implies a classification rule defined in terms of the two attributes $A_1$ and $A_2$. In general, such boxes are convex polyhedrals defined on the space of the attributes. For the single box in Figure 1.6 this is indicated by the dotted lines that define the ranges of values defined as $[a_1, a_2]$ and $[b_1, b_2]$ for the attributes $A_1$ and $A_2$, respectively. If the coordinates of a new observation fall inside these two ranges, then according to the rule depicted by the solid box in Figure 1.6, this observation belongs to the "solid dark" class. In the majority of the methods described in this book we will attempt to derive solutions like the one in Figure 1.6. That is, we will employ minimization approaches on the complexity of the derived pattern when such a pattern is expressed in terms of a compact Boolean function or, equivalently, in terms of a few classification rules.

For instance, for the case of the classification rule depicted as the solid box in Figure 1.6, this Boolean function has the form (where "$\wedge$" indicates the logical "and" operator)

$$\boldsymbol{F} = (a_2 \geq A_1 \geq a_1) \wedge (b_2 \geq A_2 \geq b_1).$$

Thus, the corresponding classification rule is

| | |
|---|---|
| **IF** | (the value of $A_1$ is between $a_1$ and $a_2$) and |
| | (the value of $A_2$ is between $b_1$ and $b_2$) |
| **THEN** | the observation belongs to the "solid dark" class. |

Similarly with the above simple case, when the data depicted in Figure 1.4 are treated according to the previous minimization objective, then a *possible* set of classification rules is the set of boxes depicted in Figure 1.7. This figure depicts rules both

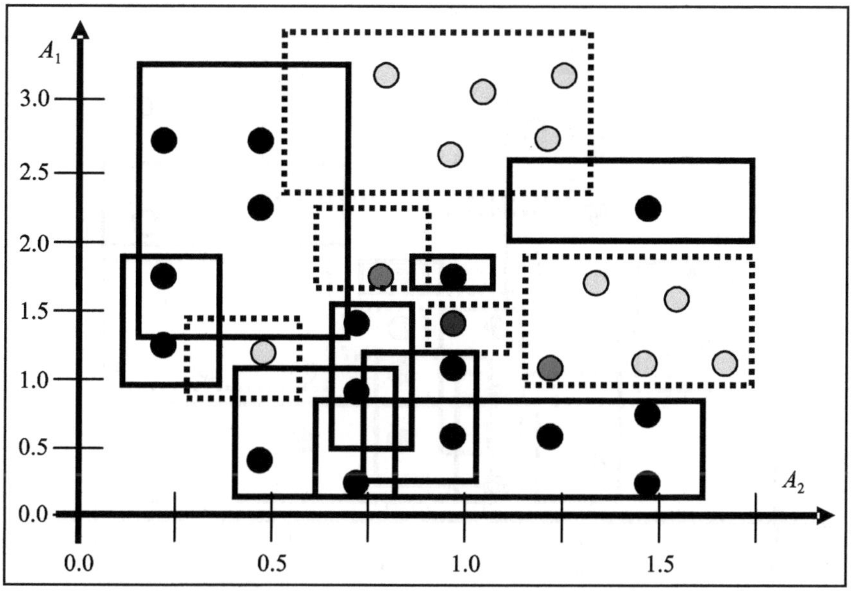


**Figure 1.7.** Some Possible Classification Rules for the Data Depicted in Figure 1.4.

for the gray and also for the solid dark sampled observations (depicted as dotted and solid-line boxes, respectively).

### 1.4.3 Which Data to Consider for Evaluation Next?

An interesting problem arises in cases in which the analyst has the capability to decide the composition of the next observation to consider, before it is sent to the expert or oracle for the determination of its class. Such a situation may arise, for instance, when one performs some kind of a test and the outcome of which may belong to two or more classes.

Let us consider Figure 1.8 which is based on the sampled data and the proposed classification rules discussed in the previous section. This figure differs from Figure 1.7 in that there are four additional observations for which we do not know their class membership (i.e., we do not know if they are gray or solid dark observations). These four observations are represented by the four plain circles near the top of the data set labeled with the question mark "?" and the numbers 1, 2, 3, and 4 for easy reference.

Should we consider new observation #1, #2, #3, or #4 for class inquiry? Let us consider each case separately and analyze these four different scenarios as follows:

*Case #1.* Suppose that we select new observation #1. This observation is covered simultaneously by a box (classification rule) from each of the two classes. This is indicated by the fact that it is covered by a solid-line box and also by a dotted-line box. This means that the current patterns (state of our knowledge about the nature of

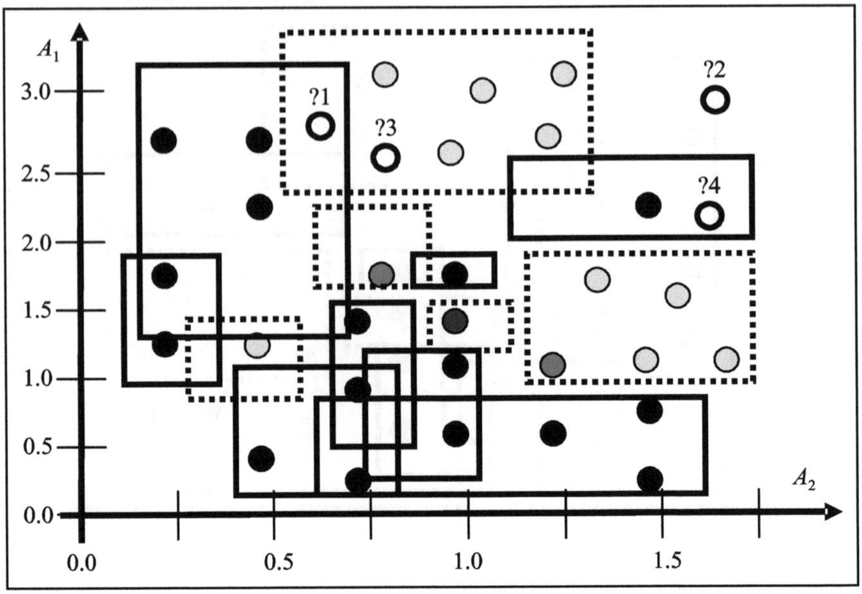


**Figure 1.8.** The Problem of Selecting a New Observation to Send for Class Determination.

the system which has classified these observations) would classify that data point as a gray and also as a solid dark point at the same time. Obviously, at least one of them is inaccurate.

When point #1 is sent to the oracle for the class inquiry, it will be classified as either a gray or a solid dark point. In the first scenario we will need to revise the solid-line box that currently covers it. In the second scenario we will need to revise the dotted-line box that currently covers it. In either scenario, one of the two sets of boxes (patterns extracted from the data) needs to be updated, and hopefully that would improve the current state of knowledge about the system of interest to the analyst.

*Case #2.* A similar situation exists with new observation #2. This observation is not covered by any of the current boxes (classification rules). That means that the solid-line boxes classify it as a gray point (since they do not cover it) and the dotted-line boxes classify it as a solid dark point for the analogous reason. Thus, if the oracle classifies it as a gray point, then we need a new or modify an existing "gray" rule (box) to cover it.

The opposite is true if the oracle classifies it as a solid dark point. Again, in any of the two possible scenarios that exist for point #2, one of the two sets of rules needs to be updated and thus, hopefully, improve the understanding of the system of interest to the analyst.

*Case #3 and #4.* The analysis now is different when we consider point #3 (or point #4). First, let us consider point #3. If that point is classified as gray by the oracle, no change to either set of rules is needed as this classification is consistent with the two sets of rules that represent our current understanding of the system of interest to the analyst.

However, if the oracle classifies it as a solid dark point, then **both** of the two sets of rules (systems) need to be modified. The set of the "gray" rules (dotted boxes) will have to reject it, while the set of the "solid dark" rules (solid boxes) will have to accept it. It should be clearly stated here that under case #3, either *none* or *both* of the two sets of rules needs (need) to be modified as a result of the classification of such a point by the oracle. An analogous reasoning can be developed for point #4 which is covered by a "solid dark" rule (box). All the above issues are part of what is known as *guided learning* and are analyzed in detail in Chapter 5.

### 1.4.4 Do Patterns Always Exist in Data?

Is it always possible to extract patterns from observations grouped into different classes? Or more accurately, do patterns always exist embedded in such groups of observations? In order to answer this question one has first to define what is a pattern.

Generally speaking, a pattern is a property or set of properties that exist in the data. According to the Merriam-Webster online dictionary, a *pattern* is a model, configuration, trait, or system. Even if data are completely random, then this by itself is a pattern. In terms of the data discussed in the previous sections, that means that the gray and solid dark points are evenly distributed in the space of the two attributes $A_1$ and $A_2$.

Thus, the answer to the above question is always a profound "yes." However, the real challenge is to be able to identify the pattern or patterns which are most useful in understanding the real nature of the system of interest to us. Otherwise, it is like looking at clouds in the sky and trying to infer any figures or images of geographic locations, plants, or even shapes of people or animals that could be imagined from them with some level of creativity. After all, many star constellations in the night sky are nothing but patterns, which, with some level of imagination, resemble such figures including those of the famous zodiac group.

Chapter 10 discusses the mathematical and algorithmic implication to DM&KD of a rather powerful property that seems to exist frequently in nature. This is the property of *monotonicity* in the data. Loosely speaking, this means that the class membership of observations is more likely to belong to a particular class as the values of certain attributes increase or decrease.

As author K.C. Cole wrote in her seminal book [1999] *The Universe and the Teacup: The Mathematics of Truth and Beauty*, "... nature bestows her blessings buried in mountains of garbage." Data mining and knowledge discovery is the development and application of a new type of computational tools for analyzing "mountains" of data for extracting useful and interesting new knowledge from them.

## 1.5 Concluding Remarks

This chapter described, in simple terms, some of the fundamental issues and challenges in DM&KD theory and applications. It described some representative application areas, although such a listing is by no means exhaustive.

It also highlighted the main steps of the entire data mining process with some of the critical issues involved at each step. Four key methodological challenges were also briefly mentioned. The same problems are the key subjects of the chapters that follow. In these chapters these problems, and many more related issues, are discussed in more detail and with the proper scientific rigor.

# Chapter 2

# Inferring a Boolean Function from Positive and Negative Examples

## 2.1 An Introduction

A central problem in data mining is how to analyze observations grouped into two categories and infer some key patterns that may be implied by these observations. As discussed in Chapter 1, these observations describe different states of nature of the system or phenomenon of interest to the analyst.

The previous chapter had a description of some possible application areas where data from observations may be used to study a variety of natural or man-made systems. Although there may be more than two classes when analyzing a system, we assume here that we have only two. Situations with more than two classes can be transformed into a set of two-class problems. Furthermore, it is assumed that these two groups (classes) of observations are exhaustive and exclusive. That is, the system has to be in only one of these two classes at any given moment.

The goal is to somehow analyze the data in these two groups of observations and try to infer some key pattern(s) that may be implied by these data. This could be important for a number of reasons. For instance, one may have the definition of a new data point (or points) but without information of its (their) class. Then, it is of interest to use the inferred patterns and assign it (them) to one of the two classes. If the available information is not adequate, then the new point(s) may not be assigned to any of the two classes and be deemed as undecidable, or as *do not know* case(s).

In the following it is assumed that the data are binary vectors (i.e., their individual fields take on 0/1 values). This is not a real limitation as nonbinary data can easily be transferred into binary ones. As the following section illustrates, this binary data and two-class problem has been studied extensively in the literature.

This chapter is organized as follows. After the following section, which reviews some key developments from the literature, a simple method is presented as to how nonbinary data can be transferred into equivalent binary data. The fourth section introduces the required terminology and notation. Sections five and six provide some formulations to this pattern inference problem. Sections seven, eight and nine

E. Triantaphyllou, *Data Mining and Knowledge Discovery via Logic-Based Methods*,
Springer Optimization and Its Applications 43, DOI 10.1007/978-1-4419-1630-3_2,

describe some developments for solving this problem by means of a branch-and-bound search. They also provide an approach for data preprocessing. Section eleven describes some computational results. This chapter concludes with section twelve.

## 2.2 Some Background Information

As mentioned above, suppose that some observations are available and they describe the behavior of a system of interest. It is also assumed that the behavior of this system is fully described by a number, say $n$, of *attributes* (also known as *parameters*, *variables*, *criteria*, *characteristics*, *predicates*, or just *features*). Thus, vectors of size $n$ define these observations. The $i$-th (for $i = 1, 2, 3, \ldots, n$) element of such a vector corresponds to the value of the $i$-th attribute. These attributes may be of any data type. For instance, they may take on continuous, discrete, or binary (i.e., $0/1$) values. Furthermore, each observation belongs to one and only one of two distinct classes. It is also assumed that the observations are *noise free*. That is, the class value associated with each observation is the correct one. In general, these two classes are called the *positive* and *negative* classes. These names are assigned arbitrarily. Thus, the examples in the positive (negative) class will be called the positive (negative) examples.

One may assume that some observations, say $m$, are already available. New observations (along with their class membership) may become available later but the analyst has no control on their composition. In addition to the previous scenario, the analyst may be able to define the composition of new observations (i.e., to set the values of the $n$ attributes) and then perform a test, or ask an expert (known as an *oracle* in the literature) to determine the class membership of a new observation. The main goal is to use the available classified observations to extract the underlying behavior of the target system in terms of a pattern. Next, this pattern is used to, hopefully, accurately infer the class membership of unclassified observations.

The extraction of new knowledge in the form of some kind of a model from collections of classified data is a particular type of *learning from examples*. Learning from examples has attracted the interest of many researchers in recent years. In the typical learning problem of this type, both positive and negative examples are available and the main goal is to determine a Boolean expression (that is, a set of logical rules or clauses) which accepts all the positive examples, while it rejects all the negative examples.

This kind of learning has been examined intensively (see, for instance, [Carbonell, *et al.*, 1983], [Dietterich and Michalski, 1983], [Kamath, *et al.*, 1992], [Kearns, *et al.*, 1987], [Pitt and Valiant, 1988], [Quinlan, 1986], and [Valiant, 1984]). Typically, the knowledge base of an intelligent system can be expressed as a Boolean function either in the conjunctive normal form (CNF) or in the disjunctive normal form (DNF) (see, for instance, [Blair, Jeroslow, and Lowe, 1985], [Cavalier, Pardalos, and Soyster, 1990], [Hooker, 1988a; 1988b], [Jeroslow, 1988; 1989], [Kamath, *et al.*, 1990], [Kamath, *et al.*, 1992], [Valiant, 1984], and [Williams, 1987]).

A considerable amount of related research is today known as the *PAC* (for *Probably Approximately Correct*) learning theory (see, for instance, [Valiant, 1984], [Angluin, 1988], and [Haussler and Warmuth, 1993]). The central idea of the PAC model is that successful learning of an unknown target concept should entail obtaining, with high probability, a hypothesis that is a good approximation of the target concept (hence the term: *probably approximately correct*). The error associated with the approximation of the target concept is defined as the probability that the proposed concept (denoted as $h$) and the target concept (denoted as $c$) will disagree on classifying a new example drawn randomly from unclassified examples. Later in this chapter this notion of error is used frequently and is related to another concept used extensively in this chapter called *accuracy rate*. The hypothesis $h$ is a good approximation of the target concept if the previous error is small (less than some quantity $\varepsilon$, where $1 > \varepsilon > 0$).

In the same framework of thought, a learning algorithm is then a computational procedure which takes a sample of random positive and negative examples of the target concept $c$ and returns a hypothesis $h$. In the literature a learning algorithm $A$ is a PAC algorithm if for all positive numbers $\varepsilon$ and $\delta$ (where $1 > \varepsilon, \delta > 0$), when $A$ runs and accesses unclassified examples, then it eventually halts and outputs a concept $h$ with probability at least $1 - \delta$ and error at most equal to $\varepsilon$ [Angluin, 1992].

Conjunctive concepts are properly PAC learnable [Valiant, 1984]. However, the class of concepts in the form of the disjunction of two conjunctions is not properly PAC learnable [Pitt and Valiant, 1988]. The same is also true for the class of existential conjunctive concepts on structural instance spaces with two objects [Haussler, 1989]. The classes of $k$-DNF, $k$-CNF, and $k$-decision lists are properly PAC learnable for each fixed $k$ (see, for instance, [Valiant, 1985], [Rivest, 1987], and [Kearns, *et al.*, 1987]), but it is unknown whether the classes of all DNF, or CNF functions are PAC learnable [Haussler and Warmuth, 1993] and [Goldman, 1990]. In [Mansour, 1992] an $n^{O(\log \log n)}$ algorithm is given for learning DNF formulas (however, *not* of minimal size) under a uniform distribution by using membership queries.

Another related issue is the sample complexity of a learning algorithm, that is, the number of examples needed to accurately approximate a target concept. The presence of bias in the selection of a hypothesis from the hypothesis space can be beneficial in reducing the sample complexity of a learning algorithm. Usually the amount of bias in the hypothesis space $H$ is measured in terms of the *Vapnik–Chernovenkis dimension*, denoted as *VCdim(H)* [Haussler, 1988].

There are many reasons why one may be interested in inferring a Boolean function with the minimum (or near minimum) number of terms. In an electronic circuit design environment, a minimum size Boolean representation is the prerequisite for a successful VLSI application. In a learning from examples environment, one may be interested in deriving a compact set of classification rules which satisfy the requirements of the input examples. As mentioned in the previous chapter, this can be motivated for achieving the maximum possible simplicity (as stated succinctly by *Occam's razor*) which could lead to easy verification and validation of the derived new knowledge.

Since the very early days it was recognized that the problem of inferring a Boolean function with a specified number of clauses is NP-complete (see, for instance, [Brayton, *et al.*, 1985] and [Gimpel, 1965]). Some related work in this area is due to [Bongard, 1970]. The classical approach for dealing with this Boolean function inference problem as a minimization problem (in the sense of minimizing the number of CNF or DNF clauses) was developed in [Quine, 1952 and 1955] and [McCluskey, 1956]. However, the exact versions of the Quine–McCluskey algorithm cannot handle large-scale problems. Thus, some heuristic approaches have been proposed. These heuristics include the systems MINI [Hong, *et al.*, 1974], PRESTO [Brown, 1981], and ESPRESSO-MV [Brayton, *et al.*, 1985]. Another widely known approach in dealing with this problem is the use of *Karnaugh maps* [Karnaugh, 1953]. However, this approach cannot be used to solve large-scale problems [Pappas, 1994]. Another application of Boolean function minimization can be found in the domain of multicast [Chang, *et al.*, 1999] where one needs a minimum number of keys.

A related method, denoted as SAT (for satisfiability), has been proposed in [Kamath, *et al.*, 1992]. In that approach one first pre-assumes an upper limit on the number of clauses to be considered, say $k$. Then a clause satisfiability (SAT) model is formed and solved using an interior point method developed by Karmakar and his associates [Karmakar, Resende, and Ramakrishnan, 1992]. If this clause satisfiability problem is feasible, then the conclusion is that it is possible to correctly classify all the examples with $k$ or fewer clauses. If the SAT problem (which essentially is an integer programming model) is infeasible, then one must increase $k$ until feasibility is reached. In this manner, the SAT approach yields a system with the minimum number of clauses.

It is important to observe at this point that from the computational point of view it is much harder to prove that a given SAT problem is infeasible than to prove that it is feasible. Therefore, trying to determine a minimum size Boolean function by using the SAT approach may be computationally too difficult. Some computational results indicate that the B&B approach proposed in [Triantaphyllou, 1994] (and as described in Chapter 3 of this book) is more efficient than the previous satisfiability-based approach. Actually, that B&B approach is on the average 5,500 times faster in those tests.

In [Felici and Truemper, 2002] the authors propose a different use of the SAT model. They formulate the problem of finding a clause with maximal coverage as a minimum cost satisfiability (MINSAT) problem and solve such problem iteratively by using the logic SAT solver *Leibniz*, which was developed by Truemper [1998]. That method is proved to be computationally feasible and effective in practice. The same authors also propose several variants and extensions to that system. Further extensions on this learning approach are also discussed in [Truemper, 2004].

A very closely related problem is to study the construction of a partially defined Boolean function (or pdBf), not necessarily of minimal size, given disjoint sets of positive and negative examples. That is, now it is required that the attributes of the function be grouped according to a given scheme (called a decomposition structure)

[Boros, *et al.*, 1994]. Typically, a pdBf may have exponentially many different extensions.

It should be stated here that there are a multitude of methods for inferring a Boolean function from two sets of training examples. A review of some recent developments of methods that infer rules (which in essence are like classification Boolean functions) can be found in [Triantaphyllou and Felici, 2006].

In summary, the most representative advances in distinguishing between observations in two or more classes can be classified into some distinct categories as follows: Some *common logic* approaches by [Zakrevskij, 1988; 1994; 1999; 2001; and 2006]. Clause *satisfiability* approaches to inductive inference such as the methods by Kamath, *et al.*, [1992, 1994] and [Felici and Truemper, 2002]. *Boolean function* (i.e., *logic*)-based approaches such as the methods in [Triantaphyllou, *et al.*, 1994], [Triantaphyllou, 1994] (these developments are described in detail later in this and the next chapter); some polynomial time and NP-complete cases of Boolean function decomposition by [Boros, *et al.*, 1994]; *association rules* [Adamo, 2000]; rough and fuzzy sets [Wang, Liu, Yao, Skowron, 2003]. *Decision tree*-based approaches [Quinlan, 1979; 1986], [Freitas, 2002] and [Witten and Eibe, 2005]. *Support vector machines* (SVM) by [Woldberg and Mangasarian, 1990], [Mangasarian, *et al.*, 1990], [Mangasarian, *et al.*, 1995], [Abe, 2005], and [Wang, 2005]. Knowledge-based learning approaches by *combining symbolic and connectionist machine* (neural networks)-based learning as proposed by Shavlik [1994], Fu [1993], Goldman and Sloan [1994] and Cohn, *et al.* [1994]. *Neural networks* [Arbib, 2002] and [Dayan and Abbot, 2001]. Various *rule induction* approaches as described in the edited book by [Triantaphyllou and Felici, 2006]; and finally, some *nearest neighbor* classification approaches [Hattori and Torri, 1993], [Kurita, 1991], [Kamgar-Parsi and Kanal, 1985], [Perner and Rosenfeld, 2003], [Berry, Kamath, and Skillicorn, 2004]. The above listing is not exhaustive as the field of data mining is still expanding rapidly, both in terms of theory and applications.

The main challenge in inferring a target set of discriminant classification rules from positive and negative examples is that the user may *never* be absolutely certain about the correctness of the classification rules, unless he/she has used the entire set of all possible examples which is of size $2^n$ in the binary case with $n$ attributes. In the general case this number is too high. Apparently, even for a small value of $n$, this task may be practically impossible to realize.

Fortunately, many real-life applications are governed by the behavior of a *monotone* system or they can be described by *a combination of a small number of monotone systems*. In data mining the property of monotonicity offers some unique computational advantages. By knowing the value of certain examples, one can easily infer the values of more examples. This, in turn, can significantly expedite the learning process. This chapter discusses the case of inferring general Boolean functions from disjoint collections of training examples. The case of inferring a monotone Boolean function is discussed in Chapter 10 of this book.

## 2.3 Data Binarization

The main idea of how to transform any data type into binary ones is best described via a simple illustrative example. Suppose that the data in Table 2.1 represent some sampled observations of the function of a system of interest. Each observation is described by the value of two *continuous* attributes denoted as $A_1$ and $A_2$. Furthermore, each observation belongs to one of two classes, denoted as Class 1 and Class 2. A number of problems can be considered at this point. The main problem is how to derive a pattern, in the form of a set of rules, which is consistent with these observations. As the set of rules we consider here logical clauses in the CNF (conjunctive normal form) or DNF (disjunctive normal form). That is, we seek the extraction of a Boolean function in CNF or DNF form. A more detailed description of the CNF and DNF forms is given in the next section.

Although, in general, many such Boolean functions can be derived, the focus of the proposed approach is on the derivation of a function of *minimal size*. By minimal size we mean a Boolean function which consists of the minimum number of CNF or DNF clauses. We leave it up to the analyst to decide whether he/she wishes to derive CNF or DNF functions. As explained in Chapter 7, Boolean functions in CNF (DNF) can easily be derived by using algorithms that initially derive Boolean functions in DNF (CNF).

Next we will demonstrate how the continuous data depicted in Table 2.1 can be represented by equivalent observations defined on only binary attributes. This is achieved as follows. We start with the first continuous attribute, i.e., attribute $A_1$ in this case, and we proceed until we cover all the continuous attributes.

From Table 2.1 it can be observed that the *ordered* set, denoted as $Val(A_1)$, with all the values of attribute $A_1$ is defined as the following ordered list:

$$\begin{aligned} Val(A_1) &= \{V_i(A_1), \text{ for } i = 1, 2, 3, \ldots, 9\} \\ &= \{0.25, 0.50, 0.75, 1.00, 1.25, 1.50, 1.75, 2.25, 2.75\}. \end{aligned}$$

That is, $V_1(A_1) = 0.25$, $V_2(A_1) = 0.50$, $V_3(A_1) = 0.75, \ldots, V_9(A_1) = 2.75$.

**Table 2.1.** Continuous Observations for Illustrative Example.

| Example No. | $A_1$ | $A_2$ | Class No. | Example No. | $A_1$ | $A_2$ | Class No. |
|---|---|---|---|---|---|---|---|
| 1 | 0.25 | 1.50 | 1 | 12 | 1.00 | 0.75 | 1 |
| 2 | 0.75 | 1.50 | 1 | 13 | 1.50 | 0.75 | 1 |
| 3 | 1.00 | 1.50 | 1 | 14 | 1.75 | 0.75 | 2 |
| 4 | 0.50 | 1.25 | 1 | 15 | 0.50 | 0.50 | 1 |
| 5 | 1.25 | 1.25 | 2 | 16 | 1.25 | 0.50 | 2 |
| 6 | 0.75 | 1.00 | 1 | 17 | 2.25 | 0.50 | 2 |
| 7 | 1.25 | 1.00 | 1 | 18 | 2.75 | 0.50 | 2 |
| 8 | 1.50 | 1.00 | 2 | 19 | 1.25 | 0.25 | 2 |
| 9 | 1.75 | 1.00 | 1 | 20 | 1.75 | 0.25 | 2 |
| 10 | 2.25 | 1.00 | 2 | 21 | 2.25 | 0.25 | 2 |
| 11 | 0.25 | 0.75 | 1 | | | | |

Obviously, the cardinality of this set (i.e., the number of elements in this set) is at most equal to the number of all available observations. In this instance, the cardinality is equal to 9. Next, we introduce 9 binary attributes $A'_{1,i}$ (for $i = 1, 2, 3, \ldots, 9$) as follows:

$$A'_{1,i} = \begin{cases} 1, & \text{if and only if } A_{1,i} \geq V_i(A_1), \text{ for } i = 1, 2, 3, \ldots, 9, \\ 0, & \text{otherwise.} \end{cases}$$

In general, the previous formula becomes for any multivalued attribute $A_j$ (where $K$ is the cardinality of the set $V_i(A_j)$):

$$A'_{j,i} = \begin{cases} 1, & \text{if and only if } A_{j,i} \geq V_i(A_j), \text{ for } i = 1, 2, 3, \ldots, K, \\ 0, & \text{otherwise.} \end{cases}$$

Using the above-introduced binary attributes, from the second observation (i.e., vector $(0.75, 1.50) = (A_{1,2}, A_{2,2})$) we get for its first attribute (please note that $A_{1,2} = 0.75$)

$$\{A'_{1,1}, A'_{1,2}, A'_{1,3}, A'_{1,4}, A'_{1,5}, A'_{1,6}, A'_{1,7}, A'_{1,8}, A'_{1,9}\} = \{1, 1, 1, 0, 0, 0, 0, 0, 0\}.$$

Similarly with the above definitions, for the second continuous attribute $A_2$ the set $Val(A_2)$ is defined as follows:

$$\begin{aligned} Val(A_2) &= \{V_i(A_2), \text{ for } i = 1, 2, 3, \ldots, 6\} \\ &= \{0.25, 0.50, 0.75, 1.00, 1.25, 1.50\}. \end{aligned}$$

Working as above, for the second observation we have

$$\{A'_{2,1}, A'_{2,2}, A'_{2,3}, A'_{2,4}, A'_{2,5}, A'_{2,6}\} = \{1, 1, 1, 1, 1, 1\}.$$

The above transformations are repeated for each of the nonbinary attributes. In this way, the transformed observations are defined on *at most* $m \times n$ binary attributes (where $m$ is the number of observations and $n$ is the original number of attributes). The precise number of the transformed attributes can be easily computed by using the following formula:

$$\sum_{i=1}^{n} |Val(A_i)|,$$

where $|s|$ denotes the cardinality of set $s$.

The binary attributed observations which correspond to the original data (as given in Table 2.1) are presented in Table 2.2 (parts (a) and (b)).

From the way the binary attributes have been defined, it follows that the two sets of observations are equivalent to each other. However, the observations in Table 2.1 are defined on continuous attributes while the observations in Table 2.2 are defined on binary ones.

**Table 2.2a.** The Binary Representation of the Observations in the Illustrative Example (first set of attributes for each example).

| Example No. | First set of attributes: $A'_{1,i}$, for $i = 1, 2, 3, \ldots, 9$ | | | | | | | | |
|---|---|---|---|---|---|---|---|---|---|
| | $A'_{1,1}$ | $A'_{1,2}$ | $A'_{1,3}$ | $A'_{1,4}$ | $A'_{1,5}$ | $A'_{1,6}$ | $A'_{1,7}$ | $A'_{1,8}$ | $A'_{1,9}$ |
| 1 | 1 | 0 | 0 | 0 | 0 | 0 | 0 | 0 | 0 |
| 2 | 1 | 1 | 1 | 0 | 0 | 0 | 0 | 0 | 0 |
| 3 | 1 | 1 | 1 | 1 | 0 | 0 | 0 | 0 | 0 |
| 4 | 1 | 1 | 0 | 0 | 0 | 0 | 0 | 0 | 0 |
| 5 | 1 | 1 | 1 | 1 | 1 | 0 | 0 | 0 | 0 |
| 6 | 1 | 1 | 1 | 0 | 0 | 0 | 0 | 0 | 0 |
| 7 | 1 | 1 | 1 | 1 | 1 | 0 | 0 | 0 | 0 |
| 8 | 1 | 1 | 1 | 1 | 1 | 1 | 0 | 0 | 0 |
| 9 | 1 | 1 | 1 | 1 | 1 | 1 | 1 | 0 | 0 |
| 10 | 1 | 1 | 1 | 1 | 1 | 1 | 1 | 1 | 0 |
| 11 | 1 | 0 | 0 | 0 | 0 | 0 | 0 | 0 | 0 |
| 12 | 1 | 1 | 1 | 1 | 0 | 0 | 0 | 0 | 0 |
| 13 | 1 | 1 | 1 | 1 | 1 | 1 | 0 | 0 | 0 |
| 14 | 1 | 1 | 1 | 1 | 1 | 1 | 1 | 0 | 0 |
| 15 | 1 | 1 | 0 | 0 | 0 | 0 | 0 | 0 | 0 |
| 16 | 1 | 1 | 1 | 1 | 1 | 0 | 0 | 0 | 0 |
| 17 | 1 | 1 | 1 | 1 | 1 | 1 | 1 | 1 | 0 |
| 18 | 1 | 1 | 1 | 1 | 1 | 1 | 1 | 1 | 1 |
| 19 | 1 | 1 | 1 | 1 | 1 | 0 | 0 | 0 | 0 |
| 20 | 1 | 1 | 1 | 1 | 1 | 1 | 1 | 0 | 0 |
| 21 | 1 | 1 | 1 | 1 | 1 | 1 | 1 | 1 | 0 |

**Table 2.2b.** The Binary Representation of the Observations in the Illustrative Example (second set of attributes for each example).

| Example No. | First set of attributes: $A'_{2,i}$, for $i = 1, 2, 3, \ldots, 6$ | | | | | | Class No. |
|---|---|---|---|---|---|---|---|
| | $A'_{2,1}$ | $A'_{2,2}$ | $A'_{2,3}$ | $A'_{2,4}$ | $A'_{2,5}$ | $A'_{2,6}$ | |
| 1 | 1 | 1 | 1 | 1 | 1 | 1 | 1 |
| 2 | 1 | 1 | 1 | 1 | 1 | 1 | 1 |
| 3 | 1 | 1 | 1 | 1 | 1 | 1 | 1 |
| 4 | 1 | 1 | 1 | 1 | 1 | 0 | 1 |
| 5 | 1 | 1 | 1 | 1 | 1 | 0 | 2 |
| 6 | 1 | 1 | 1 | 1 | 0 | 0 | 1 |
| 7 | 1 | 1 | 1 | 1 | 0 | 0 | 1 |
| 8 | 1 | 1 | 1 | 1 | 0 | 0 | 2 |
| 9 | 1 | 1 | 1 | 1 | 0 | 0 | 1 |
| 10 | 1 | 1 | 1 | 1 | 0 | 0 | 2 |
| 11 | 1 | 1 | 1 | 0 | 0 | 0 | 1 |
| 12 | 1 | 1 | 1 | 0 | 0 | 0 | 1 |
| 13 | 1 | 1 | 1 | 0 | 0 | 0 | 1 |
| 14 | 1 | 1 | 1 | 0 | 0 | 0 | 2 |
| 15 | 1 | 1 | 0 | 0 | 0 | 0 | 1 |
| 16 | 1 | 1 | 0 | 0 | 0 | 0 | 2 |
| 17 | 1 | 1 | 0 | 0 | 0 | 0 | 2 |
| 18 | 1 | 1 | 0 | 0 | 0 | 0 | 2 |
| 19 | 1 | 0 | 0 | 0 | 0 | 0 | 2 |
| 20 | 1 | 0 | 0 | 0 | 0 | 0 | 2 |
| 21 | 1 | 0 | 0 | 0 | 0 | 0 | 2 |

Given the above considerations, it follows that the original problem has been transformed to the binary problem depicted in Table 2.2 (parts (a) and (b)). This problem has the following two sets of positive and negative examples, denoted as $E^+$ and $E^-$, respectively.

$$E^+ = \begin{bmatrix} 1&0&0&0&0&0&0&0&0&1&1&1&1&1&1 \\ 1&1&1&0&0&0&0&0&0&1&1&1&1&1&1 \\ 1&1&1&1&0&0&0&0&0&1&1&1&1&1&1 \\ 1&1&0&0&0&0&0&0&0&1&1&1&1&1&0 \\ 1&1&1&0&0&0&0&0&0&1&1&1&1&0&0 \\ 1&1&1&1&1&0&0&0&0&1&1&1&1&0&0 \\ 1&1&1&1&1&1&1&0&0&1&1&1&1&0&0 \\ 1&0&0&0&0&0&0&0&0&1&1&1&0&0&0 \\ 1&1&1&1&0&0&0&0&0&1&1&1&0&0&0 \\ 1&1&1&1&1&1&0&0&0&1&1&1&0&0&0 \\ 1&1&0&0&0&0&0&0&0&1&1&0&0&0&0 \end{bmatrix} \text{ and}$$

$$E^- = \begin{bmatrix} 1&1&1&1&1&0&0&0&0&1&1&1&1&1&0 \\ 1&1&1&1&1&1&0&0&0&1&1&1&1&0&0 \\ 1&1&1&1&1&1&1&1&0&1&1&1&1&0&0 \\ 1&1&1&1&1&1&1&0&0&1&1&1&0&0&0 \\ 1&1&1&1&1&0&0&0&0&1&1&0&0&0&0 \\ 1&1&1&1&1&1&1&1&0&1&1&0&0&0&0 \\ 1&1&1&1&1&1&1&1&1&1&1&0&0&0&0 \\ 1&1&1&1&1&0&0&0&0&1&0&0&0&0&0 \\ 1&1&1&1&1&1&1&0&0&1&0&0&0&0&0 \\ 1&1&1&1&1&1&1&1&0&1&0&0&0&0&0 \end{bmatrix}.$$

Finally, it should be stated here that [Bartnikowski, *et al.*, 2006] present a detailed study of the general binarization problem.

## 2.4 Definitions and Terminology

Let $\{A_1, A_2, A_3, \ldots, A_n\}$ be a set of $n$ Boolean *attributes*. Each attribute $A_i$ (for $i = 1, 2, 3, \ldots, n$) can be either true (denoted by 1) or false (denoted by 0). Let $F$ be a *Boolean function* defined on these attributes. For instance, the expression $(A_1 \vee A_2) \wedge (A_3 \vee \bar{A}_4)$ is such a Boolean function, where "$\vee$" and "$\wedge$" stand for the logical "OR" and "AND" operators, respectively. That is, $F$ is a mapping from $\{0, 1\}^n \rightarrow \{0, 1\}$ which determines for each combination of truth values of the attributes $A_1, A_2, A_3, \ldots, A_n$ of $F$, whether $F$ is true or false (denoted as 1 or 0, respectively).

For each Boolean function $F$, the *positive examples* are the vectors $v \in \{0, 1\}^n$ such that $F(v) = 1$. Similarly, the *negative examples* are the vectors $v \in \{0, 1\}^n$ such that $F(v) = 0$. Therefore, given a function $F$ defined on the $n$ attributes

$\{A_1, A_2, A_3, \ldots, A_n\}$, then a vector $v \in \{0, 1\}^n$ is either a positive or a negative example.

Equivalently, we say that a vector $v \in \{0, 1\}^n$ is *accepted* (or *rejected*) by a Boolean function $F$ if and only if the vector $v$ is a positive (or a negative) example of $F$. For instance, let $F$ be the Boolean function $(A_1 \vee A_2) \wedge (A_3 \vee \bar{A}_4)$. Consider the two vectors $v_1 = (1, 0, 0, 0)$ and $v_2 = (1, 0, 0, 1)$. Then, it can be easily verified that $F(v_1) = 1$. That is, the vector $v_1$ is a positive example of the function $F$. However, the vector $v_2$ is a negative example of $F$ (since $F(v_2) = 0$).

The motivation for the following developments is best illustrated via a simple illustrative example. Consider a system of interest (represented by some Boolean function) that involves the following four attributes: $A_1$, $A_2$, $A_3$, and $A_4$. We do not know its structure yet. In any situation each attribute can either be *true* (denoted by 1) or *false* (denoted by 0). For instance, in example $(0, 1, 1, 0)$ the attributes $A_2, A_3, \bar{A}_1, \bar{A}_4$, are true or, equivalently, $A_1, A_4, \bar{A}_2, \bar{A}_3$, are false. There are $2^4 = 16$ possible examples (also known as states of nature) for this system (Boolean function). If a Boolean function is specified, then each of these 16 examples could be categorized either as *positive* or as *negative*.

For systems in CNF (to be formally defined below) a state of nature corresponds to a positive example if and only if it is satisfied by each clause in the system (i.e., Boolean function to be inferred). For instance, the state $(0, 1, 0, 1)$ satisfies the clause $(A_1 \vee A_2 \vee \bar{A}_3)$ and thus it corresponds to a positive example (in terms of that clause). Similarly, a state is a negative example if it violates at least one of the clauses in the system (Boolean function). Next consider the following three Boolean clauses:

$$(\bar{A}_1 \vee \bar{A}_2 \vee A_3 \vee A_4), (A_1 \vee A_2), \text{ and } (\bar{A}_1 \vee A_2 \vee A_3).$$

Then, all the 16 possible states are characterized as $(1, 1, 1, 1)$ positive, $(1, 0, 0, 0)$ negative, $(1, 1, 0, 0)$ negative, and so on.

The terms Boolean function, Boolean expression, and system would be used to denote the same concept. Also, in this chapter it is assumed, unless otherwise stated, that any Boolean expression (and consequently any clause) is expressed in the conjunctive normal form (CNF). An example of a Boolean expression in CNF is $(A_1 \vee A_3 \vee A_4) \wedge (A_2 \vee \bar{A}_7) \wedge (A_1 \vee \bar{A}_6)$, which simply is a conjunction of disjunctions. The above expression evaluates to true value if and only if all three disjunctions evaluate to true value. It evaluates to false value if and only if at least one of the disjunctions evaluates to false value. More formally, a Boolean expression is in the *conjunctive normal form* (CNF) if it is in the form (where $a_i$ is either $A_i$ or $\bar{A}_i$ and $\rho_j$ is the set of indices)

$$\bigwedge_{j=1}^{k} \left( \bigvee_{i \in \rho_j} a_i \right). \tag{2.1}$$

Similarly, a Boolean expression is in the *disjunctive normal form* (DNF) if it is in the form

$$\bigvee_{j=1}^{k} \left( \bigwedge_{i \in \rho_j} a_i \right). \tag{2.2}$$

An example of a Boolean function in DNF is $(A_1 \wedge A_2) \vee (A_3 \wedge \bar{A}_4 \wedge A_5)$, which is a disjunction of conjunctions. Such an expression evaluates to true value if and only if at least one of the conjunctions evaluates to true value. It evaluates to false value if and only if all the conjunctions evaluate to false value. In other words, a CNF expression is a conjunction of disjunctions, while a DNF expression is a disjunction of conjunctions. It should be stated here that the "boxes" (rules) discussed in some of the figures in Chapter 1 correspond to DNF systems. This is true because a single box can be viewed as a conjunction of a set of conditions (see also Section 1.4.2).

It is known [Peysakh, 1987] that any Boolean function can be transformed into the CNF or the DNF form. Chapter 7 of this book provides a simple approach of how any algorithm that derives a Boolean function in CNF (DNF) can also derive a function in DNF (CNF) by performing some simple transformations.

In summary, a set of positive examples (to be denoted as $E^+$ in this book) and a set of negative examples (to be denoted as $E^-$ in this book) are assumed to be available. These data will be used as the training data to infer a Boolean function. Given these two sets of positive and negative examples, the constraints to be satisfied by a Boolean function are as follows. In the CNF case, each positive example should be accepted by all the disjunctions in the CNF expression and each negative example should be rejected by at least one of the disjunctions. In the case of DNF systems, any positive example should be accepted by at least one of the conjunctions in the DNF expression, while each negative example should be rejected by all the conjunctions.

The general problem we analyze in this chapter is the construction of a set of Boolean expressions (clauses in CNF form) which correctly classify a set of sampled examples. We assume that each of these examples can be correctly classified (by an "oracle" or "expert") either as a *positive example* or as a *negative example*. The "expert" somehow knows the correct identification of any example. Such an expert opinion could be the result of a test, or a series of tests, which could be used to classify examples of the way the system operates. Furthermore, the underlying system of interest is not explicitly known. As illustrated in the first chapter, this can be a common and very important problem in many and diverse application areas.

The "expert" somehow can identify (probably through experience or special tests) the nature of any particular example but lacks the ability to characterize the classification rules to be used for such classifications. Thus, an important challenge is to develop methods to approximate the hidden system in situations in which the nature of finite numbers of examples is known.

We will consider this problem in a practical and applied context. Instead of four attributes, consider the scenario in which we may have, say, 50 attributes. Here the number of all the possible states (examples) is $2^{50} = 1{,}125{,}899{,}906{,}842{,}624$. This is more than one quadrillion. It would be impractical to generate all possible examples. However, one may be able to generate and categorize a few hundreds or even thousands of sampled examples. From this *partial* set of examples, we will determine a particular set of CNF clauses which correctly classify all the sampled examples and, hopefully, a large proportion of the remaining ones.

## 2.5 Generating Clauses from Negative Examples Only

Consider any example $\alpha$ defined on $n$ binary attributes. For instance, if $n = 5$, then consider an example such as $(1, 0, 1, 1, 0)$. Next, observe that the CNF clause $(\bar{A}_1 \vee A_2 \vee \bar{A}_3 \vee \bar{A}_4 \vee A_5)$ is satisfied by all examples $(d_1, d_2, d_3, d_4, d_5)$, where $d_i \in \{0, 1\}$, except $(1, 0, 1, 1, 0)$. The previous observation leads to the realization that a clause $C_\alpha$ can always be constructed which rejects any single example $\alpha$ while it accepts all other possible examples in the binary space of dimension $n$. In order to formalize this, let *ATTRIBUTES* $(\alpha)$ be the set of indices of the attributes which are true in example $\alpha$. For instance, *ATTRIBUTES* $((1, 0, 1, 1, 0)) = \{1, 3, 4\}$. If the clause $C_\alpha$ is defined as

$$C_\alpha = (\beta_1 \vee \beta_2 \vee \beta_3 \vee \cdots \vee \beta_N),$$

where

$$\beta_i = \left\{ \begin{array}{l} \bar{A}_i, \text{ if and only if } i \in \textit{ATTRIBUTES}(\alpha) \\ A_i, \text{ otherwise} \end{array} \right\},$$

$$\text{for each } \ i = 1, 2, 3, \ldots, n,$$

then the clause $C_\alpha$ will reject only example $\alpha$ and it will accept any other example. For instance, for the vector $\alpha = (1, 0, 1, 1, 0)$ the $C_\alpha$ clause is the previous CNF clause $(\bar{A}_1 \vee A_2 \vee \bar{A}_3 \vee \bar{A}_4 \vee A_5)$.

Suppose that $m$ examples are somehow generated. Define $E^+$ as the set of $m_1$ examples which have been classified as positive examples and $E^-$ as the set of the examples which have been classified as negative examples. These are the training examples to be used to generate a Boolean function.

For each of the $m_2$ (where $m_2 = m - m_1$) examples in $E^-$, we generate the unique clause as defined above. Each of these clauses rejects one and only one example and, hence, accepts all the examples in $E^+$. This set of $m_2$ clauses precisely satisfies the first objective of inferring a Boolean function which would accept all positive examples and reject all the negative ones. However, this approach would be *impractical* for large selections of negative examples, since it would result in large numbers of clauses.

More importantly, the above approach would suffer immensely of the *overfitting* problem. That is, the pattern (set of Boolean clauses) that would be generated as described previously, would fit perfectly well the negative data and nothing else. Its generalization capability would be almost nil. In terms of the solid-line and dotted-line boxes depicted in Figure 1.7 in Chapter 1, the above situation would be like generating solid-line boxes that cover each of the solid dark points in Figure 1.7 (assuming that the solid dark points in that figure are the negative examples). It should be mentioned here that the other extreme situation is to have *overgeneralization* of the data. Having a way to control the overfitting and overgeneralization properties of the inferred system is of high priority in data mining and the subject of ongoing research activity.

From this discussion it becomes clear that it is important to have an approach that constructs a rather small (relative to $m_1$ and $m_2$) number of clauses. This would also be desirable for the reasons of simplicity as described earlier in Chapter 1 (based on Occam's razor, etc.). The methods described in the following sections are such approaches.

## 2.6 Clause Inference as a Satisfiability Problem

In [Karmakar, *et al.*, 1991] it is shown that given two collections of positive and negative examples, then a DNF system can be inferred to satisfy the requirements of these examples. This is achieved by formulating a satisfiability (SAT) problem, which essentially is an integer programming (IP) problem, and then solve this IP problem by using the interior point method of Karmakar and his associates [Karmakar, *et al.*, 1991] as the solution strategy. This approach requires the specification of the number of conjunctions in the DNF system. The SAT problem uses the following Boolean variables [Karmakar, *et al.*, 1991]:

$$s_{ji} = \begin{cases} 0, & \text{if } A_i \text{ is in the } j\text{-th conjunction} \\ 1, & \text{if } A_i \text{ is not in the } j\text{-th conjunction} \end{cases}$$

$$s'_{ji} = \begin{cases} 0, & \text{if } \bar{A}_i \text{ is in the } j\text{-th conjunction} \\ 1, & \text{if } \bar{A}_i \text{ is not in the } j\text{-th conjunction} \end{cases}$$

$$\sigma^{\alpha}_{ji} = \begin{cases} s_{ji}, & \text{if } A_i = 1 \text{ in the positive example } \alpha \in E^+ \\ s'_{ji}, & \text{if } A_i = 0 \text{ in the positive example } \alpha \in E^+ \end{cases}$$

$$z^{\alpha}_{j} = \begin{cases} 1, & \text{if the positive example } \alpha \text{ is accepted by the } j\text{-th conjunction} \\ 0, & \text{otherwise} \end{cases}$$

Then, the clauses of this SAT problem are as follows:

$$s_{ji} \vee s'_{ji}, \quad \text{for } i = 1, \ldots, n, \text{ and } j = 1, \ldots, k, \tag{2.1a}$$

$$\left( \bigvee_{i \in P_r} \bar{s}'_{ji} \right) \vee \left( \bigvee_{i \in \bar{P}_r} \bar{s}_{ji} \right), \quad \text{for } i = 1, \ldots, k, \text{ and } r = 1, \ldots, m_2, \tag{2.2a}$$

$$\bigvee_{j=1}^{k} z^{\alpha}_{j}, \quad \text{for } \alpha = 1, \ldots, m_1, \tag{2.3a}$$

$$\sigma^{\alpha}_{ji} \vee \bar{z}^{\alpha}_{j}, \quad \text{for } i = 1, \ldots, n, j = 1, \ldots, k, \text{ and } \alpha = 1, \ldots, m_1, \tag{2.4a}$$

where $P_r$ is the set of indices of $A$ for which $A_i = 1$ in the negative example $r \in E^-$. Similarly, $\bar{P}_r$ is the set of indices of $A$ for which $A_i = 0$ in the negative example $r \in E^-$.

Clauses of type (2.1a) ensure that both $A_i$ and $\bar{A}_i$ will never appear in any conjunction. Clauses of type (2.2a) ensure that each negative example is rejected by all conjunctions. Clauses of type (2.3a) ensure that each positive example is accepted by at *least one* conjunction. Finally, clauses of type (2.4a) ensure that $z_i^{\alpha} = 1$ if and only if the positive example $\alpha$ is accepted by the $j$-th disjunction. In general, this SAT problem has $k(n(m_1+1)+m_2)+m_1$ clauses, and $k(2n(1+m_1)+nm_2+m_1)$ Boolean variables. A detailed example of this formulation can be found in [Karmakar, *et al.*, 1991].

## 2.7 An SAT Approach for Inferring CNF Clauses

The SAT formulation for deriving CNF systems is based on the original SAT formulation for deriving DNF systems as described in the previous section. The variables used in the new formulation are similar to the ones used in the DNF case. They are defined in a similar way as in the previous section as follows:

$$s_{ji} = \begin{cases} 0, & \text{if } A_i \text{ is in the } j\text{-th disjunction} \\ 1, & \text{if } A_i \text{ is not in the } j\text{-th disjunction} \end{cases}$$

$$s'_{ji} = \begin{cases} 0, & \text{if } \bar{A}_i \text{ is in the } j\text{-th disjunction} \\ 1, & \text{if } \bar{A}_i \text{ is not in the } j\text{-th disjunction} \end{cases}$$

$$\sigma_{ji}^{\beta} = \begin{cases} s_{ji}, & \text{if } A_i = 1 \text{ in the negative example } \beta \in E^- \\ s'_{ji}, & \text{if } A_i = 0 \text{ in the negative example } \beta \in E^- \end{cases}$$

$$z_j^{\beta} = \begin{cases} 1, & \text{if the negative example } \beta \text{ is accepted by the } j\text{-th disjunction} \\ 0, & \text{otherwise} \end{cases}$$

The clauses of the SAT formulation for deriving a CNF system which has up to $k$ disjunctions are as follows (where $n$ is the number of attributes):

$$s_{ji} \vee s'_{ji}, \quad \text{for } i = 1, \ldots, n, \text{ and } j = 1, \ldots, k, \tag{2.1b}$$

$$\left( \bigvee_{i \in P_r} \bar{s}_{ji} \right) \vee \left( \bigvee_{i \in \bar{P}_r} \bar{s}'_{ji} \right), \quad \text{for } i = 1, \ldots, k, \text{ and } r = 1, \ldots, m_1, \tag{2.2b}$$

$$\bigvee_{j=1}^{k} z_j^{\beta}, \quad \text{for } \beta = 1, \ldots, m_2, \tag{2.3b}$$

$$\sigma_{ji}^{\beta} \vee \bar{z}_j^{\beta}, \quad \text{for } i = 1, \ldots, n, \ j = 1, \ldots, k, \text{ and } \beta = 1, \ldots, m_2, \tag{2.4b}$$

where $P_r$ is the set of indices of $A$ for which $A_i = 1$ in the positive example $r \in E^+$. Similarly, $\bar{P}_r$ is the set of indices of $A$ for which $A_i = 0$ in the positive example $r \in E^+$.

Clauses of type (2.1b) ensure that *both* $A_i$ and $\bar{A}_i$ will *never* appear in any disjunction at the same time. Clauses of type (2.2b) ensure that each positive example will be accepted by *all* $k$ disjunctions. Clauses of type (2.3b) ensure that each negative example will be rejected by at *least one* of the $k$ disjunctions. Finally, clauses of type (2.4b) ensure that $z_i^\beta = 1$ if and only if the negative example is rejected by the $j$-th conjunction. In general, this problem has $k(n(m_2+1)+m_1)+m_2$ clauses, and $k(2n+m_2)$ binary variables.

Next, suppose that the following are the two sets (it is assumed that $n = 4$) $E^+$ and $E^-$ with the positive and negative examples of cardinality $m_1$ and $m_2$, respectively. These data were used to derive the integer programming (IP) model of the SAT formulation and its solution, for the CNF case, shown in the Appendix of this chapter. This IP model is written for the LINDO integer programming solver and can be easily adapted to fit many other IP solvers.

$$E^+ = \begin{bmatrix} 0 & 1 & 0 & 0 \\ 1 & 1 & 0 & 0 \\ 0 & 0 & 1 & 1 \\ 1 & 0 & 0 & 1 \end{bmatrix} \quad \text{and} \quad E^- = \begin{bmatrix} 1 & 0 & 1 & 0 \\ 0 & 0 & 0 & 1 \\ 1 & 1 & 1 & 1 \\ 0 & 0 & 0 & 0 \\ 1 & 0 & 0 & 0 \\ 1 & 1 & 1 & 0 \end{bmatrix}.$$

## 2.8 The One Clause At a Time (OCAT) Concept

The simple approach for creating clauses (disjunctions for a CNF expression) from negative examples described in Section 2.5, is very inefficient and ineffective. However, it provides some interesting insights. First of all, it is clear that in inferring a Boolean function from training data, one could start with a single data point and then move to another one and so on until all the points are covered. In the method described in Section 2.5 that strategy took place for the negative examples only. That is why that method is very inefficient as the derived systems suffer from extreme overfitting. The above idea provides the foundation for a sequential strategy. This is also the notion of the sequential covering algorithm discussed in [Tan, Steinbach, and Kumara, 2005].

Given a single clause, defined as before one could alleviate its problem of overfitting by removing items (i.e., attributes or their negations) off the definition of the clause. In that way, the clause would become less specific and more generalizing. That is, the modified clause would cover more than just a single negative example. This is how those clauses could be expanded in a gradual manner. In terms of the "boxes" (rules) idea discussed in Figure 1.7, this means that now the boxes would cover more than just a single negative example. One could keep removing such items off the definition of the clause until positive examples are covered too or some threshold value is reached. This idea can be explored from various implementation points of view but the end result is usually a rather large number of clauses.

Even after the above modification, the derived system of clauses may not be the best. There is no attention paid to the number of clauses derived to be small or, ideally, minimal. The strategy discussed next was first proposed in [Triantaphyllou, *et al.*, 1994] and [Triantaphyllou, 1994] and provides a greedy algorithm for achieving this goal. That approach is conceptually very close to Occam's razor and computationally superior to the SAT approach discussed in the previous two sections.

**Input:** Training data sets $E^+$ and $E^-$
$i = 0$; $C = \varnothing$; {initializations}
**DO WHILE** ($E^- \neq \varnothing$)
**Step 1:** $i \leftarrow i + 1$;
**Step 2:** Find a clause $c_i$ which accepts all members of $E^+$ while it rejects as many members of $E^-$ as possible;
**Step 3:** Let $E^-(c_i)$ be the set of members of $E^-$ which are rejected by $c_i$;
**Step 4:** Let $C \leftarrow C \wedge c_i$;
**Step 5:** Let $E^- \leftarrow E^-—E^-(c_i)$;
**REPEAT;**
**Output:** A CNF expression $C$ which accepts all examples in set $E^+$ while it rejects all examples in set $E^-$

**Figure 2.1.** The One Clause At a Time (OCAT) Approach (for the CNF case).

As mentioned in the previous section, the problem of deriving a Boolean function from sets of observations has been extensively studied in the literature. In our setting each example is a binary vector of size $n$ (number of binary attributes). The proposed *One Clause At a Time (or OCAT)* approach is based on a greedy algorithm. It uses as input data the two collections of disjoint positive and negative examples. It determines a set of CNF clauses that, when taken together, reject all the negative examples and each of them accepts all the positive examples.

The OCAT approach is sequential. In the first iteration it determines a clause in CNF form (in the current implementation) that accepts all the positive examples in the $E^+$ set while it rejects as many negative examples in the current $E^-$ set as possible. In the second iteration it performs the same task using the original $E^+$ set but the current $E^-$ set has only those negative examples that have not been rejected by any clause so far. The iterations continue until a set of clauses is constructed which reject all the negative examples. Figure 2.1 summarizes the iterative nature of the OCAT approach.

The core of the OCAT approach is Step 2 in Figure 2.1. The way this step is defined in Figure 2.1, implies the solution of an optimization problem. This is how the number of the inferred clauses could be controlled and not allowed to increase too much. In the next section a branch-and-bound (B&B)-based algorithm is presented that solves the problem posed in Step 2. Another faster B&B algorithm is presented in Chapter 3. However, the first B&B algorithm is presented to motivate the introduction of the second one. A fast heuristic for solving the problem posed in Step 2 of the OCAT approach is described in Chapter 4. The OCAT approach returns the set of desired clauses as set $C$.

For the DNF case one needs to modify Step 2 by deriving a clause which rejects all the negative examples while it accepts as many positive examples in the current version of the $E^+$ set as possible. Next, one needs to update the set of the positive examples (in modified Steps 3 and 5) by keeping only those examples which have not been accepted so far and repeat the loop. Step 4 needs to be modified too so the

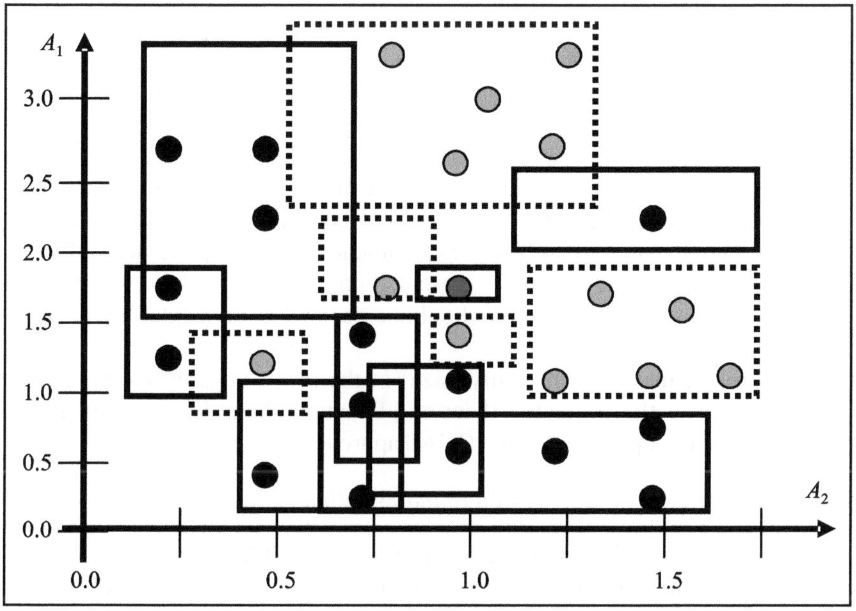


**Figure 2.2.** Some Possible Classification Rules for the Data Depicted in Figure 1.4.

derived system is a disjunction of conjunctions. The process is terminated when no positive examples are left in the $E^+$ set.

Next, we will try to get an idea about the difficulty of this Boolean function inference problem. Suppose that a learning problem involves a total of $m$ training examples each of which is a binary vector of size $n$. Then, as Section 5.4 proves, there are $2^L$, where $L = 2^n - m$, different Boolean functions which satisfy the requirements of these training data. This is an astronomically large number of possible solutions. The illustrative example later in Section 5.5 attempts to give a practical feeling of how incredibly large this solution space can be even for trivially small size learning problems. Thus, which Boolean function should one try to determine out of this huge number of possible solutions? The OCAT approach uses the greedy algorithm described in Figure 2.1 which is based on what is called the *principal of maximum simplicity*. This principle is best expressed by Occam's razor described in Section 1.4.2 as OCAT tries to infer a Boolean function of minimal or near-minimal number of clauses.

If one revisits the ideas behind Figure 1.7 (which is repeated here as Figure 2.2) about the two sets of boxes (rules) which cover the two groups of observations, then the OCAT approach solves a type of a *set covering problem.* Furthermore, it has the following interpretation.

Suppose that we start with the task of first covering the solid black points in Figure 2.2. We would like to derive a set of solid boxes. Such boxes correspond to classification rules which, as was shown in Section 1.4.2, are nothing but DNF expressions, although not defined on binary data. Then, according to the OCAT algorithm, and for the DNF case, the first step is to determine a box (rule) which covers the largest concentration of solid black points without covering any of the gray points. Next, determine a second box which covers the second largest concentration of solid black points, also without covering any of the gray points. We repeat this step successively, until a set of boxes is derived which, when are taken together, cover all the solid black points without covering any of the gray ones.

This set of boxes is the "solid black" set of classification rules. In an analogous manner, the "gray" (or "dotted") set of classification rules can be derived as well. In the following section a strategy is presented on how to classify new examples as being in either class or whether they should be deemed as undecidable (i.e., unclassifiable) cases.

Next, suppose that the cardinality (size) of the set of negative examples $E^-$ is equal to $m_2$. Then, the following theorem [Triantaphyllou, Soyster, and Kumara, 1994] states a critical property of the OCAT approach.

**Theorem 2.1.** *The OCAT approach terminates within $m_2$ iterations.*

*Proof.* From Section 2.5 it follows that it is always possible to construct a clause $C_\alpha$ that rejects only one negative example while it accepts any other possible example. At worst, Step 2 of the OCAT approach could propose a clause that rejects only one negative example at a given iteration. Therefore, the maximum number of iterations of the OCAT approach is $m_2$. ■

In Sections 2.6 and 2.7 we discussed a Boolean inference algorithm based on a satisfiability (SAT) formulation. In the above version of the OCAT approach, Boolean functions are derived in CNF. The two approaches have a major difference.

The OCAT approach, as defined in Figure 2.1, attempts to minimize the number of disjunctions in the proposed CNF system. However, the SAT approach *preassumes* a given number, say $k$, of conjunctions in the DNF (or disjunctions in the CNF) system to be inferred and solves an SAT problem. If this SAT problem is infeasible, then the conclusion is that there is no DNF system which has $k$ or fewer conjunctions and satisfies the requirements imposed by the examples. It should be emphasized here that it is not very critical whether an inference algorithm determines a CNF or DNF system (i.e., CNF or DNF Boolean function). As shown in [Triantaphyllou and Soyster, 1995b] and also presented in detail in Chapter 7, either a CNF or DNF system can be derived by using *either algorithm.*

## 2.9 A Branch-and-Bound Approach for Inferring a Single Clause

Branch-and-bound (B&B) is a search strategy which can be used to solve a wide spectrum of problems. It takes different forms depending on the specific problem under consideration. For Step 2 of the OCAT approach (for the CNF case), a B&B approach is given in [Triantaphyllou, Soyster, and Kumara, 1994]. It can be best described via an illustrative example. Suppose that the following are the two sets (it is assumed that $n = 4$, i.e., the system involves 4 attributes) $E^+$ and $E^-$ with the positive and negative examples of cardinality $m_1$ and $m_2$, respectively.

$$E^+ = \begin{bmatrix} 0 & 1 & 0 & 0 \\ 1 & 1 & 0 & 0 \\ 0 & 0 & 1 & 1 \\ 1 & 0 & 0 & 1 \end{bmatrix} \quad \text{and} \quad E^- = \begin{bmatrix} 1 & 0 & 1 & 0 \\ 0 & 0 & 0 & 1 \\ 1 & 1 & 1 & 1 \\ 0 & 0 & 0 & 0 \\ 1 & 0 & 0 & 0 \\ 1 & 1 & 1 & 0 \end{bmatrix}.$$

We number the positive examples as $(1, 2, 3, 4)$ and the negative examples as $(1, 2, 3, 4, 5, 6)$. For instance, the set of the negative examples $\{1, 3\}$ means the set of the first and the third negative examples (i.e., vectors $(1, 0, 1, 0)$ and $(1, 1, 1, 1)$, respectively). The B&B approach will determine a single clause (in CNF) that accepts all the positive examples in the $E^+$ set, while rejecting as many negative examples from the current $E^-$ set as possible. Before proceeding with the description of the B&B approach, it is instructive to compare it with a complete enumeration methodology (or brute force approach).

Consider the first positive example $(0, 1, 0, 0)$. One can observe that in order to accept this positive example at least one of the four attributes $A_1, A_2, A_3, A_4$ must be specified as follows: ($A_1 =$ false, i.e., $\bar{A}_1 =$ true), ($A_2 =$ true), ($A_3 =$ false, i.e., $\bar{A}_3 =$ true), and ($A_4 =$ false, i.e., $\bar{A}_4 =$ true). Hence, any valid CNF clause must include at least one of the following attributes: $\bar{A}_1$, $A_2$, $\bar{A}_3$, or $\bar{A}_4$. Similarly, the second positive example $(1, 1, 0, 0)$ implies that any valid CNF clause must include at least one of the following attributes: $A_1$, $A_2$, $\bar{A}_3$, or $\bar{A}_4$. In this manner, it can be concluded that any valid CNF clause must include at least one attribute as specified from each of the following four sets:

$$\{\bar{A}_1, A_2, \bar{A}_3, \bar{A}_4\},$$
$$\{A_1, A_2, \bar{A}_3, \bar{A}_4\},$$
$$\{\bar{A}_1, \bar{A}_2, A_3, A_4\}, \text{ and}$$
$$\{A_1, \bar{A}_2, \bar{A}_3, A_4\}.$$

As mentioned in the previous section, this is a special case of the set covering problem which we denote as the *minimum cardinality problem* (or MCP). Let $|s|$ denote the cardinality of a set $s$. For the clause inference problem, the corresponding MCP problem takes the following general form:

Problem MCP (the initial formulation):

$$\text{minimize} \left| \bigcup_{i=1}^{m_1} \beta_i \right|$$

Subject to:

$$\beta_i \in B_i, \text{ for } i = 1, 2, 3, \ldots, m_1,$$

where the sets $B_i$ are defined next.

**Table 2.3.** The $NEG(A_k)$ Sets for the Illustrative Example.

| Attribute | Set of Negative Examples | Attribute | Set of Negative Examples |
|---|---|---|---|
| $A_1$ | $NEG(A_1) = \{1, 3, 5, 6\}$ | $\bar{A}_1$ | $NEG(\bar{A}_1) = \{2, 4\}$ |
| $A_2$ | $NEG(A_2) = \{3, 6\}$ | $\bar{A}_2$ | $NEG(\bar{A}_2) = \{1, 2, 4, 5\}$ |
| $A_3$ | $NEG(A_3) = \{1, 3, 6\}$ | $\bar{A}_3$ | $NEG(\bar{A}_3) = \{2, 4, 5\}$ |
| $A_4$ | $NEG(A_4) = \{2, 3\}$ | $\bar{A}_4$ | $NEG(\bar{A}_4) = \{1, 4, 5, 6\}$ |

The MCP formulation for the current clause inference problem (in CNF) is developed as follows. Define as $NEG(A_k)$ the set of the negative examples which are accepted by a clause when the attribute $A_k$ is included in that clause. For the illustrative example in this section the $NEG(A_k)$ sets are presented in Table 2.3.

In the light of the definition of the $NEG(A_k)$ set and the $ATTRIBUTES(\alpha)$ set (as defined in Section 2.5), the sets $B_i$ in problem MCP are defined as follows:

$$B_i = \{NEG(A_k), \text{ for each } A_k \in ATTRIBUTES(\alpha_i)\},$$

where $\alpha_i$ is the $i$-th positive example in $E^+$.

Therefore, the previous minimization problem takes the following more precise form:

Problem MCP (more detailed formulation):

$$\text{Minimize} \left| \bigcup_{i=1}^{m_1} \beta_i \right| \tag{2.3}$$

Subject to:

$$\beta_i \in B_i, \text{ for } i = 1, 2, 3, \ldots, m_1,$$

where $B_i = \{NEG(A_k), \text{ for each } A_k \in ATTRIBUTES(\alpha_i)\}$, and $\alpha_i$ is the $i$-th positive example in $E^+$.

By using the data presented in Table 2.3, formulation (2.3) takes the following form for the case of the current illustrative example:

$$\text{Minimize} \left| \bigcup_{i=1}^{4} \beta_i \right|$$

Subject to:

$$\beta_1 \in B_1, \text{ where } B_1 = \{\{2, 4\}, \{3, 6\}, \{2, 4, 5\}, \{1, 4, 5, 6\}\},$$

$$\beta_2 \in B_2, \text{ where } B_2 = \{\{1, 3, 5, 6\}, \{3, 6\}, \{2, 4, 5\}, \{1, 4, 5, 6\}\},$$

$$\beta_3 \in B_3, \text{ where } B_3 = \{\{2, 4\}, \{3, 6\}, \{1, 2, 4, 5\}, \{1, 3, 6\}, \{2, 3\}\},$$

$$\beta_4 \in B_4, \text{ where } B_4 = \{\{1, 3, 5, 6\}, \{1, 2, 4, 5\}, \{2, 4, 5\}, \{2, 3\}\}.$$

An *exhaustive enumeration* approach to solve this MCP problem is to construct a tree that has nodes arranged in $4(= m_1)$ levels. In the description of the search that follows, we call these levels *stages*. These levels correspond to the four positive examples enumerated as $\{1, 2, 3, 4\}$ in $E^+$. Each interior node (i.e., a node with descendents), say at level $h$ (where $1 \leq h < 4$), is connected to $n$ nodes in the next higher level via $n$ arcs. These $n$ arcs represent the attributes that are true at the $h$-th positive example (i.e., the members of the set $ATTRIBUTES(\alpha_h)$, where $\alpha_h$ is the $h$-th positive example), as described in Section 2.5. The nodes (or *search states*) in this tree represent sets of negative examples. In our illustrative example these are subsets of the set $\{1, 2, 3, 4, 5, 6\}$.

For instance, the state $\{2, 3, 5\}$ refers to the second, third, and fifth negative examples in the set $E^-$. The set of negative examples that corresponds to a node (state) is the set of all the negative examples accepted by the attributes that correspond to the arcs that connect that node with the root node. That is, if one is at node (search state) $Y_k$ and one follows the arc that corresponds to attribute $A_i$, then the resulting state, say $Y_L$, is

$$Y_L = Y_K \cup NEG(A_i).$$

If the above strategy is followed, then the current illustrative example would create $4 \times 4 \times 4 \times 4 = 256$ terminal nodes and, in the general case, $n^{m_1}$ terminal nodes (where $m_1 = |E^+|$). Then, a clause which accepts all the positive examples and rejects as many negative examples as possible can be found by simply selecting a terminal node that corresponds to a search state with the minimum cardinality. This is true because such a state accepts the *minimum* number (or equivalently, rejects the *maximum* number) of negative examples.

Apparently, an exhaustive enumeration strategy is impractical. This is true because an exhaustive enumeration would require one to construct a search tree with $n^{m_1}$ different terminal nodes (final states). However, this B&B approach, which is based on the previous tree, is much faster because it is capable of *pruning* this tree rather efficiently. As is explained next, each node of the tree is examined in terms of two tests. If any of these two tests succeeds, then that node is *fathomed* and it is not expanded further.

The tree of this search is shown in Figure 2.3. Consider the two nodes which correspond to the two states $\{2, 4\}$ and $\{2, 4, 5\}$ in the second stage of the search tree (see also Figure 2.3). Clearly, the states that correspond to the leaves (terminal nodes) that have the state $\{2, 4, 5\}$ as an ancestor are going to have *at least as* many members (i.e., negative examples) as the states of the leaves (terminal nodes) that have as ancestor the state $\{2, 4\}$. This is true because subsequent states are derived by performing union operations on these two states with the same sets. Therefore, if at any stage of building the search tree there is a state that has another state (in the current stage) as a subset, then that state (node) can be fathomed without eliminating any optimal solutions. This characterization of the states is formalized by the following definitions of dominated and undominated states, which is derived from the above discussion.

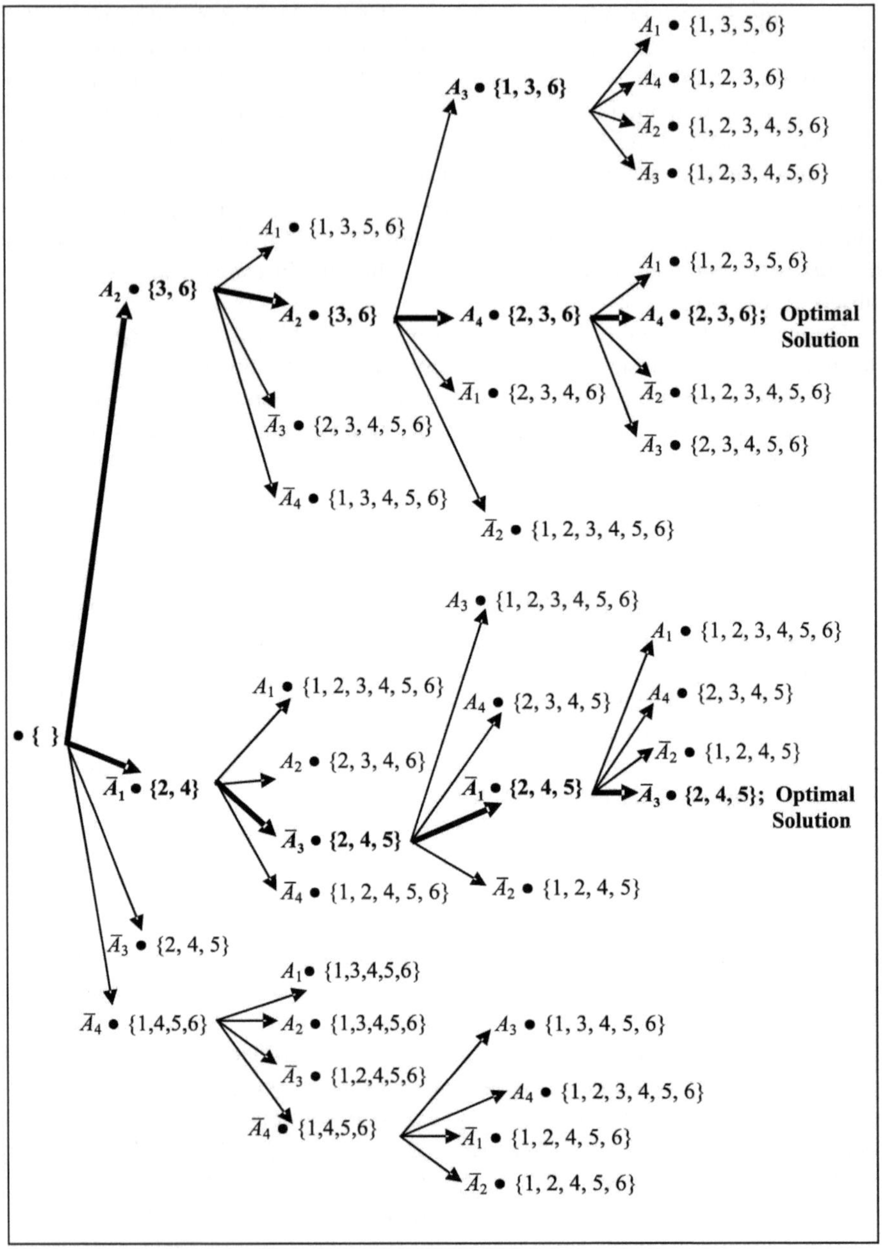


**Figure 2.3.** The Branch-and-Bound Search for the Illustrative Example.

**Definition 2.1.** *A state $S_k$ is a* **dominated state** *if there is another state $S_j$ in the same stage which is a* **proper** *subset of $S_k$, i.e., if $S_i \subset S_k$. Otherwise, the state $S_k$ is an* **undominated** *state.*

The notion of dominated states leads to an important simplification of the MCP problem. Define as MCP$'$ the problem derived from MCP when all dominated states are eliminated.

Problem MCP$'$:

$$\text{Minimize } \left| \bigcup_{i=1}^{m_1} \beta_i \right|$$

Subject to:

$$\beta_i \in B_i', \text{ for } i = 1, 2, 3, \ldots, m_1,$$

where $B_i'$ (for $i = 1, 2, 3, \ldots, m_1$) is the set that has as members only the *undominated* members of the set $B_i$.

Then, the previous definitions and discussion about dominated and undominated states lead to the following theorem [Triantaphyllou, Soyster, and Kumara, 1994]:

**Theorem 2.2.** *An optimal solution to MCP$'$ is also optimal to MCP.*

The following corollary is a direct implication of Theorem 2.2:

**Corollary 2.1.** *The optimal solutions of the original MCP problem, given as (2.3), and the previous MCP$'$ problem are identical.*

The previous corollary can be used for *problem preprocessing*. That is, when an MCP problem formulated as (2.3) is given, then it is beneficial to first transform it to the problem MCP$'$. In this way, the number of options (arcs in the B&B search graph) available at each node (search state) of the search graph will be the same or *smaller* than in the original MCP problem. Clearly, this means that the search can be done faster than in the original MCP problem.

The states in the last stage (i.e., the leaves of the tree) with the minimum number of negative examples indicate an optimal solution (see also Figure 2.3). In this example there are two such minimum size states. These are the states {2, 3, 6} and {2, 4, 5}. The first optimal state (i.e., {2, 3, 6}) is derived from the clause $(A_2 \vee A_4)$. This is true because the attributes $A_2$ and $A_4$ are the only attributes (as indicated by the B&B search) which are involved in the decisions that generate the state {2, 3, 6}. Similarly, the second optimal state (i.e., {2, 4, 5}) is derived from the clause $(\bar{A}_1 \vee \bar{A}_3)$.

Next we discuss some other ways for making this B&B search (and possibly other B&B algorithms which share similar principles) even more efficient. One way to do so for this B&B formulation is to keep in memory only the nodes (states) of the current level (stage). Then, when an optimal state $S$ is determined at the last stage, the optimal clause can be found by simply including in the definition of the current clause all the attributes along the path of the arcs which connect the optimal node with the root node.

Note that the optimal solution $(A_2 \vee A_4)$ does not reject the second, third, and sixth of the current negative examples in $E^-$. Hence, the remaining negative examples are

$$E^- = \begin{bmatrix} 0 & 0 & 0 & 1 \\ 1 & 1 & 1 & 1 \\ 1 & 1 & 1 & 0 \end{bmatrix}.$$

Similarly, the second OCAT iteration, when applied to the $E^+$ set and the new $E^-$ set, yields the clause $(\bar{A}_2 \vee \bar{A}_3)$. Now the remaining negative examples are

$$E^- = [0 \ \ 0 \ \ 0 \ \ 1].$$

Iterating further, the third OCAT iteration yields the clause $(A_1 \vee A_3 \vee \bar{A}_4)$. That is, the CNF clauses which are generated from the original $E^+$ and $E^-$ training examples are as follows:

$$\begin{aligned} &\textit{Clause } 1 : (A_2 \vee A_4) \\ &\textit{Clause } 2 : (\bar{A}_2 \vee \bar{A}_3) \\ &\textit{Clause } 3 : (A_1 \vee A_3 \vee \bar{A}_4). \end{aligned}$$

Thus, the inferred Boolean function is

$$(A_2 \vee A_4) \wedge (\bar{A}_2 \vee \bar{A}_3) \wedge (A_1 \vee A_3 \vee \bar{A}_4).$$

It can be easily verified that the previous three clauses, when taken together, reject all the negative examples in $E^-$. Moreover, each of the three clauses accepts all the positive examples in $E^+$. That is, this function satisfies the desired requirements.

This is the Boolean function derived from the original positive and negative training examples. Thus, we will call it the "positive" Boolean function or just the "positive" system. Next, one can treat the original negative examples as positive and the original negative examples as positive and apply the OCAT approach with the previous B&B algorithm on this reversed set of data. Then, a "negative" system can be derived in a similar manner. These two systems, that is, the "positive" and the "negative" system, correspond to the idea of the "solid" and "dotted" rules depicted in Figure 1.7 (or, equivalently, in Figure 2.2). They together can be used to classify new examples of unknown class value.

Given these two systems, and a new example of unknown class value, then the following four scenarios are possible when the new example is classified by these two systems:

1) It is accepted by the "positive" system and rejected by the "negative" system. Then, this new example would be characterized as a positive one.
2) It is accepted by the "negative" system and rejected by the "positive" system. Then, this new example would be characterized as a negative one.
3) It is accepted by both the "positive" and the "negative" system. Then, this new example would be characterized as a *do not know* case (i.e., as undecidable/ unclassifiable due to limited information).
4) It is rejected by both the "positive" and the "negative" system. Then, this new example would be characterized as a *do not know* case (i.e., as undecidable/ unclassifiable due to limited information).

Please recall that the rules derived this way correspond to CNF expressions.

Regarding the algorithmic steps of the previous B&B approach, there is another observation that allows for further reduction on the number of states in the B&B search. Suppose that it is known (possibly via a heuristic) that one of the terminal states in the B&B search (not necessarily an optimal one) has $k$ elements. Then, at any stage of the B&B approach, all states which have more than $k$ elements can be deleted from further consideration. This is a valid step because any descendent of a state may only get larger at subsequent stages. This observation is summarized in the following theorem [Triantaphyllou, Soyster, and Kumara, 1994]:

**Theorem 2.3.** *Suppose some feasible solution to MCP (or MCP′) has cardinality k. Then, an optimal solution to a modified B&B search in which all states that have more than k members are deleted, is also optimal for MCP (or MCP′).*

**Corollary 2.2.** *The optimal solutions of the original MCP problem, given as (2.3), and the following problem are identical.*

$$\text{Minimize } \left| \bigcup_{i=1}^{m_1} \beta_i \right|$$

Subject to:

$$\beta_i \in B_i', \text{ for } i = 1, 2, 3, \ldots, m_1,$$

where $B_i'$ (for $i = 1, 2, 3, \ldots, m_1$) is the set that has as members only the members of the original set $B_i$ which have *less than or equal* to $k$ members (defined as above).

## 2.10 A Heuristic for Problem Preprocessing

The last corollary can be used for *problem preprocessing*. That is, when an MCP problem is formulated as (2.3), then it is a good idea first to run a heuristic (as will be described next) that very quickly yields a good feasible solution of size $k$ (i.e., $k$ is small) to the original MCP problem. When a value for $k$ is available, the B&B search does not need to expand nodes in the search graph that have cardinality greater than $k$. This is true even for nodes that correspond to undominated states. In this way, the number of nodes to be expanded in the B&B search tree will, in general, be *smaller* than those in the original MCP problem. This step has the potential to expedite the B&B search.

Theorem 2.3 can further improve the performance of the proposed B&B search. When the B&B search is performed, the number of states at each stage (i.e., level of the search tree) may also increase dramatically. Therefore, the time and memory requirements of the search may increase dramatically. An efficient way to overcome this complication is to run the B&B search in two or more phases. In the first phase the B&B search is applied by allowing up to a small number, say 5, of states (i.e., nodes in the search tree) to be considered at any stage (i.e., level of the search tree).

These 5 states are the ones with the smallest cardinalities. That is, if more than 5 states (nodes) are formed at any stage, then only the 5 states with the smallest cardinalities will be considered for the next stage. This type of search is used in the AI literature often and is called *beam search* (see, for instance, [Dietterich and Michalski, 1981]).

Since up to 5 states are allowed to be considered at any stage of the B&B search and the number of stages is equal to the number of positive examples, it follows that the first phase will terminate quickly. Furthermore, the terminal nodes (final states) of the search tree will tend to represent states which have a tendency to have small cardinalities. This is expected to be the case because at each stage only the 5 states with the smallest cardinalities are considered (any ties are broken arbitrarily).

Suppose that in the first phase of the B&B process more than 5 states were generated at some stage. Let $k$ be the cardinality of the smallest state that is dropped from further consideration due to the upper limit of 5 states per stage. Then, if one of the terminal nodes has cardinality *less than* $k$, then one can conclude that this node (state) represents an optimal solution. This is true because in this case none of the deleted states could lead to a terminal state with cardinality less than $k$. If there is no terminal state with cardinality less than $k$, then a terminal node (search state) with the minimal cardinality represents a potentially good feasible solution which may or may not be optimal. It should be emphasized here that by an optimal solution we mean the one that represents a single clause in CNF (i.e., a single disjunction) which accepts all the positive examples in $E^+$ while it rejects as many negative examples in the current $E^-$ set as possible.

If after the first phase optimality is not provable, then the second phase is initiated. In the second phase, the B&B process is repeated with a higher limit, say 20, states per stage. As in the first phase, these 20 states are the states with the 20 smallest cardinalities. Suppose that $L$ is the cardinality of the best solution obtained in the first phase. Then in the second phase, Theorem 2.3 is applied by eliminating any state that has cardinality greater than $L$. However, memory limitations may prohibit this B&B search from reaching an optimal solution. It should be stated here that if a too large number of states were allowed to be considered at any stage, then the B&B approach would take excessive time in ranking these states. The previous limit of 20 states was empirically found to be a reasonable choice.

As was done in the first phase, if more than 20 states are generated at any stage, then only 20 states are allowed at each stage. Similarly to the first phase, let $k$ be the cardinality of the smallest state that was dropped from further consideration due to the upper limit of 20 states per stage. Then, if one of the terminal nodes has cardinality *less than* $k$, one can conclude that this node (state) represents an optimal solution. Otherwise optimality is not provable. In this case one may want to proceed with a third phase, or a fourth phase until optimality is eventually reached.

Some computational experiments indicate that Theorems 2.2 and 2.3 provide a rather efficient way for keeping the states at each stage in a manageable number and the resulting CPU requirements are dramatically reduced. For instance, a case with $n$ equal to 10, 50 positive examples, and 170 negative examples required more than 1,100 CPU seconds on an IBM ES/3090-600S machine (Penn State's mainframe computer in the 1980s and 1990s) running an integer programming implementation of the OCAT approach by using the MPSX software. However, the same problem took less than 30 CPU seconds with the proposed B&B formulation. Other similar comparisons also demonstrated significant improvement in time performance for this B&B approach.

## 2.11 Some Computational Results

In order to gain some computational experience with the OCAT approach and this B&B formulation, some random problems were generated and tested. The derived computational results are depicted in Table 2.4. For these problems, $n$, the number of attributes, was set equal to 30. First a set of 40 random clauses (disjunctions) was generated (the number 40 is arbitrary). Each such clause included, on the average, 5 attributes (as was the case with the experiments reported in [Hooker, 1988b]). The range of the number of variables per clause was from 1 to 10. Next, a collection $E^o$ of random examples was generated. In these experiments we generated groups of 100, 200, 300, ..., 1,000 random examples.

Each such random example was classified, according to the previous 40 clauses, either as a positive or as a negative example. With 40 clauses, this process resulted in more negative than positive examples. Because the stages in the B&B algorithm correspond to positive examples, problems with higher percentages of positive examples would demand more CPU time.

Next, the OCAT approach was applied on the previous positive and negative examples. The computational results are shown in Table 2.4. In this table the number of clauses derived by OCAT is denoted as $S$. The CPU time of the OCAT approach was recorded as well. This simulation program was written in the PL/I programming language and run on an IBM ES/3090-600S computer.

Each entry in Table 2.4 represents the performance of a single test problem, rounded to the nearest integer. Recall that $|s|$ indicates the *size* (or cardinality) of a set $s$. The computational results in Table 2.4 strongly suggest that the B&B approach is computationally tractable. For instance, no test problem took more than 836 CPU seconds (with an overage of 96.17 CPU seconds). As was anticipated, the number of clauses created by this B&B search increases with the number of input examples.

It is also interesting to observe the behavior of the CPU time used by OCAT under the B&B formulation. Since the number of stages in the B&B search is equal to the number of positive examples, the CPU time increases with the size of the set of the positive examples. Furthermore, the total number of examples $|E^o|$ is critical too.

In these test problems the B&B formulation was applied as follows. During the first phase up to 5 states were allowed. If after the final stage optimality was not proved, then the best (i.e., the one with the smallest cardinality) solution available at this point was kept and the B&B approach was repeated by allowing up to 20 B&B states per stage (20 was an upper limit for memory considerations). These 20 states were selected as follows. If more than 20 B&B states were generated at some

**Table 2.4.** Some Computational Results When $n = 30$ and the OCAT Approach Is Used.

| $\|E^o\|$ | $\|E^+\|$ | $\|E^-\|$ | $S$ | Time | $\|E^0\|$ | $\|E^+\|$ | $\|E^-\|$ | $S$ | Time |
|---|---|---|---|---|---|---|---|---|---|
| 100 | 9 | 91 | 4 | 2 | 400 | 10 | 390 | 6 | 10 |
| 100 | 5 | 95 | 4 | 2 | 400 | 7 | 393 | 6 | 10 |
| 100 | 15 | 85 | 4 | 7 | 400 | 36 | 364 | 13 | 282 |
| 100 | 7 | 93 | 4 | 6 | 400 | 47 | 353 | 6 | 97 |
| 100 | 3 | 97 | 4 | 1 | 400 | 49 | 351 | 12 | 400 |
| 100 | 8 | 92 | 4 | 2 | 400 | 15 | 385 | 5 | 7 |
| 100 | 7 | 93 | 4 | 2 | 400 | 5 | 395 | 5 | 3 |
| 100 | 1 | 99 | 4 | 1 | 400 | 17 | 383 | 6 | 23 |
| 100 | 7 | 93 | 4 | 3 | 400 | 16 | 384 | 6 | 8 |
| 100 | 5 | 95 | 4 | 3 | 500 | 35 | 465 | 12 | 194 |
| 200 | 5 | 195 | 4 | 2 | 500 | 16 | 484 | 5 | 38 |
| 200 | 2 | 198 | 5 | 1 | 500 | 7 | 493 | 6 | 15 |
| 200 | 18 | 182 | 5 | 18 | 500 | 34 | 466 | 7 | 73 |
| 200 | 6 | 194 | 5 | 2 | 500 | 13 | 487 | 6 | 8 |
| 200 | 1 | 199 | 4 | 1 | 500 | 20 | 480 | 5 | 19 |
| 200 | 11 | 189 | 7 | 38 | 500 | 6 | 494 | 5 | 13 |
| 200 | 19 | 181 | 4 | 4 | 600 | 83 | 517 | 6 | 300 |
| 200 | 51 | 149 | 2 | 212 | 600 | 49 | 551 | 15 | 315 |
| 200 | 10 | 190 | 4 | 6 | 600 | 44 | 556 | 5 | 41 |
| 200 | 4 | 196 | 5 | 2 | 600 | 8 | 592 | 6 | 16 |
| 300 | 22 | 278 | 8 | 70 | 600 | 23 | 577 | 12 | 184 |
| 300 | 14 | 286 | 6 | 25 | 600 | 11 | 589 | 6 | 15 |
| 300 | 14 | 286 | 7 | 29 | 700 | 56 | 644 | 16 | 467 |
| 300 | 2 | 298 | 5 | 1 | 700 | 18 | 682 | 6 | 30 |
| 300 | 22 | 278 | 1 | 102 | 700 | 19 | 681 | 6 | 15 |
| 300 | 36 | 264 | 1 | 243 | 700 | 19 | 681 | 9 | 60 |
| 300 | 24 | 276 | 4 | 12 | 700 | 13 | 687 | 6 | 26 |
| 300 | 71 | 229 | 4 | 524 | 800 | 64 | 736 | 18 | 739 |
| 300 | 3 | 297 | 5 | 2 | 900 | 72 | 828 | 17 | 836 |
| 300 | 17 | 283 | 1 | 107 | 1,000 | 47 | 953 | 14 | 80 |

NOTE: The time is in seconds.

stage, then these states were ranked in descending order according to the number of elements (negative examples) per state and the top 20 states were selected.

In this second phase of the B&B search, the best solution found at the end of the first phase was used to reduce the state space at each stage (i.e., Theorem 2.3 was applied to reduce the memory requirements). The process was terminated after this second phase (in which the 20 states per stage limit was imposed) regardless of whether the current best solution could be confirmed as optimal or not. It should be mentioned here that if a higher limit of states was used, then the B&B approach takes more time because at each stage more states need to be considered. Some computational tests indicated that the previous limits (i.e., 5 and 20 states) seem to be reasonable. In 83% of the problems examined, confirmation of optimality could

**Table 2.5.** Some Computational Results When $n = 16$ and the SAT Approach Is Used.

| $\|E^0\|$ | Problem ID | $K$ | Vars | Clauses | Time |
|---|---|---|---|---|---|
| 100 | 16A1 | 15 | 1,650 | 19,368 | 2,039 |
| 100 | 16C1 | 20 | 1,580 | 16,467 | 758 |
| 200 | 16D1 | 10 | 1,230 | 15,901 | 1,547 |
| 200 | 16E1 | 15 | 1,245 | 14,766 | 2,156 |
| 300 | 16A2 | 6 | 1,602 | 23,281 | 608 |
| 300 | 16B1 | 8 | 1,728 | 24,792 | 78 |
| 400 | 16B2 | 4 | 1,076 | 16,121 | 236 |
| 400 | 16C2 | 4 | 924 | 13,803 | 521 |
| 400 | 16D2 | 4 | 836 | 12,461 | 544 |
| 400 | 16E2 | 4 | 532 | 7,825 | 376 |

NOTE: The time is in seconds.

**Table 2.6.** Some Computational Results When $n = 32$ and the SAT Approach Is Used.

| $\|E^0\|$ | Problem ID | $k$ | Vars | Clauses | Time |
|---|---|---|---|---|---|
| 50 | 32B1 | 3 | 228 | 1,374 | 5 |
| 50 | 32C1 | 3 | 225 | 1,280 | 24 |
| 50 | 32D1 | 4 | 332 | 2,703 | 66 |
| 50 | 32E1 | 3 | 222 | 1,186 | 8 |
| 100 | 32B2 | 3 | 261 | 2,558 | 57 |
| 100 | 32C2 | 3 | 249 | 2,182 | 9 |
| 100 | 32D2 | 4 | 404 | 5,153 | 178 |
| 100 | 32E2 | 3 | 267 | 2,746 | 10 |
| 150 | 32C3 | 3 | 279 | 3,272 | 14 |
| 200 | 32E3 | 3 | 330 | 5,680 | 133 |
| 250 | 32A1 | 3 | 459 | 9,212 | 177 |
| 250 | 32B3 | 3 | 348 | 5,734 | 190 |
| 300 | 32B4 | 3 | 381 | 6,918 | 259 |
| 300 | 32E4 | 3 | 387 | 7,106 | 277 |
| 400 | 32D3 | 4 | 824 | 19,478 | 1,227 |
| 400 | 32E5 | 3 | 450 | 9,380 | 390 |
| 1,000 | 32C4 | 3 | 759 | 20,862 | 155 |

NOTE: The time is in seconds.

be made. The low CPU times indicate that this B&B approach is rather efficient both in terms of CPU time and memory requirements.

Tables 2.5 and 2.6 present some computational results when the SAT approach is used. These results are the ones originally reported in [Kamath, Karmakar, *et al.*, 1992]. The CPU times are approximated to the closest integer value (in seconds). Those experiments were performed on a VAX 8700 running UNIX and that computer program was written in a combination of FORTRAN and C codes. The strategy of generating and testing the random problems is similar to the one mentioned in the

OCAT case. The only difference is that now the "hidden system" is in DNF form and consists of a few conjunctions (three to four). Please recall that in the OCAT case the "hidden logic" was a system in CNF form consisting of 40 randomly generated disjunctions.

The main point with the SAT results is that even for a small number of (positive and negative) examples the CPU times are rather high. This happens because the resulting SAT problems (as was indicated in formulas presented in Section 2.6) require many variables and clauses (as is shown under the "Vars" and "Clauses" columns in Tables 2.5 and 2.6). In Table 2.6 the test problems considered 32 attributes. The CPU times are smaller than the ones with 16 attributes (in Table 2.5) because now $k$ was allowed to take much smaller values (3 or 4). In the 16-attribute case, however, $k$ was allowed to take relatively speaking larger values (4 to 20).

In other words, the CPU requirements increase dramatically with the number of conjunctions assumed in the SAT formulation (denoted as $k$). This behavior is in direct agreement with the formulas mentioned in Section 2.6. However, if the original $k$ value is *too small*, then infeasibility will be reached and the SAT problem needs to run again (with a larger $k$ value) until a feasible solution is reached. This situation may increase the *actual* CPU requirements even more dramatically than the numbers shown in Tables 2.5 and 2.6.

## 2.12 Concluding Remarks

This chapter examined the problem of inferring a Boolean function from two sets of disjoint binary data. This is a fundamental problem in data mining and knowledge discovery and thus has received lots of attention by the scientific community. It may be hard to determine what is the best way to solve this problem. A computationally demanding approach is to formulate this problem as a satisfiability (SAT) problem. In this way a Boolean function of minimal size (in terms of the number of CNF or DNF clauses that comprise it) can be inferred. However, the computational cost may make the SAT approach impractical for large size problems.

We have chosen to work with CNF or DNF because any Boolean function can be transferred into these two forms [Blair, Jeroslow, and Lowe, 1986]. DNF expressions can be visualized easily as convex polyhedral shapes in the space of the attributes, while CNF expressions offer more intuitive formulation capabilities.

The approach proposed in this chapter helps to quickly infer a Boolean function in CNF or DNF. It is termed OCAT (for *One Clause At a Time*). It is a greedy approach for inferring a Boolean function by means of one clause at a time. A key step of the OCAT approach involves the solution of an optimization problem and thus the OCAT approach may lead to systems comprised of a few clauses. That would make it consistent with the desire to infer the system of *maximum simplicity* as Occam's razor would dictate.

The previous optimization problem, as part of the OCAT approach, was solved according to a branch-and-bound (B&B) algorithm. This B&B algorithm can be expedited by exploiting certain key properties of the problem. These properties could be used with other B&B algorithms in the future. Solving this problem is the foundation to inferring a Boolean function from training data.

At this point it should be pointed out that one may consider different approaches besides the one which tries to minimize the inferred CNF expression. One such reasonable approach would be to derive a Boolean function which would minimize a weighted average of the false-positive, false-negative, and undecidable rates. More on this idea is discussed later in Chapter 4, Section 4.5.

As stated earlier, this Boolean function inference problem is open-ended. One will always have a strong incentive to develop new methods that would be faster and methods to partition large-scale inference problems. The most important aspect is to have methods which would indeed capture the real essence of the input data, and thus the actual nature of the system or phenomenon that generated these data. Thus, this problem will always be one of keen interest to the research and practitioners communities in the field of data mining and knowledge discovery from data.

# Appendix

## The SAT Formulation and Solution for the Illustrative Example in Section 2.7 (for the CNF case)

```
! **************************************************
!    This is the integer IP formulation (to run
! on LINDO) for the illustrative example presented
! in this chapter (for the CNF case).
!    The variable names are not identical, but they
! closely reflect the notation used in this chapter.
! **************************************************
! Note:
! We are interested in checking for feasibility.
! Thus, any objective function is applicable here.
! **************************************************

MIN S11

ST
!...................
S11 + SP11 >= 1
S21 + SP21 >= 1
S12 + SP12 >= 1
S22 + SP22 >= 1
S13 + SP13 >= 1
S23 + SP23 >= 1
!...................
SS11 + SS12 + SSP13 >= 1
SS21 + SS22 + SSP23 >= 1
SSP11 + SSP12 + SSP13 >= 1
SSP21 + SSP22 + SSP23 >= 1
!...................
! NEXT ARE THE NEGATIONS
S11 + SS11 <= 1
S21 + SS21 <= 1
S12 + SS12 <= 1
S22 + SS22 <= 1
S13 + SS13 <= 1
S23 + SS23 <= 1
SP11 + SSP11 <= 1
SP21 + SSP21 <= 1
SP12 + SSP12 <= 1
SP22 + SSP22 <= 1
SP13 + SSP13 <= 1
```

```
SP23 + SSP23 <= 1
!...................
Z11 + Z12 >= 1
Z21 + Z22 >= 1
Z31 + Z32 >= 1
!...................
! NEXT ARE MORE NEGATIONS
Z11 + ZZ11 <= 1
Z12 + ZZ12 <= 1
Z13 + ZZ13 <= 1
Z21 + ZZ21 <= 1
Z22 + ZZ22 <= 1
Z23 + ZZ23 <= 1
Z32 + ZZ32 <= 1
Z31 + ZZ31 <= 1
!...................
SP11 + ZZ11 >= 1
S12 + ZZ11 >= 1
SP13 + ZZ11 >= 1
!.........
SP21 + ZZ12 >= 1
S22 + ZZ12 >= 1
SP23 + ZZ12 >= 1
!.........
SP11 + ZZ21 >= 1
S12 + ZZ21 >= 1
S13 + ZZ21 >= 1
!.........
SP21 + ZZ22 >= 1
S22 + ZZ22 >= 1
S23 + ZZ22 >= 1
!......
S11 + ZZ31 >= 1
SP12 + ZZ31 >= 1
SP13 + ZZ31 >= 1
!.........
S21 + ZZ32 >= 1
SP22 + ZZ32 >= 1
SP23 + ZZ32 >= 1
!......
END
INTEGER 40
```

## This is the corresponding solution as generated by LINDO

```
NEW INTEGER SOLUTION OF .000000000 AT BRANCH 0 PIVOT 46
LP OPTIMUM FOUND AT STEP 46
OBJECTIVE VALUE = .000000000
ENUMERATION COMPLETE. BRANCHES= 0 PIVOTS= 46

LAST INTEGER SOLUTION IS THE BEST FOUND
RE-INSTALLING BEST SOLUTION...

OBJECTIVE FUNCTION VALUE

1) .000000000

   VARIABLE        VALUE        REDUCED COST
        S11        .000000         1.000000
       SP11       1.000000          .000000
        S21       1.000000          .000000
       SP21        .000000          .000000
        S12       1.000000          .000000
       SP12        .000000          .000000
        S22        .000000          .000000
       SP22       1.000000          .000000
        S13       1.000000          .000000
       SP13       1.000000          .000000
        S23        .000000          .000000
       SP23       1.000000          .000000
       SS11       1.000000          .000000
       SS12        .000000          .000000
      SSP13        .000000          .000000
       SS21        .000000          .000000
       SS22       1.000000          .000000
      SSP23        .000000          .000000
      SSP11        .000000          .000000
      SSP12       1.000000          .000000
      SSP21       1.000000          .000000
      SSP22        .000000          .000000
       SS13        .000000          .000000
       SS23        .000000          .000000
        Z11       1.000000          .000000
        Z12        .000000          .000000
        Z21       1.000000          .000000
        Z22        .000000          .000000
        Z31        .000000          .000000
        Z32       1.000000          .000000
```

| | | |
|---|---|---|
| ZZ11 | .000000 | .000000 |
| ZZ12 | 1.000000 | .000000 |
| Z13 | .000000 | .000000 |
| ZZ13 | .000000 | .000000 |
| ZZ21 | .000000 | .000000 |
| ZZ22 | 1.000000 | .000000 |
| Z23 | .000000 | .000000 |
| ZZ23 | .000000 | .000000 |
| ZZ32 | .000000 | .000000 |
| ZZ31 | 1.000000 | .000000 |

| ROW | SLACK OR SURPLUS | DUAL PRICES |
|---|---|---|
| 2) | .000000 | .000000 |
| 3) | .000000 | .000000 |
| 4) | .000000 | .000000 |
| 5) | .000000 | .000000 |
| 6) | 1.000000 | .000000 |
| 7) | .000000 | .000000 |
| 8) | .000000 | .000000 |
| 9) | .000000 | .000000 |
| 10) | .000000 | .000000 |
| 11) | .000000 | .000000 |
| 12) | .000000 | .000000 |
| 13) | .000000 | .000000 |
| 14) | .000000 | .000000 |
| 15) | .000000 | .000000 |
| 16) | .000000 | .000000 |
| 17) | 1.000000 | .000000 |
| 18) | .000000 | .000000 |
| 19) | .000000 | .000000 |
| 20) | .000000 | .000000 |
| 21) | .000000 | .000000 |
| 22) | .000000 | .000000 |
| 23) | .000000 | .000000 |
| 24) | .000000 | .000000 |
| 25) | .000000 | .000000 |
| 26) | .000000 | .000000 |
| 27) | .000000 | .000000 |
| 28) | .000000 | .000000 |
| 29) | 1.000000 | .000000 |
| 30) | .000000 | .000000 |
| 31) | .000000 | .000000 |
| 32) | 1.000000 | .000000 |
| 33) | .000000 | .000000 |
| 34) | .000000 | .000000 |

```
35)        .000000      .000000
36)        .000000      .000000
37)        .000000      .000000
38)        .000000      .000000
39)        .000000      .000000
40)       1.000000      .000000
41)        .000000      .000000
42)        .000000      .000000
43)        .000000      .000000
44)        .000000      .000000
45)        .000000      .000000
46)        .000000      .000000
47)        .000000      .000000
48)        .000000      .000000
49)       1.000000      .000000
50)        .000000      .000000
51)        .000000      .000000
52)        .000000      .000000

NO. ITERATIONS= 46
BRANCHES= 0 DETERM.= -1.000E 0
```

# Chapter 3

# A Revised Branch-and-Bound Approach for Inferring a Boolean Function from Examples

## 3.1 Some Background Information

This chapter discusses a revised branch-and-bound (B&B) algorithm for inferring a single clause (in CNF or DNF) from two disjoint sets of binary training examples. This algorithm is an extension of the B&B algorithm described in the previous chapter. Now the states of the search space are described by using more information and this seems to be critical in leading to good search results faster. This chapter is based on the developments first presented in [Triantaphyllou, 1994].

This chapter is organized as follows. The next section describes the revised B&B algorithm. Besides the CNF version, it also describes a version of it in which the inferred clauses are in DNF. We present some extensive computational results that indicate that the revised B&B algorithm has good performance characteristics. The chapter ends with a brief conclusions section.

## 3.2 The Revised Branch-and-Bound Algorithm

The revised B&B algorithm will also be demonstrated on the examples presented in Chapter 2 (Section 2.9). These examples are repeated here as follows:

$$E^+ = \begin{bmatrix} 0 & 1 & 0 & 0 \\ 1 & 1 & 0 & 0 \\ 0 & 0 & 1 & 1 \\ 1 & 0 & 0 & 1 \end{bmatrix} \quad \text{and} \quad E^- = \begin{bmatrix} 1 & 0 & 1 & 0 \\ 0 & 0 & 0 & 1 \\ 1 & 1 & 1 & 1 \\ 0 & 0 & 0 & 0 \\ 1 & 0 & 0 & 0 \\ 1 & 1 & 1 & 0 \end{bmatrix}.$$

The above positive and negative examples are numbered as in Section 2.9. First, it will be shown how the algorithm can derive a single CNF clause (i.e., a single disjunction). Next, the basic algorithm will be modified to derive a single DNF clause (i.e., a conjunction). Please recall that according to the OCAT approach (Step 2 in

E. Triantaphyllou, *Data Mining and Knowledge Discovery via Logic-Based Methods*, Springer Optimization and Its Applications 43, DOI 10.1007/978-1-4419-1630-3_3,

**Table 3.1.** The $NEG(A_k)$ Sets for the Illustrative Example.

| Attribute | Set of Negative Examples | Attribute | Set of Negative Examples |
|---|---|---|---|
| $A_1$ | $NEG(A_1) = \{1, 3, 5, 6\}$ | $\bar{A}_1$ | $NEG(\bar{A}_1) = \{2, 4\}$ |
| $A_2$ | $NEG(A_2) = \{3, 6\}$ | $\bar{A}_2$ | $NEG(\bar{A}_2) = \{1, 2, 4, 5\}$ |
| $A_3$ | $NEG(A_3) = \{1, 3, 6\}$ | $\bar{A}_3$ | $NEG(\bar{A}_3) = \{2, 4, 5\}$ |
| $A_4$ | $NEG(A_4) = \{2, 3\}$ | $\bar{A}_4$ | $NEG(\bar{A}_4) = \{1, 4, 5, 6\}$ |

**Table 3.2.** The $POS(A_k)$ Sets for the Illustrative Example.

| Attribute | Set of Positive Examples | Attribute | Set of Positive Examples |
|---|---|---|---|
| $A_1$ | $POS(A_1) = \{2, 4\}$ | $\bar{A}_1$ | $POS(\bar{A}_1) = \{1, 3\}$ |
| $A_2$ | $POS(A_2) = \{1, 2\}$ | $\bar{A}_2$ | $POS(\bar{A}_2) = \{3, 4\}$ |
| $A_3$ | $POS(A_3) = \{3\}$ | $\bar{A}_3$ | $POS(\bar{A}_3) = \{1, 2, 4\}$ |
| $A_4$ | $POS(A_4) = \{3, 4\}$ | $\bar{A}_4$ | $POS(\bar{A}_4) = \{1, 2\}$ |

Figure 2.1) for the CNF case, the requirement is for the clause to accept all the positive examples, while rejecting as many negative examples as possible.

### 3.2.1 Generating a Single CNF Clause

Define as $POS(A_k)$ the set of the positive examples which are accepted by a CNF clause when the attribute $A_k$ is included in that clause. The new B&B algorithm also uses the concepts of the $NEG(A_k)$ and *ATTRIBUTES(v)* sets, as they were defined in the previous chapter and they are repeated again for the current data as Table 3.1. The $POS(A_k)$ sets for the current illustrative example are presented in Table 3.2.

Now a typical search state is described in terms of *two sets* (as opposed to only one set as was the case with the B&B algorithm presented in the previous chapter). The first set refers to the *positive examples* which are accepted by the attributes which correspond to the arcs which connect that state (node) with the root node. Similarly, the second set refers to the *negative examples* which are accepted by the attributes which correspond to the arcs which connect that state with the root node. Suppose that we are at state $S_i = [P_i, N_i]$ (where $P_i$, $N_i$ correspond to the previous two sets of positive and negative examples, respectively). Now assume that the search considers the state (node) which is derived by following the arc which corresponds to the attribute $A_k$. Then, the new state is $S_j = [P_j, N_j]$, where the new sets $P_j$ and $N_j$ are defined as follows:

$$P_j = P_i \cup POS(A_k) \quad \text{and} \quad N_j = N_i \cup NEG(A_k).$$

Therefore, the search continues until terminal states are reached. A state $S_i = [P_i, N_i]$ is a *terminal state* if and only if the set $P_i$ refers to *all positive* examples, that is, if and only if $P_i = \{1, 2, 3, \ldots, m_1\}$, or equivalently, if and only if $P_i = E^+$. Apparently, a terminal state with a *minimum cardinality* of the set $N_i$ is *optimal* (in the OCAT sense). In the light of the previous considerations, the central problem to be solved by the revised B&B search can be summarized as follows (where $a_i$ is either $A_i$ or $\bar{A}_i$):

*Find a set of attributes S such that the following two conditions are true:*

$$\left| \bigcup_{a_i \in S} NEG(a_i) \right| = minimum$$

and

$$\bigcup_{a_i \in S} POS(a_i) = E^+.$$

In other words, the above problem statement calls for the search (by whatever means such a search may be implemented) to identify a state which corresponds to a set of attributes (given as set $S$). These attributes are used to form a single clause in CNF. Then this clause would accept all the positive examples (this is indicated by the second condition which calls for the union of the $POS(a_i)$ sets to be equal to the set of positive examples $E^+$), while the same clause would also reject as many of the negative examples as possible. The later requirement is indicated by the need to have the union of the $NEG(a_i)$ sets of minimum cardinality (i.e., with as few elements as possible). Then, the general form of such a clause in CNF (i.e., a single disjunction) would be as follows:

$$\left( \bigvee_{a_i \in S} a_i \right).$$

Given the above definitions some useful derivations are possible. We say that a state $S_i$ *absorbs* another state $S_j$ if by expanding the state $S_j$ we cannot reach any better terminal state than the ones derived by expanding the state $S_i$. In such a case we call state $S_j$ an *absorbed state*. From the previous considerations it becomes obvious that once a state can be identified to be an absorbed state, then it can be dropped from further consideration. Then the following two theorems are applicable only when a CNF clause is to be generated and they provide some conditions for identifying absorbed states. The proofs of these two theorems [Triantaphyllou, 1994] follow directly from the previous definitions and discussion.

**Theorem 3.1.** *A state $S_i = [P_i, N_i]$* **absorbs** *a state $S_j = [P_j, N_j]$ if and only if the following two conditions are true: $P_j \subseteq P_i$ and $N_i \subseteq N_j$.*

**Theorem 3.2.** *Suppose that state $S_i = [P_i, N_i]$ is a* **terminal** *state. Then, any state $S_j = [P_j, N_j]$, such that $|N_j| \geq |N_i|$, is* **absorbed** *by state $S_i$.*

The search tree for the current illustrative example is depicted in Figure 3.1. Observe that the arcs starting from a given node (search state) *are not* in the order $A_1, A_2, \ldots, A_n, \bar{A}_1, \bar{A}_2, \ldots, \bar{A}_n$ as was the case with the original B&B search but, instead, now they are *ranked*. They are ranked in terms of two criteria as follows.

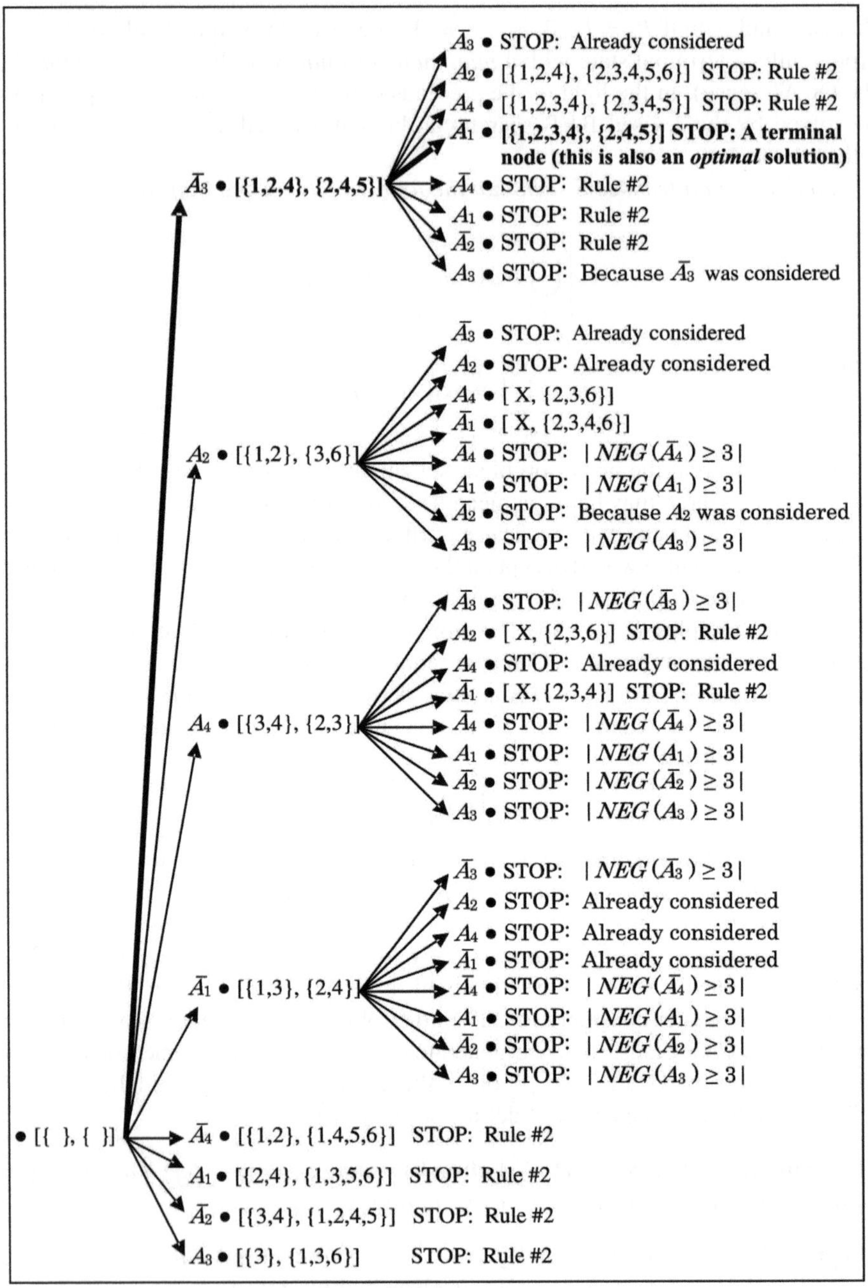


**Figure 3.1.** The Search Tree for the Revised Branch-and-Bound Approach.

The first criterion is to rank the attributes in descending order of the size of the corresponding $POS(A_k)$ set (in Table 3.2). If there is a tie, then they are ranked in ascending order in terms of the size of the corresponding $NEG(A_k)$ set (in Table 3.1). Therefore, the resulting ranking is as follows: $\bar{A}_3$, $A_2$, $A_4$, $\bar{A}_1$, $\bar{A}_4$, $A_1$, $\bar{A}_2$, $A_3$. In this way, it is more likely to reach terminal states quickly, and thus the fathoming test of Theorem 3.1 can be utilized more frequently.

This search tree indicates that fathoming of states may occur very often. In this figure, Rule #2 refers to the application of Theorem 3.2. For instance, the state $[\{1, 2\}, \{1, 4, 5, 6\}]$ (i.e., the fifth *state* in the second *stage*) does not need to be expanded. That is, it is an absorbed state because there is a terminal state whose size of its $N_i$ set is equal to 3 (this absorbed state has a corresponding value of 4). As terminal state we consider the fourth state from the top of the third *stage*. That is, state $[\{1, 2, 3, 4\}, \{2, 4, 5\}]$. At this point it is not known that this is also an optimal state (solution). Since this search process was successful in determining a terminal state very early, new states are generated by considering *at first* their $N_i$ sets. If the $N_i$ sets have a size greater or equal to 3, then they are fathomed.

This is the reason why their $P_i$ sets *do not even need* to be considered. For this reason they are indicated with the symbol **X** in Figure 3.1. It is interesting to observe that any arc $A_k$ for which the size of the $NEG(A_k)$ set is greater or equal to 3 necessarily leads to an absorbed state. Therefore, all states derived by such arcs are fathomed. Now consider the state $[\{1, 2\}, \{3, 6\}]$. This state was created by following the arc of attribute $A_2$ from the root state. If we follow the arc which corresponds to the negation of $A_2$ (i.e., arc $\bar{A}_2$) on any of its descendent states, then the new state would correspond to *both* attributes $A_2$ and $\bar{A}_2$ (possibly among other attributes). That is, the partial clause would include the subexpression $(A_2 \vee \bar{A}_2)$. However, if that happens, then the new state would accept all the negative (and also all the positive) examples. Clearly, such a clause would have no discriminatory power and thus would be totally useless. This is the reason why the seventh child state of $[\{1, 2\}, \{3, 6\}]$ is fathomed in the B&B search. In this illustrative example Theorem 3.1 was never utilized.

Since the state $[\{1, 2, 3, 4\}, \{2, 4, 5\}]$ is the only unfathomed terminal state, this is also an optimal one. Thus, the derived CNF clause is $(\bar{A}_3 \vee \bar{A}_1)$. This clause accepts all the positive examples and also accepts the negative examples $\{2, 4, 5\}$ (or, equivalently, it rejects the negative examples $\{1, 3, 6\}$).

From the previous considerations it follows that there is a great advantage to reach terminal nodes early in the search process. In this way, the minimum size of their $N_i$ sets can be used to effectively fathom search states. This situation suggests the application of the B&B search in *two search phases* (as was the case with the original B&B search discussed in Chapter 2). During the first phase only a very small number (say, 10) of active states is allowed. If there are more than 10 active states, then they are ranked according to their $P_i$ and $N_i$ sizes (i.e., in a manner similar to the ranking of the attributes). In this way, the states with the highest potential of being optimal are kept in memory. Recall that this is the principle of *beam search* in artificial intelligence (see, for instance, [Dietterich and Michalski, 1983]). At the end of phase one, a terminal state of small cardinality becomes available. Next, phase two is initiated. During the second phase a larger number (say, 50) of active states is allowed. However, states now can be fathomed more frequently because the size of a small $N_i$ set of a terminal state is known. Also note that the new B&B search does not have to follow a fixed number of stages (as was the case with the original B&B search as discussed in Chapter 2).

An important issue with the previous two phases is to be able to decide when a terminal state is optimal (in the OCAT sense). As was mentioned above, memory limitations may force the search to *drop* states which are *not* absorbed by any other state. Therefore, *there is a possibility to drop a state which could have led to an optimal state (and thus to an optimal clause).*

Suppose that $L$ nonabsorbed states had to be dropped because of memory limitations. Let $K_1, K_2, K_3, \ldots, K_L$ represent the cardinalities of their corresponding $N_i$ sets. Next, let us define the quantity $K_{\text{MIN}}$ as the minimum of the previous $L$ numbers. Similarly, suppose that the B&B search has identified $N$ terminal states. Let $Y_1, Y_2, Y_3, \ldots, Y_N$ represent the cardinalities of their corresponding $N_i$ sets. Also, define as $Y_{\text{MIN}}$ the minimum of the previous $N$ cardinalities. Then, the previous considerations lead to the proof of the following theorem [Triantaphyllou, 1994] which states a condition for establishing optimality.

**Theorem 3.3.** *A terminal state* $S_i = [P_i, N_i]$ *is also an* **optimal state** *if the following two conditions are true:* $|N_i| = Y_{\text{MIN}}$ *and* $K_{\text{MIN}} \geq Y_{\text{MIN}}$.

Note that this theorem can be applied after each of the two search phases. Obviously, if it is applicable after the first phase, then the second phase does not need to be initiated. The following lemma states a fact when optimality is not provable.

**Lemma 3.1.** *Suppose that the following relation is true:* $K_{\text{MIN}} < Y_{\text{MIN}}$. *Then, an optimal clause accepts no less than* $K_{\text{MIN}}$ *negative examples.*

This lemma indicates that if optimality cannot be proven, then it is still possible to establish a *lower limit* on the number of negative examples which can be accepted by an optimal clause (or, equivalently, an upper limit on the number of negative examples which can be *rejected* by an optimal clause).

Finally, it should be stated here that the CNF version of the B&B algorithm was implemented in FORTRAN and used in the computational experiments described in Section 3.3. The next section briefly describes a modification to this B&B approach which can derive DNF clauses. An alternative approach to deriving DNF systems, which is based on some data transformations, is discussed in Chapter 7.

### 3.2.2 Generating a Single DNF Clause

The previously described B&B search can be easily modified to generate DNF clauses (i.e., conjunctions). Please recall that the requirements now are as follows. Every positive example should be accepted by *at least one* of the conjunctions (DNF clauses). Also, every negative example should be rejected by *all* the conjunctions. Therefore, the DNF clause proposed by the B&B approach during a single OCAT iteration should reject all the negative examples and accept as many positive examples as possible.

The search states are defined in a manner analogous to that of the CNF implementation. Each state is described in terms of two sets. The first set (denoted by $P_i$) refers to the *positive examples* which are accepted by the attributes which correspond to the arcs which connect that state (node) with the root node. Similarly, the second set (denoted by $N_i$) refers to the *negative examples* which are accepted by the attributes which correspond to the arcs which connect that state with the root node. Suppose that we are at state $S_i = [P_i, N_i]$. Now assume that the search considers the state (node) which is derived by following the arch which corresponds to the attribute $A_k$. Then, the new state is: $S_j = [P_j, N_j]$, where the new sets $P_j$ and $N_j$ are defined as follows:

$$P_j = P_i \cap POS(A_k), \quad \text{and}$$
$$N_j = N_i \cap NEG(A_k).$$

In other words, instead of the union operator now there is the *intersection* operator for sets. A state $S_i = [P_i, N_i]$ is a *terminal state* if and only if the set $N_i$ is equal to the *empty set* (i.e., $N_i = \varnothing$). That is, a state is terminal if and only if it rejects all the negative examples. A terminal state with the *maximum cardinality* of the set $P_i$ is *optimal* (in the OCAT sense). The concept of *absorbed* states is the same as in the CNF case. However, in the light of the previous definitions Theorem 3.2 takes the following form:

**Theorem 3.4.** *Suppose that state $S_i = [P_i, N_i]$ is a* **terminal** *state. Then, any state $S_j = [P_j, N_j]$, such that $|P_j| \leq |P_i|$, is* **absorbed** *by state $S_i$.*

The attributes in the arcs now are ranked in an analogous manner as in the previous section. The first criterion is to rank the attributes in ascending (instead of descending) order of the size of the corresponding $NEG(A_k)$ set. If there is a tie, then they are ranked in descending (instead of ascending) order in terms of the size of the corresponding $POS(A_k)$ set.

As was the case with the CNF version of the B&B algorithm, suppose that $L$ nonabsorbed states had to be dropped because of memory limitations. Let $K_1, K_2, K_3, \ldots, K_L$ represent the cardinalities of their corresponding $P_i$ sets (note that earlier we had considered the $N_i$ sets). Next, define the quantity $K_{\text{MAX}}$ as the *maximum* of the previous $L$ numbers (before we had considered the minimum values). Similarly, suppose that the B&B search has identified $N$ terminal states. Let $Y_1, Y_2, Y_3, \ldots, Y_N$ represent the cardinalities of their corresponding $P_i$ sets. Define as $Y_{\text{MAX}}$ the *maximum* of the previous $N$ cardinalities. Then, the previous considerations lead to the proof of a new theorem, which is analogous to Theorem 3.4, regarding optimality.

**Theorem 3.5.** *A terminal state $S_i = [P_i, N_i]$ is also an* **optimal state** *if the following two conditions are true: $|P_i| = Y_{\text{MAX}}$ and $K_{\text{MAX}} \leq Y_{\text{MAX}}$.*

### 3.2.3 Some Computational Results

Several computational experiments were conducted in order to gain a better understanding of the performance of the revised branch-and-bound approach. These experiments were conducted in a manner similar to the experiments reported in [Kamath, *et al.*, 1992]. At first, a Boolean function was chosen to be the *hidden logic/system.* This system is *hidden* in the sense that we try to infer it from limited sets of positive and negative examples. The Boolean functions used as *hidden logic/system* are exactly the same as the ones also used in [Kamath, *et al.*, 1992]. These are the 15 Boolean expressions depicted in Table 3.3. The solution statistics are shown in Table 3.4. The notation of the Problem ID column in Table 3.4 is derived from the notation in Table 3.3 where the second index denotes the difference in the number of the input training examples. For instance, in Table 3.4 the problem IDs which are equal to 8A1, 8A2, 8A3, and 8A4 indicate that the *hidden system/logic* is system 8A (as defined in Table 3.3) with 10, 25, 50, and 100 training examples, respectively. A similar interpretation holds for the rest of the problem IDs.

Once a *hidden logic/system* is selected, a number of random training examples were generated. Each random example was classified according to the *hidden logic* as either positive or negative. The random examples used in these experiments are identical to the ones used in [Kamath, *et al.*, 1992]. After the sets of the $E^+$ and $E^-$ examples were generated this way, the OCAT approach with the CNF version of the revised B&B algorithm was applied.

Since the SAT approach had derived Boolean functions in DNF, a simple transformation on the data was first performed in order the previous updated B&B approach would yield DNF functions (and not CNF). The transformation is to first derive the complements of the data. That is, every 0 becomes 1 and vice versa. Next, the complemented positive examples are treated as the *negative* examples, while the complemented negative examples are treated as the *positive* examples. More on this data transformation and some other related issues can be found in Chapter 7 of this book.

We also studied some lower bounds related to the number of clauses regarding the derivation of compact Boolean functions. The details of the approach for deriving such lower bounds are described in Chapter 8 of this book. That chapter describes how one can partition a problem of inferring a Boolean function from training data into smaller problems by first constructing a special graph, called the *Rejectability Graph* (or *R-Graph*) from the two sets of examples. Building such a graph is independent of the inference algorithm used and its analysis provides many insights into the two sets of the training examples. This is how the lower limits on the number of clauses reported in Table 3.4 were calculated.

When a Boolean function was inferred, it was compared in terms of 10,000 random examples with the *hidden logic/system.* That is, 10,000 random examples were generated and then they were classified according to these two Boolean functions. The percentage of the times the two Boolean functions agreed, was reported as the *accuracy* of the inferred Boolean function.

**Table 3.3.** Description of the Boolean Functions Used as Hidden Logic/System in the Computational Experiments.

| System ID | System Description | System ID | System Description |
|---|---|---|---|
| 8A | $(A_4 \vee \bar{A}_7) \wedge (\bar{A}_3 \vee A_4) \wedge (A_1 \vee A_2 \vee \bar{A}_6)$ | 16D | $(\bar{A}_5 \vee \bar{A}_8 \vee \bar{A}_{10} \vee A_{16}) \wedge (\bar{A}_2 \vee \bar{A}_{12} \vee \bar{A}_{16}) \wedge (\bar{A}_1 \vee \bar{A}_{12}) \wedge (A_3 \vee \bar{A}_5 \vee A_6)$ |
| 8B | $(\bar{A}_1 \vee \bar{A}_4 \vee A_6) \wedge (A_2) \wedge (\bar{A}_2 \vee A_8)$ | 16E | $(A_1 \vee \bar{A}_2 \vee A_3 \vee \bar{A}_4) \wedge (A_5 \vee \bar{A}_6 \vee \bar{A}_7 \vee A_8) \wedge (A_9 \vee \bar{A}_{10} \vee \bar{A}_{11} \vee \bar{A}_{12}) \wedge (\bar{A}_{13} \vee A_{14} \vee \bar{A}_{15} \vee A_{16})$ |
| 8C | $(A_5) \wedge (A_6 \vee \bar{A}_8) \wedge (A_7)$ | 32A | $(A_1 \vee \bar{A}_{12}) \wedge (A_2 \vee \bar{A}_5 \vee A_{32}) \wedge (A_{19} \vee \bar{A}_{23} \vee A_{26})$ |
| 8D | $(\bar{A}_6) \wedge (\bar{A}_2) \wedge (\bar{A}_3 \vee \bar{A}_7)$ | 32B | $(A_1 \vee A_2 \vee \bar{A}_9 \vee \bar{A}_{12} \vee A_{31}) \wedge (A_{19} \vee \bar{A}_{23} \vee A_{26}) \wedge (A_2 \vee \bar{A}_5 \vee \bar{A}_{20} \vee A_{32})$ |
| 8E | $(A_8) \wedge (A_2 \vee A_5) \wedge (\bar{A}_3 \vee A_5)$ | 32C | $(A_1 \vee A_2 \vee \bar{A}_9 \vee \bar{A}_{12} \vee A_{31}) \wedge (A_2 \vee \bar{A}_{20} \vee A_{32}) \wedge (A_1 \vee A_2 \vee A_{19} \vee \bar{A}_{23} \vee A_{26})$ |
| 16A | $(A_1 \vee \bar{A}_{12}) \wedge (A_2 \vee A_3 \vee \bar{A}_5)$ | 32D | $(A_4 \vee A_{11} \vee \bar{A}_{22}) \wedge (A_2 \vee A_{12} \vee \bar{A}_{15} \vee \bar{A}_{29}) \wedge (\bar{A}_3 \vee A_9 \vee A_{20}) \wedge (\bar{A}_{10} \vee A_{11} \vee \bar{A}_{29} \vee A_{32})$ |
| 16B | $(A_3 \vee A_{12} \vee A_{15}) \wedge (\bar{A}_3 \vee \bar{A}_{11}) \wedge (\bar{A}_2 \vee \bar{A}_{10} \vee \bar{A}_{16}) \wedge (A_1 \vee A_2)$ | 32E | $(A_9 \vee A_{10} \vee A_{23}) \wedge (A_2 \vee A_{29} \vee \bar{A}_{31}) \wedge (A_2 \vee \bar{A}_4 \vee A_6 \vee \bar{A}_7 \vee A_{19} \vee \bar{A}_{32})$ |
| 16C | $(A_4 \vee \bar{A}_7 \vee A_{11}) \wedge (A_4 \vee A_{10} \vee A_{14}) \wedge (\bar{A}_9 \vee \bar{A}_{14} \vee A_{15}) \wedge (\bar{A}_3 \vee A_8)$ | | |

For the case of the systems which were defined on 8 attributes (i.e., systems 8A1, 8A2, 8A3, . . ., 8E2), all possible 256 examples were considered. The active list in the B&B search contained up to 10 search states at any time. However, this restriction did not prevent the search from reaching optimal solutions in any OCAT iteration. It should be stated here that only the first search (i.e., only one phase) was enough to derive an optimal clause.

Table 3.4, with the computational results, also depicts the results derived by using the SAT approach. The SAT results were derived by using a VAX 8700 computer running 10th Edition UNIX. The code for the SAT tests was written in the FORTRAN and C programming languages. The SAT results were originally

**Table 3.4.** Solution Statistics.

| Problem | | SAT Solution Characteristics | | | | OCAT Solution Characteristics | | | | |
|---|---|---|---|---|---|---|---|---|---|---|
| **Problem ID** | **No. of Examples** | ***k* value** | **Clauses** | **CPU time** | **Accuracy** | **Clauses** | **Lower Limit** | **CPU time** | **Accuracy** | **CPU time comparison with SAT** |
| 8A1 | 10 | 3 | 3 | 0.42 | 0.86 | 3 | 2 | 0.004 | 0.81 | 105 |
| 8A2 | 25 | 6 | 3 | 21.37 | 0.87 | 3 | 2 | 0.004 | 0.86 | 5,343 |
| 8A3 | 50 | 6 | 3 | 29.77 | 0.92 | 3 | 3 | 0.007 | 0.92 | 4,253 |
| 8A4 | 100 | 6 | 3 | 9.33 | 1.00 | 3 | 3 | 0.015 | 1.00 | 622 |
| 8B1 | 50 | 3 | 2 | 2.05 | 0.97 | 2 | 2 | 0.002 | 0.97 | 1,025 |
| 8B2 | 100 | 6 | 3 | 97.07 | 0.97 | 3 | 3 | 0.006 | 0.99 | 16,178 |
| 8B3 | 150 | 10 | 3 | 167.07 | 0.96 | 3 | 3 | 0.007 | 0.96 | 23,867 |
| 8B4 | 200 | 6 | 3 | 122.62 | 1.00 | 3 | 3 | 0.009 | 0.97 | 13,624 |
| 8C1 | 50 | 10 | 2 | 8.02 | 0.94 | 2 | 2 | 0.002 | 0.94 | 4,010 |
| 8C2 | 100 | 10 | 3 | 84.8 | 1.00 | 3 | 3 | 0.005 | 0.94 | 16,960 |
| 8D1 | 50 | 10 | 3 | 116.98 | 0.94 | 3 | 3 | 0.004 | 1.00 | 29,245 |
| 8D2 | 100 | 10 | 3 | 45.8 | 1.00 | 3 | 3 | 0.006 | 1.00 | 7,620 |
| 8E1 | 50 | 10 | 3 | 122.82 | 1.00 | 3 | 3 | 0.005 | 1.00 | 24,564 |
| 8E2 | 100 | 10 | 3 | 16.68 | 1.00 | 3 | 3 | 0.007 | 1.00 | 2,383 |
| 16A1 | 100 | 15 | 4 | 2,038.55 | 1.00 | 4 | 3 | 0.065 | 1.00 | 31,362 |
| 16A2 | 300 | 6 | 4 | 607.8 | 1.00 | 4 | 4 | 0.134 | 1.00 | 4,536 |
| 16B1 | 200 | 8 | 5 | 78.27 | 0.99 | 4 | 4 | 0.402 | 1.00 | 195 |
| 16B2 | 400 | 4 | 4 | 236.07 | 1.00 | 4 | 4 | 0.284 | 1.00 | 831 |

**Table 3.4.** Continued.

| | | | | | | | | | |
|---|---|---|---|---|---|---|---|---|---|
| 16C1 | 100 | 20 | 5 | 757.55 | 0.87 | 4 | 4 | 0.314 | 1.00 | 2,413 |
| 16C2 | 400 | 4 | 4 | 520.6 | 1.00 | 4 | 4 | 0.477 | 1.00 | 1,091 |
| 16D1 | 200 | 10 | 4 | 1,546.78 | 1.00 | 4 | 4 | 0.089 | 1.00 | 17,380 |
| 16D2 | 400 | 4 | 4 | 544.25 | 1.00 | 4 | 4 | 0.105 | 1.00 | 5,183 |
| 16E1 | 200 | 15 | 5 | 2,156.42 | 0.99 | 5 | 4 | 1.545 | 1.00 | 1,396 |
| 16E2 | 400 | 4 | 4 | 375.83 | 1.00 | 4 | 4 | 2.531 | 1.00 | 149 |
| 32A1 | 250 | 3 | 3 | 176.73 | 1.00 | 3 | 3 | 0.134 | 1.00 | 1,319 |
| 32B1 | 50 | 3 | 3 | 5.02 | 0.83 | 2 | 1 | 0.044 | 0.93 | 114 |
| 32B2 | 100 | 3 | 3 | 56.65 | 0.96 | 3 | 2 | 0.058 | 0.96 | 977 |
| 32B3 | 250 | 3 | 3 | 189.8 | 0.97 | 3 | 2 | 0.143 | 0.97 | 1,327 |
| 32B4 | 300 | 3 | 3 | 259.43 | 1.00 | 3 | 2 | 0.198 | 0.99 | 1,310 |
| 32C1 | 50 | 3 | 3 | 23.85 | 0.80 | 2 | 1 | 0.019 | 0.92 | 1,255 |
| 32C2 | 100 | 3 | 3 | 9.38 | 0.88 | 2 | 1 | 0.02 | 0.91 | 469 |
| 32C3 | 150 | 3 | 3 | 14.27 | 0.92 | 3 | 1 | 1.969 | 0.91 | 7 |
| 32C4 | 1000 | 3 | 3 | 154.62 | 1.00 | 3 | 3 | 0.654 | 1.00 | 236 |
| 32D1 | 50 | 4 | 4 | 65.65 | 0.74 | 3 | 1 | 0.197 | 0.81 | 333 |
| 32D2 | 100 | 4 | 4 | 178.1 | 0.91 | 3 | 2 | 0.308 | 0.92 | 578 |
| 32D3 | 400 | 4 | 4 | 1,227.40 | 1.00 | 4 | 2 | 1.103 | 1.00 | 1,113 |
| 32E1 | 50 | 3 | 2 | 8.33 | 0.86 | 2 | 1 | 0.015 | 0.83 | 555 |
| 32E2 | 100 | 3 | 3 | 9.67 | 0.97 | 2 | 1 | 0.096 | 0.99 | 101 |
| 32E3 | 200 | 3 | 3 | 132.83 | 0.98 | 3 | 2 | 0.105 | 0.98 | 1,265 |
| 32E4 | 300 | 3 | 3 | 276.93 | 0.98 | 3 | 2 | 0.209 | 0.99 | 1,325 |
| 32E5 | 400 | 3 | 3 | 390.22 | 0.98 | 3 | 2 | 0.184 | 0.98 | 2,121 |

reported in [Kamath, *et al.*, 1992]. The OCAT results were derived by using an IBM 3090-600S computer (the supercomputer at Penn State in the 1980s and 1990s) and the code was written in FORTRAN. As can be seen from this table the OCAT approach outperformed the SAT approach in an order of many times. If we exclude the case of the test with ID 32C3, then for all the other cases OCAT was 101 to 31,362 times faster than the SAT approach. On the average, OCAT was 5,579 times faster.

One may observe at this point that a direct comparison would have required that the computer codes for the SAT (which was written in FORTRAN and C) formulation and the OCAT approach (which was written in FORTRAN) had to be written in the same programming language. However, this was impractical at the time of those studies. Thus, the current results can convey only a flavor on the relative performance of the two methods, and by no means should be considered as a direct comparison of the two approaches.

In terms of the accuracy rate, both the SAT and OCAT approaches performed considerably well. It should also be stated here that when the lower limit is equal to the number of clauses derived by OCAT, then we can conclude that OCAT derived a minimum size (in terms of the number of clauses) function. Note that this situation occurred in these experiments 51.2% of the time. However, if the lower limit is less than the number of clauses, then this *does not necessarily imply* that OCAT failed to derive a minimum size function. More on this bound on the number of inferred clauses is discussed in Chapter 8 of this book.

It should be recalled that if optimality (i.e., the minimum number of inferred clauses) is not proven in the OCAT case, then the SAT approach can be applied with *successively decreasing* $k$ values. When an infeasible SAT problem is reached, the conclusion is that the *last feasible* SAT solution yielded a Boolean function with an *optimal* (i.e., minimum) number of clauses.

Finally, it is interesting to emphasize here that in these computational experiments it was found that *most of the time* the OCAT approach derived a minimum size system. Therefore, it is anticipated that under the proposed strategy the SAT approach (which is very CPU time consuming) does not have to be used very often. Optimality (i.e., the minimum number of clauses) can be checked by comparing the number of clauses derived by OCAT with the lower limit established via the graph theoretic approach given as Theorem 8.3 in Section 8.2.2 of this book. Actually, Table 8.1 is complementary to Table 3.4.

Table 3.5 presents the results of solving some *large problems*. In these tests the number of attributes is equal to 32, the total number of examples is equal to 1,000, and each *hidden logic/system* was assumed to have 30 clauses. The strategy followed in these experiments is the same as in the previous tests. The numbers of positive and negative examples are shown as well.

Next, Table 3.6 presents the exact structure of the *hidden* and inferred Boolean functions of the first of these test problems (i.e., for problem 32H1). In Table 3.6 only the indexes of the attributes are depicted (in order to save space). For instance,

**Table 3.5.** Solution Statistics of Some Large Test Problems (the Number of Attributes Is Equal to 32).

| Problem Characteristics | | | | | OCAT Solution Characteristics | | | |
|---|---|---|---|---|---|---|---|---|
| **Problem ID** | **Clauses** | **Total No. of Examples** | **$\|E^+\|$** | **$\|E^-\|$** | **CPU Time** | **Clauses** | **Lower Limit** | **Accuracy** |
| 32H1 | 30 | 1,000 | 943 | 57 | 135.71 | 17 | 3 | 84.13% |
| 32H2 | 30 | 1,000 | 820 | 180 | 45.18 | 10 | 3 | 93.83% |
| 32H3 | 30 | 1,000 | 918 | 18 | 175.50 | 7 | 2 | 95.85% |
| 32H4 | 30 | 1,000 | 944 | 56 | 64.16 | 20 | 2 | 82.84% |
| 32H5 | 30 | 1,000 | 988 | 12 | 13.41 | 5 | 2 | 97.83% |

the first clause of the *inferred system* is represented by the list $[13, 15, 25, -6, -19, -30, -32]$ which implies the CNF clause:

$$(A_{13} \vee A_{15} \vee A_{25} \vee \bar{A}_6 \vee \bar{A}_{19} \vee \bar{A}_{30} \vee \bar{A}_{32}).$$

It can be observed that now the CPU times are considerably (with the OCAT standards) higher. However, relatively speaking these times are still kept in low levels. The lower limit, however, is not tight enough. Furthermore, determining the maximum clique of the complemented rejectability graph (as discussed in Chapter 8) took considerably more time than determining the inferred system.

It should also be stated here that the *hidden* Boolean functions were not defined in terms of a minimum representation. That is, it might be possible to represent an equivalent system with less than 30 clauses. The OCAT approach always returned, compared to the original *hidden* system, a very compact system.

Finally, the accuracy of the inferred Boolean function was rather high. The size of the population of all possible examples is $2^{32} = 4.29496 \times 10^9$. Out of these examples, the tests considered only 1,000 random inputs as the training data. This represents an extremely small sample of the actual population and, therefore, the corresponding accuracy values can be considered rather high. The computational results in Tables 3.2 and 3.5 suggest that the OCAT approach, when it is combined with the revised B&B algorithm, constitutes an efficient and effective strategy for inferring a Boolean function from examples.

## 3.3 Concluding Remarks

The results of the previous computational experiments suggest that the proposed OCAT approach, when combined with the revised B&B algorithm, provides a fast way for inferring logical clauses from positive and negative examples. It is interesting to observe that OCAT also derived systems for which very often it could be proved (by using the idea of the rejectability graph as discussed in detail in Chapter 8) to be

**Table 3.6.** Descriptions of the Boolean Function in the First Test Problem *(i.e., when ID = 32H1)*.

| The "Hidden Logic" System | | | | | | | | | | | | | | | | |
|---|---|---|---|---|---|---|---|---|---|---|---|---|---|---|---|---|
| CLAUSE: 1 = | 2 | 6 | 11 | 14 | 16 | 21 | 25 | 27 | −10 | −13 | −17 | −18 | −29 | | | |
| CLAUSE: 2 = | 15 | 17 | 19 | 26 | 31 | −4 | −8 | −20 | | | | | | | | |
| CLAUSE: 3 = | 24 | 25 | −1 | −10 | −17 | −18 | −20 | −26 | | | | | | | | |
| CLAUSE: 4 = | 1 | 9 | 11 | 17 | 32 | −10 | −15 | −16 | −21 | −22 | | | | | | |
| CLAUSE: 5 = | 4 | 8 | 14 | 15 | 17 | 18 | 24 | 32 | −11 | −16 | −21 | −22 | −28 | | | |
| CLAUSE: 6 = | 9 | 10 | 12 | 16 | 19 | −4 | −5 | −8 | −14 | −25 | −30 | −31 | | | | |
| CLAUSE: 7 = | 1 | 12 | −8 | −9 | −11 | −13 | −15 | −17 | −26 | −27 | −28 | −31 | | | | |
| CLAUSE: 8 = | 1 | 12 | 17 | 26 | −2 | −5 | −8 | −20 | −22 | −24 | −28 | −31 | | | | |
| CLAUSE: 9 = | 1 | 9 | 17 | 26 | −2 | −4 | −6 | −8 | | | | | | | | |
| CLAUSE: 10 = | 5 | 7 | 19 | −2 | −12 | −17 | −18 | −20 | −23 | −26 | −28 | | | | | |
| CLAUSE: 11 = | 20 | 25 | 31 | 32 | −2 | −3 | −8 | −12 | −13 | −15 | −22 | −28 | | | | |
| CLAUSE: 12 = | 13 | 25 | −6 | −19 | −30 | −32 | | | | | | | | | | |
| CLAUSE: 13 = | 11 | 19 | 21 | 23 | 24 | 27 | 31 | −4 | −8 | −9 | −12 | −13 | −15 | −26 | −28 | |
| CLAUSE: 14 = | 5 | 8 | 20 | 21 | 23 | −2 | −9 | −16 | | | | | | | | |
| CLAUSE: 15 = | 3 | 4 | 13 | 16 | −6 | −10 | −12 | −26 | −28 | −29 | −31 | | | | | |
| CLAUSE: 16 = | 4 | 17 | 19 | 24 | 26 | 32 | −1 | −5 | −10 | −13 | −18 | −21 | −28 | | | |
| CLAUSE: 17 = | 14 | 25 | 28 | −11 | −12 | −17 | −22 | −29 | | | | | | | | |
| CLAUSE: 18 = | 2 | 3 | 7 | 11 | 23 | 24 | 27 | 31 | 32 | −5 | −8 | −12 | −14 | −26 | −29 | |
| CLAUSE: 19 = | 2 | 12 | 22 | 29 | −8 | −18 | −27 | −30 | −31 | | | | | | | |
| CLAUSE: 20 = | 3 | 17 | 25 | −4 | −6 | −24 | −26 | −30 | | | | | | | | |
| CLAUSE: 21 = | 8 | 9 | 12 | 22 | 25 | −3 | −18 | −24 | −26 | −29 | | | | | | |
| CLAUSE: 22 = | 1 | 2 | 8 | 9 | 10 | 11 | 22 | 24 | 27 | 32 | −3 | −7 | −14 | −16 | −29 | −30 |
| CLAUSE: 23 = | 15 | 18 | 21 | 23 | −2 | −17 | | | | | | | | | | |
| CLAUSE: 24 = | 10 | 13 | 16 | 17 | 18 | −23 | −26 | −27 | | | | | | | | |
| CLAUSE: 25 = | 3 | 6 | 16 | 19 | 20 | 21 | 25 | −1 | −12 | −14 | −15 | −18 | −23 | −26 | −32 | |
| CLAUSE: 26 = | 1 | 2 | 8 | 9 | 13 | 20 | 27 | 28 | −5 | −16 | −23 | −25 | −26 | | | |
| CLAUSE: 27 = | 5 | 7 | 15 | 19 | 25 | 28 | −6 | −8 | −10 | −11 | | | | | | |
| CLAUSE: 28 = | 1 | 8 | 9 | 11 | 15 | 20 | 21 | 22 | 29 | −13 | −24 | −28 | −31 | −32 | | |
| CLAUSE: 29 = | 1 | 5 | 9 | 15 | 25 | 29 | −13 | −26 | −32 | | | | | | | |
| CLAUSE: 30 = | 17 | 18 | 22 | 23 | 30 | 32 | −6 | −8 | −9 | −10 | −12 | −24 | −27 | −29 | 31 | |

**Table 3.6.** Continued.

| The Inferred System | | | | | | | | | | | | |
|---|---|---|---|---|---|---|---|---|---|---|---|---|
| CLAUSE: 1 = | 13 | 15 | 25 | −6 | −19 | −30 | −32 | | | | | |
| CLAUSE: 2 = | 15 | 18 | 20 | 21 | 23 | 25 | −2 | −32 | | | | |
| CLAUSE: 3 = | 15 | 18 | 21 | 23 | 25 | −2 | −4 | −17 | −30 | | | |
| CLAUSE: 4 = | 5 | 14 | 15 | 21 | 25 | −1 | −2 | −8 | −10 | −28 | −32 | |
| CLAUSE: 5 = | 5 | 17 | 21 | 25 | −2 | −4 | −6 | −28 | −30 | −32 | | |
| CLAUSE: 6 = | 18 | 25 | 29 | −6 | −8 | −19 | −30 | −32 | | | | |
| CLAUSE: 7 = | 15 | 18 | 21 | 23 | −2 | −9 | −28 | | | | | |
| CLAUSE: 8 = | 1 | 9 | 17 | −2 | −4 | −6 | −8 | −27 | | | | |
| CLAUSE: 9 = | 2 | 5 | 11 | 15 | 17 | 29 | −23 | −26 | −27 | −28 | −32 | |
| CLAUSE: 10 = | 3 | 11 | 12 | 17 | 22 | 25 | 29 | −4 | −8 | −28 | | |
| CLAUSE: 11 = | 3 | 12 | 15 | 17 | 29 | −4 | −8 | −13 | −18 | −20 | −32 | |
| CLAUSE: 12 = | 10 | 12 | 25 | −14 | −23 | −24 | −26 | −28 | −32 | | | |
| CLAUSE: 13 = | 3 | 7 | 11 | 14 | 15 | 17 | 18 | 25 | 28 | −8 | −20 | −26 |
| CLAUSE: 14 = | 3 | 5 | 6 | 15 | 24 | 25 | 27 | −4 | −13 | −18 | | |
| CLAUSE: 15 = | 10 | 16 | 21 | 23 | 31 | −4 | −8 | −14 | −20 | −28 | | |
| CLAUSE: 16 = | 7 | 16 | 30 | −9 | −13 | −15 | −17 | −26 | −28 | −29 | | |
| CLAUSE: 17 = | 12 | 25 | 19 | | | | | | | | | |

of minimum size. Furthermore, the OCAT and the SAT approaches can be combined into a single strategy in order to efficiently derive a minimum size CNF and DNF system.

The high CPU time efficiency and effectiveness of the proposed combination of methods make it a practical approach for inferring clauses from examples. Future work could focus on inferring *Horn clauses* (and not just general CNF or DNF systems as is the case currently). Another interesting expansion of the B&B search is to apply these concepts on *partially defined* examples. That is, examples now are not defined in the domain $\{0, 1\}^n$, but instead in the domain $\{0, 1, *\}^n$, where "$*$" indicates an unknown value. In Chapter 4, a heuristic approach is presented which can process data with missing entries. However, a B&B approach that pursues the goal of inferring a minimal size CNF or DNF expression would be of interest here. Related here is also the discussion presented in Section 4.5 regarding the minimization of a weighted average of the three error rates which occur when one uses a Boolean function on new data of unknown class values. Another extension is to develop methods (such as B&B and/or heuristics) that would work directly on nonbinary data. As binarization usually results in a high number of binary variables, such methods may be faster if they work directly on nonbinary data.

The problem of learning a Boolean function from past experience (training examples) is the keystone in building truly intelligent systems and thus constitutes the keystone of data mining and knowledge discovery from data sets. Thus, more efficient decomposition and clause inference approaches are required in order to make learning feasible for large-scale applications. More research in this area has the potential of making more contributions in this vital area of data mining and knowledge discovery.

# Chapter 4

# Some Fast Heuristics for Inferring a Boolean Function from Examples

## 4.1 Some Background Information

The previous two chapters discussed the development and key mathematical properties of some branch-and-bound (B&B) approaches for inferring a Boolean function in the form of a compact (i.e., with as few clauses as possible) CNF or DNF expression from two collections of disjoint examples. As was described in Chapters 2 and 3, the B&B approaches may take a long time to run (actually, they are of exponential time complexity).

This chapter presents a simple heuristic approach which may take much less time (of polynomial time complexity) than the B&B approaches. This heuristic attempts to return a small (but not necessarily minimum or near minimum) number of clauses of the CNF or DNF expressions inferred from two disjoint collections of examples. Some variants of this heuristic are also discussed. This chapter is based on the developments first reported in [Deshpande and Triantaphyllou, 1998].

In this chapter we consider two closely related problems. The first one is how to infer a Boolean function fast from two disjoint collections of positive and negative examples. All the examples are again assumed to be binary vectors and we also assume that we know them precisely. The second problem considers cases in which we have *partial* knowledge of some of the examples.

In order to help fix ideas, consider the following two disjoint sets of positive and negative examples (which were also used in the previous two chapters):

$$E^{+} = \begin{bmatrix} 0 & 1 & 0 & 0 \\ 1 & 1 & 0 & 0 \\ 0 & 0 & 1 & 1 \\ 1 & 0 & 0 & 1 \end{bmatrix} \quad \text{and} \quad E^{-} = \begin{bmatrix} 1 & 0 & 1 & 0 \\ 0 & 0 & 0 & 1 \\ 1 & 1 & 1 & 1 \\ 0 & 0 & 0 & 0 \\ 1 & 0 & 0 & 0 \\ 1 & 1 & 1 & 0 \end{bmatrix}.$$

Given these data we want to derive a single Boolean function (in CNF or DNF form) that satisfies the requirements implied in the previous examples. For the CNF

E. Triantaphyllou, *Data Mining and Knowledge Discovery via Logic-Based Methods*,
Springer Optimization and Its Applications 43, DOI 10.1007/978-1-4419-1630-3_4,

case, we would like each clause (i.e., each disjunction of the CNF expression) to be accepted by each of the positive examples, while each negative example to be rejected by at least one of the clauses. For instance, the following CNF expression satisfies these requirements:

$$(A_2 \vee A_4) \wedge (\bar{A}_2 \vee \bar{A}_3) \wedge (A_1 \vee A_3 \vee \bar{A}_4).$$

This, in essence, is *Problem 1*, that is, how to construct a set (of hopefully small size) of clauses which would correctly classify all the available positive and negative examples and hopefully classify new examples with high accuracy.

Next, in order to illustrate *Problem 2*, we consider the following hypothetical sample of input data:

$$E^+ = \begin{bmatrix} 0 & * & 0 & 0 \\ 1 & 0 & * & 1 \\ 0 & 1 & 0 & 1 \end{bmatrix}, \quad E^- = \begin{bmatrix} 0 & 0 & * & 1 \\ * & 0 & 1 & 0 \\ 0 & 1 & 0 & 0 \end{bmatrix}, \quad \text{and}$$

$$E^U = \begin{bmatrix} 0 & 1 & 1 & * \\ 1 & 1 & * & * \end{bmatrix},$$

In this chapter the $E^U$ set will denote the set with the *undecidable* (not to be confused with *unclassified*) examples. The symbol "$*$" in the three data sets represents attributes whose binary values are missing.

In the previous $E^+$ and $E^-$ data sets it is assumed that the missing data values (indicated by "$*$") did not prohibit the "*oracle*" (i.e., the hidden function) from classifying the corresponding examples as positive or negative. For instance, the first positive example in $E^+$ (i.e., $(0, *, 0, 0)$) implies that this example should be positive *regardless* of the actual nature of the "$*$" element. This observation indicates that the following two examples (note that $2 = 2^1$, where 1 is the number of the missing elements in that example): $(0, 0, 0, 0)$ and $(0, 1, 0, 0)$ are also positive examples. That is, for positive and negative examples the missing values can be treated as *do not care* cases (i.e., they can be either 1 or 0 without changing the classification of that example). The notion of the *do not care* concept was first introduced by Kamath, *et al.*, in [1993] in order to condense the information representation in this type of learning problems.

An obvious restriction for the data in the two sets to be valid is that every possible pair of a positive and a negative example should have at least one of their common fixed attributes with a different value. For instance, when $n = 8$, the examples $(1, 1, *, *, 0, 0, *, *)$ and $(1, 1, *, 0, 0, 0, *, 1)$ *cannot* belong to different classes (i.e., one to be positive and the other to be negative). This is true because the example $(1, 1, 0, 0, 0, 0, 0, 1)$ is implied by either of the previous two examples which have missing (i.e., *do not care*) elements.

To further illustrate the concept of the undecidable examples consider the following Boolean function defined on five attributes (i.e., $n = 5$):

$$(A_1 \vee A_4) \wedge (A_2 \vee \bar{A}_3 \vee A_5).$$

The above Boolean function would make an example such as $(0, 1, 1, *, 1)$ undecidable chiefly because the value of $A_4$ is missing in this example.

A naive way for dealing with data which have missing values would be to ignore all the undecidable examples and concentrate attention only on the positive and negative examples, that is, to ignore all undecidable examples, and expand all the positive and negative examples, and thus transform this truncated version of Problem 2 into an instance of Problem 1. Recall that if a positive or negative (but not undecidable) example has $k$ (where $k < n$) *do not care* values, then it can be expanded into $2^k$ positive or negative examples defined in $\{1, 0\}^n$. However, if this were done, then one would have ignored the information present in the undecidable examples. That is, by knowing that the inferred system should neither accept nor reject any of the undecidable examples, the search for an accurate Boolean function may be better directed.

## 4.2 A Fast Heuristic for Inferring a Boolean Function from Complete Data

The aim of the clause inference strategies in this chapter is to derive a very small (hopefully minimum) number of disjunctions (for the CNF case). Although the Boolean functions derived in the proposed approach are in CNF form, DNF functions can also be derived from the same data set and vice versa (see also Chapter 7 of this book).

As was stated earlier, a Boolean function in CNF must satisfy the following requirements:

(i) *Each clause in the derived expression should accept all the examples in the $E^+$ set; and*
(ii) *All the clauses, when they are taken together, should reject all the examples in the $E^-$ set.*

In order to offset the drawback of the exponential time complexity of the B&B algorithms used to implement Step 2 of the OCAT approach (as described in Figure 2.1, in Section 2.8), the clauses formed by the heuristics are built as follows. Each clause accepts all the examples in the $E^+$ set while it rejects *many* (as opposed to *as many as possible* in the B&B approaches) examples in the $E^-$ set. Note that this is the main procedural difference between the B&B algorithms and the proposed heuristics.

In the first heuristic of this chapter this is achieved by choosing the attributes to form a clause based on an evaluative function (to be described later in this section). Only attributes with high values in terms of this evaluative function are included in the current clause. A single clause is completely derived when all the examples in the $E^+$ set are accepted. The clause forming procedure is repeated until all the examples in the $E^-$ set are rejected by the proposed set of clauses. This is the essence of the OCAT approach. As some computational results (presented in a later section) indicate, this strategy may often result in Boolean functions with a rather small number of clauses.

One may observe at this point that if the attribute with the *highest value* of the evaluative function is always included in the clause, then there is an inherent danger of being trapped at a local optimal point. To prevent the Boolean expression from being degenerated as a result of being trapped at a local optimal point, a *randomized* approach is used. In this randomized approach, instead of a single attribute being included in a clause due to its highest value of the evaluative function, a *candidate list* of attributes is formed. The attributes in this candidate list are selected based on their values in terms of the evaluative function. These values are close to the highest value derived from the evaluative function. Next, an attribute is *randomly* chosen out of the candidate list and is included in the CNF clause being derived.

It is also possible for a CNF clause to reject as many negative examples as possible (and, of course, to accept all positive examples) but the entire Boolean expression not to have a small (ideally minimum) number of clauses. Recall that the proposed heuristics are components based on the OCAT approach (i.e., they are not run as procedures of the OCAT approach as was the case with the B&B algorithms). That is, sometimes it may be more beneficial to have a less "effective" clause which does not reject a large number of negative examples, and still derive a Boolean function with very few clauses. Such Boolean functions are possible to derive with the use of randomized algorithms. A randomized algorithm, with a sufficiently large number of iterations, is difficult to be trapped at a local optimum.

The first heuristic approach, termed *RA1* (for *Randomized Algorithm 1*), is proposed next to solve the first problem considered in this chapter. Before the RA1 heuristic is formally presented, some new definitions and terminology are needed to be introduced next.

**Definitions.**

$C$ = *The set of attributes in the current clause (disjunction).*

$A_k$ = *An attribute such that* $A_k \in A$, *where A is the set of all attributes* $A_1, \ldots, A_n$ *and their negations.*

$POS(A_k)$ = *The number of all positive examples in* $E^+$ *which would be accepted if attribute* $A_k$ *is included in the current clause.*

$NEG(A_k)$ = *The number of all negative examples in* $E^-$ *which would be accepted if attribute* $A_k$ *is included in the current clause.*

$l$ = *The size of the candidate list.*

$ITRS$ = *The number of times the clause forming procedure is repeated.*

As an illustrative example of the above definitions, consider the following sets of positive and negative examples (which were also given in earlier chapters):

$$E^+ = \begin{bmatrix} 0 & 1 & 0 & 0 \\ 1 & 1 & 0 & 0 \\ 0 & 0 & 1 & 1 \\ 1 & 0 & 0 & 1 \end{bmatrix} \quad \text{and} \quad E^- = \begin{bmatrix} 1 & 0 & 1 & 0 \\ 0 & 0 & 0 & 1 \\ 1 & 1 & 1 & 1 \\ 0 & 0 & 0 & 0 \\ 1 & 0 & 0 & 0 \\ 1 & 1 & 1 & 0 \end{bmatrix}.$$

The set $A$ of all attributes for the above set of examples is

$$A = \{A_1, A_2, A_3, A_4, \bar{A}_1, \bar{A}_2, \bar{A}_3, \bar{A}_4\}.$$

Therefore, the $POS(A_k)$ and the $NEG(A_k)$ values are

$$POS(A_1) = 2 \quad NEG(A_1) = 4 \quad POS(\bar{A}_1) = 2 \quad NEG(\bar{A}_1) = 2$$

$$POS(A_2) = 2 \quad NEG(A_2) = 2 \quad POS(\bar{A}_2) = 2 \quad NEG(\bar{A}_2) = 4$$

$$POS(A_3) = 1 \quad NEG(A_3) = 3 \quad POS(\bar{A}_3) = 3 \quad NEG(\bar{A}_3) = 3$$

$$POS(A_4) = 2 \quad NEG(A_4) = 2 \quad POS(\bar{A}_4) = 2 \quad NEG(\bar{A}_4) = 4$$

The problem now is to derive a small set of clauses (disjunctions) which would correctly classify all the above examples. Suppose that there exists a *hidden* system given by the following Boolean function:

$$(A_2 \vee A_4) \wedge (\bar{A}_2 \vee \bar{A}_3) \wedge (A_1 \vee A_3 \vee \bar{A}_4).$$

It can be easily seen that the above Boolean function correctly classifies all the previous examples. Therefore, the first problem is to accurately estimate the above *hidden* system. This is accomplished by using heuristic RA1, which is described in Figure 4.1.

Let $m_1$ and $m_2$ be the cardinalities (sizes) of the sets of examples $E^+$ and $E^-$, respectively. Then, the following theorem [Deshpande and Triantaphyllou, 1998] states an upper bound on the number of clauses which can be inferred by RA1, and it is similar to Theorem 2.1 in Chapter 2.

**Theorem 4.1.** *The RA1 approach terminates within at most $m_2$ iterations.*

*Proof.* A clause $C_k$ can always be formed which rejects only the $k$-th negative example while accepting all other examples. For instance, if the $k$-th negative example to be rejected is $(1, 0, 1, 0)$, then the clause which rejects the $k$-th example, while accepting all other examples, is $(\bar{A}_1 \vee A_2 \vee \bar{A}_3 \vee A_4)$. Therefore, in Step 2 of the RA1 procedure, a clause which at best rejects only a single example from the $E^-$ set could be formed. As a result, the maximum number of clauses required to reject all the $E^-$ examples is $m_2$. ■

Next, let $n$ be the number of attributes in the data set. Then Theorem 4.2 states the time complexity of the RA1 algorithm [Deshpande and Triantaphyllou, 1998].

**Theorem 4.2.** *The RA1 algorithm has a polynomial time complexity of order $O(n(m_1 + m_2)m_1 m_2\ ITRS)$.*

*Proof.* Calculating the values of the ratios $POS(a_i)/NEG(a_i)$, for $i = 1$ to $n$, requires $n(m_1 + m_2)$ simple computations. In order to sort out the attributes in descending order of their $POS(a_i)/NEG(a_i)$ value, we can use the "quick sort" procedure (see, for instance, [Aho, *et al.*, 1974]) which has time complexity of order $O(n \log n)$.

**DO** for *ITRS* number of iterations
**BEGIN** { Reset the $E^+$ and $E^-$ sets };
  **DO WHILE** ($E^- \neq \varnothing$)
  $C = \varnothing$; {*initialization*}
    **DO WHILE** ($E^+ \neq \varnothing$)
      **Step 1:** Rank in descending order all attributes $a_i \in A$ (*where* $a_i$ *is either* $A_i$ *or* $\bar{A}_i$) according to their $POS(a_i)/NEG(a_i)$ value. If $NEG(a_i) = 0$, then use as an alternative scoring function the product of an arbitrarily large number times $POS(a_i)$. We call this the $ALT(a_i)$ value;
      **Step 2:** Form a candidate list of the attributes which have the $l$ top highest $POS(a_i)/NEG(a_i)$ ratios or $ALT(a_i)$ values (when $NEG(a_i) = 0$);
      **Step 3:** Randomly choose an attribute $a_k$ from the candidate list;
      **Step 4:** Let the partial current clause be $C \leftarrow C \vee a_k$;
      **Step 5:** Let $E^+(a_k)$ be the set of members of $E^+$ accepted when $a_k$ is included in the current CNF clause;
      **Step 6:** Let $E^+ \leftarrow E^+ - E^+(a_k)$;
      **Step 7:** Let $A \leftarrow A - a_k$;
      **Step 8:** Calculate the new $POS(a_i)$ values for all $a_i \in A$;
    **REPEAT**
      **Step 9:** Let $E^-(C)$ be the set of members of $E^-$ which are rejected by $C$;
      **Step 10:** Let $E^- \leftarrow E^- - E^-(C)$;
      **Step 11:** Reset $E^+$;
  **REPEAT;**
**END;**
**CHOOSE** the final Boolean system among the previous *ITRS* systems which has the smallest number of clauses.

**Figure 4.1.** The RA1 Heuristic.

Each clause is completely formed when all the $m_1$ examples in the positive set are accepted. Each Boolean function is completely formed when all the $m_2$ negative examples are rejected. The entire clause forming procedure is repeated *ITRS* number of times. Therefore, the time complexity of the RA1 algorithm is $O((n(m_1 + m_2) + n \log n)m_1 m_2 ITRS) = O(n(m_1 + m_2)m_1 m_2 ITRS)$. ■

From the way the $POS(a_k)$ and $NEG(a_k)$ values are defined, some critical observations can be made. When an attribute with a rather high value of the *POS* function is included in the CNF clause being formed, then chances are that some additional positive examples will be accepted by that clause as a result of the inclusion of that attribute. Similarly, attributes which correspond to low *NEG* values are likely not to cause many new negative examples to be accepted as a result of the inclusion of that attribute in the current clause. Therefore, it makes sense to include as attributes in the CNF clause under formation, the ones which correspond to high *POS* values and, at the same time, to low *NEG* values.

For the current illustrative example, the values of the $POS(a_k)/NEG(a_k)$ ratios are

$$POS(A_1)/NEG(A_1) = 0.50, \quad POS(\bar{A}_1)/NEG(\bar{A}_1) = 1.00,$$

$$POS(A_2)/NEG(A_2) = 1.00, \quad POS(\bar{A}_2)/NEG(\bar{A}_2) = 0.50,$$

$$POS(A_3)/NEG(A_3) = 0.33, \quad POS(\bar{A}_3)/NEG(\bar{A}_3) = 1.00,$$

$$POS(A_4)/NEG(A_4) = 1.00, \quad POS(\bar{A}_4)/NEG(\bar{A}_4) = 0.50.$$

The above discussion illustrates the motivation for considering as possible candidates for the evaluative function, the functions *POS/NEG*, *POS-NEG*, or some type of a weighted version of the previous two expressions. Some exploratory computational experiments indicated that the evaluative function *POS/NEG* was the most effective one. That is, it led to the formation of Boolean functions with less clauses than when other alternative evaluative functions were considered.

The randomization of the RA1 algorithm is done as follows. In Step 2, the first $l$ attributes with the highest value of the $POS(a_k)/NEG(a_k)$ ratio are chosen as the members of the candidate list and an attribute in the list is randomly chosen out of the candidate list in Step 3. This is done in order to obtain different solutions at each iteration and prevent the system from being trapped by a locally optimal point.

In choosing a fixed value for the size $l$ of the candidate list, there is a possibility that an attribute with a very low value of the ratio $POS(a_k)/NEG(a_k)$ could be selected if the value of $l$ is large enough (how large depends on the particular data). That could occur if there are not $l$ attributes with a sufficiently high value of the ratio $POS(a_k)/NEG(a_k)$. If an attribute with a low value of the ratio $POS(a_k)/NEG(a_k)$ is chosen to be included in the clause, then the clause would accept less examples from the $E^+$ set or accept more examples from the $E^-$ set, or both. All of these three situations should be avoided as they would lead to an increase in the number of attributes in a clause (if it accepts less examples from the $E^+$ set) or to an increase in the number of clauses (if the attribute accepts more examples from the $E^-$ set), or both. To prevent the above situation from happening, a candidate list is formed of attributes, each of whose $POS(a_k)/NEG(a_k)$ value is within a certain percentage, say $\alpha\%$, of the highest value of the $POS(a_k)/NEG(a_k)$ value in the current candidate list. This ensures that the attribute (randomly chosen out of the candidate list) to be included in the clause has a value close to the highest value of the $POS(a_k)/NEG(a_k)$ ratios. The alternative (evaluative) scoring function $ALT(a_k)$ was introduced in Step 1 in order to avoid degenerative cases when $NEG(a_k) = 0$ and, at the same time, to give a high priority to such attributes. In case of multiple maximum values, ties are broken arbitrarily.

The above idea of using randomization in a search algorithm has been explored by other researchers as well. For instance, Feo and Resende in [1995] have successfully used randomization to solve clause satisfiability (SAT) problems. They called their approach GRASP (for *Greedy Randomized Adaptive Search Procedures*) and the previous developments are an implementation of the GRASP approach. Also, in a book Motwani and Raghavan [1995] provide a comprehensive presentation of the theory on randomized algorithms. Randomization also offers a natural and intuitive way for implementing *parallelism* in algorithms.

To obtain a Boolean function with a very small number of clauses, the whole procedure is subjected to a certain number of iterations (denoted by the value of the *ITRS* parameter) and the system which has the least number of disjunctions is chosen as the final inferred Boolean system.

Referring to the previous illustrative example, if $l = 3$, then the values of the three best $POS(a_k)/NEG(a_k)$ ratios are {1.0, 1.0, 1.0} (note that it is a coincidence that all three values are identical) which correspond to the attributes $\bar{A}_1$, $A_2$, and $A_4$, respectively. Let attribute $A_2$ be the randomly selected attribute from the candidate list. Note that attribute $A_2$ accepts examples number 2 and 3 from the current $E^+$ set. Therefore, at least one more attribute is required to complete the formation of the current clause. Thus, the entire process of finding a new attribute (other than attribute $A_2$ which has already been selected) with a very high value of *POS/NEG* is repeated. Now, suppose that the attribute with a high *POS/NEG* value happened to be $A_4$. It can be observed now that when attributes $A_2$ and $A_4$ are combined together, they accept all the elements in the $E^+$ set. Therefore, the first clause is $(A_2 \vee A_4)$.

This clause fails to reject examples number 2, 3, and 6 in the $E^-$ set. Therefore, examples number 2, 3, and 6 in the original $E^-$ set constitute the reduced (and thus new) $E^-$ set. The above process is repeated until a set of clauses are formed which, when taken together, reject all the examples in the original $E^-$ set. Therefore, a final Boolean function for this problem could be as follows (recall that the algorithm is a randomized one and thus it may not return the same solution each time it is run):

$$(A_2 \vee A_4) \wedge (\bar{A}_2 \vee \bar{A}_3) \wedge (A_1 \vee A_3 \vee \bar{A}_4).$$

A very important factor in deriving a Boolean function from positive and negative examples is the number of examples needed to infer the logic. This is also known as the *sample complexity* of a given approach. The problem of inferring a pattern with a *sequence of very few new examples* has been examined by Bshouty, *et al.*, [1993], Goldman and Sloan [1994], and Triantaphyllou and Soyster [1995b]. This is also known as the *guided learning* approach. A guided learning approach is described later in Chapter 5.

## 4.3 A Fast Heuristic for Inferring a Boolean Function from Incomplete Data

The second heuristic algorithm deals with the case in which some of the examples contain missing values. That is, now the examples are defined in the $\{0, 1, *\}^n$ space. Some algorithmic consequences of having missing elements in the positive, negative, or undecidable examples (denoted as $E^U$) have already been discussed in Section 4.1.

If an example is determined as undecidable by the "*hidden*" system, then it has also to remain undecidable by the derived Boolean function. In other words, the property for inferring Boolean functions when undecidable examples are also considered (along with positive and negative examples), is that the examples in the $E^U$ set should neither be accepted nor rejected by the derived system.

To help fix ideas, consider the Boolean function which we have also seen in the previous section:

$$(A_2 \vee A_4) \wedge (\bar{A}_2 \vee \bar{A}_3) \wedge (A_1 \vee A_3 \vee \bar{A}_4).$$

If any example in $E^U$ has $A_2 = A_4 = 0$, then the first clause would reject that example. This, however, should not be permissible and thus the above function *cannot* be a candidate solution, *regardless* of what are the positive and negative examples.

On the other hand, if in a particular undecidable example $A_2 = 0$ and $A_4 = *$, then that example is neither rejected nor accepted. Therefore, in the later scenario, the previous function is a possible candidate as far as that single undecidable example is concerned.

The second algorithm, termed *RA2* (for *Randomized Algorithm 2*), is also based on the OCAT approach (as described in Figure 2.1, in Section 2.8). That is, the RA2 generates a sequence of clauses as shown in Figure 4.2. Each clause accepts all the positive examples in the $E^+$ set, while it does not reject any of the examples in the undecidable set $E^U$ and it also attempts to reject a large number (but not necessarily as many as possible) of the negative examples in the $E^-$ set.

Subsequent clauses are formed with a reduced $E^-$ set (which is comprised by the negative examples which have not been rejected so far). When all the examples in the $E^-$ set have been rejected, then the RA2 algorithm enters its second phase (see also Figure 4.2). In the second phase the entire set of clauses is tested against the $E^U$ set to satisfy the necessary Boolean function forming condition for the undecidable examples. That is, all the clauses when grouped together should not accept any example from the $E^U$ set.

Recall that after Phase I the clauses may not reject any of the undecidable examples. Any undecidable examples which are accepted by the clauses which have been formed during Phase I, are grouped into the set $E^A$ (that is: $E^A \subseteq E^U$). The clauses which were formed during Phase I are appended with a new set of clauses which are formed in the second phase. The new clauses, when grouped together with the first set of clauses, do not accept any example in the set $E^A$ (i.e., now $E^A = \varnothing$). This is accomplished in a sequential manner as shown in Figure 4.2.

Let the set $E^A$ contain examples from the original $E^U$ set which the derived Boolean function has accepted. Therefore, the maximum number of examples in the set $E^A$ is the same as the cardinality of the $E^U$ set (i.e., equal to $m_3$). Suppose that there are still $m'_3$ (where $m'_3 \leq m_3$) undecidable examples in the $E^A$ set after all the examples from the $E^-$ set are rejected and all the clauses are tested against the examples in the $E^U$ set for nonacceptance (i.e., now $E^A \neq \varnothing$ and $E^- = \varnothing$). Then, how does the RA2 heuristic terminate? The upper bound for the terminating condition is given in the following theorem [Deshpande and Triantaphyllou, 1998]:

**DO** for *ITRS* number of iterations

**BEGIN** { Reset the $E^+$ and $E^-$ sets };

**Phase I: DO WHILE** $(E^- \neq \varnothing)$

$C = \varnothing$; {*initialization* }

**DO WHILE** ($C$ does not reject any examples in $E^U$)

**DO WHILE** $(E^+ \neq \varnothing)$

**Step 1:** Rank in descending order all attributes $a_i \in A$ (*where* $a_i$ *is either* $A_i$ *or* $\bar{A}_i$) according to their $POS(a_i)/NEG(a_i)$ value. If $NEG(a_i) = 0$, then use as an alternative scoring function the product of an arbitrarily large number times $POS(a_i)$. We call this the $ALT(a_i)$ value;

**Step 2:** Form a candidate list of the attributes which have the $l$ top highest $POS(a_i)/NEG(a_i)$ ratios or $ALT(a_i)$ values (when $NEG(a_i) = 0$);

**Step 3:** Randomly choose an attribute $a_k$ from the candidate list;

**Step 4:** Let $E^+(a_k)$ be the set of members of $E^+$ accepted when $a_k$ is included in the current CNF clause;

**Step 5:** Let $E^+ \leftarrow E^+ - E^+(a_k)$; also, let $A \leftarrow A - a_k$;

**Step 6:** Let the current partial clause be $C \leftarrow C \vee a_k$;

**REPEAT**

**Step 7:** Let the current clause be $C \leftarrow C \vee a_j$, where $a_j$ is any attribute with a value of "*" in each of the examples in $E^U$ which were rejected by the clause $C$;

**Step 8:** Let $E^-(C)$ be the set of members of $E^-$ which are rejected by $C$;

**Step 9:** Let $E^- \leftarrow E^- - E^-(C)$;

**Step 10:** Reset $E^+$;

**REPEAT;**

**Phase II:** Denote as $E^A$ the updated set of undecidable examples in $E^U$ accepted by the current set of clauses $C_1 \vee C_2 \cdots \vee C_m$ where $m$ is the total number of clauses formed so far;

**DO WHILE** $(E^A \neq \varnothing)$

**Step 11:** $m \leftarrow m + 1$;

**Step 12:** Form (according to the proof of Theorem 4.3) a clause $C_m$ which does not accept the first undecidable example from the current $E^A$ set ; Update the $E^A$ set;

**REPEAT;**

**END;**

**CHOOSE** the final Boolean system among the previous *ITRS* systems which has the smallest number of clauses.

**Figure 4.2.** The RA2 Heuristic.

**Theorem 4.3.** *The maximum number of* **additional** *conjunctions to be formed during Phase II in heuristic RA2 is equal to* $m'_3$.

*Proof.* A clause $C_k$ can always be formed which does not accept the $k$-th undecidable example (i.e., it can either reject or make undecidable this example) from the $E^A$ set while accepting *all* other examples (refer to the proof of Theorem 4.1 for the validity of the previous statement). Since all the negative examples have already been rejected, a maximum of $m'_3$ new clauses can be formed which do not accept the $m'_3$ undecidable examples from the $E^A$ set. For instance, if the $k$-th example in $E^A$ is $(1, *, 0, *)$, then the clause $C_k$ is $(\bar{A}_1 \vee A_2 \vee A_3 \vee A_4)$ and it would fail to accept or reject the $k$-th example. ■

If $n$ is the number of attributes and $m_1$, $m_2$, and $m_3$ are the cardinalities of the $E^+$, $E^-$, and $E^U$ sets, respectively, then the complexity of the RA2 algorithm is stated in the following theorem [Deshpande and Triantaphyllou, 1998]:

**Theorem 4.4.** *The time complexity of the RA2 heuristic is* $O(n(m_1 + m_2)m_1m_2m_3\,ITRS)$.

*Proof.* Calculating the values of the $POS(a_i)/NEG(a_i)$ ratios (for $i = 1$ to $n$) requires $n(m_1 + m_2)$ computations. To sort out the attributes in descending order of their $POS(a_i)/NEG(a_i)$ value we can use the "quick sort" approach which is of time complexity of order $O(n \log n)$ [Aho, *et al.*, 1974]. To form a Boolean expression in which each clause accepts all the $m_1$ positive examples, rejects none of the $m_3$ undecidable examples and the whole set of clauses rejecting all the $m_2$ negative examples is of order $m_1m_2m_3$. The complexity of Phase II is $m_3n$. This is indicated in the second loop in Figure 4.2. Therefore, the complexity of the RA2 heuristic is of order $O((((n(m_1+m_2)+n \log n)m_1m_2m_3)+m_3n)ITRS) = O(n(m_1+m_2)m_1m_2m_3ITRS)$. ■

Next, for demonstrative purposes of the above issues consider the following illustrative example. Let the three classes of data be as follows:

$$E^+ = \begin{bmatrix} 0 & 1 & 0 & 0 \\ 1 & 1 & 0 & 1 \\ 0 & 0 & 1 & 0 \end{bmatrix}, \quad E^- = \begin{bmatrix} 1 & 0 & 1 & 0 \\ 0 & 0 & 0 & 1 \\ 1 & 1 & 1 & 1 \\ 0 & 0 & 0 & 0 \end{bmatrix}, \quad \text{and}$$

$$E^U = \begin{bmatrix} 1 & * & 0 & * \\ 1 & 0 & * & 0 \\ * & 1 & * & 0 \end{bmatrix}.$$

In the previous positive and negative examples there are no examples with missing fields. If there were such missing data, then one could first expand them and replace them with the expanded data. For instance, an example such as $(*, 1, 0, *)$ could be expanded and then be replaced by the following four ($4 = 2^2$) examples with no missing data:

$$(0,\ 1,\ 0,\ 0)$$
$$(0,\ 1,\ 0,\ 1)$$
$$(1,\ 1,\ 0,\ 0)$$
$$(1,\ 1,\ 0,\ 1)$$

Next, the $POS(a_k), NEG(a_k), POS(a_k)/NEG(a_k)$ values are as follows:

$$\begin{aligned}
POS(A_1) &= 1, & NEG(A_1) &= 2, & POS(A_1)/NEG(A_1) &= 0.50,\\
POS(A_2) &= 2, & NEG(A_2) &= 1, & POS(A_2)/NEG(A_2) &= 2.00,\\
POS(A_3) &= 1, & NEG(A_3) &= 2, & POS(A_3)/NEG(A_3) &= 0.25,\\
POS(A_4) &= 1, & NEG(A_4) &= 2, & POS(A_4)/NEG(A_4) &= 0.50,\\
POS(\bar{A}_1) &= 2, & NEG(\bar{A}_1) &= 2, & POS(\bar{A}_1)/NEG(\bar{A}_1) &= 1.00,\\
POS(\bar{A}_2) &= 1, & NEG(\bar{A}_2) &= 3, & POS(\bar{A}_2)/NEG(\bar{A}_2) &= 0.33,\\
POS(\bar{A}_3) &= 2, & NEG(\bar{A}_3) &= 2, & POS(\bar{A}_3)/NEG(\bar{A}_3) &= 0.50,\\
POS(\bar{A}_4) &= 2, & NEG(\bar{A}_4) &= 2, & POS(\bar{A}_4)/NEG(\bar{A}_4) &= 1.00.
\end{aligned}$$

If $l = 3$, then the three highest ratios of the $POS(a_k)/NEG(a_k)$ values are $\{2.0, 1.0, 1.0\}$ which correspond to attributes $A_2$, $\bar{A}_1$, and $\bar{A}_4$, respectively. Let attribute $A_2$ be the randomly selected attribute from the current candidate list. If attribute $A_2$ is introduced into the current clause, then as a result this clause will accept the first and the second examples in the $E^+$ set. The whole process of finding the values of $POS(a_k)/NEG(a_k)$ (with $k \neq 2$) is repeated. For the next iteration suppose that attribute $A_3$ is chosen. When attributes $A_2$ and $A_3$ are introduced into the clause, then this clause accepts all examples in the $E^+$ set. This set of attributes does not reject any example in the $E^U$ set. Therefore, the first clause is $(A_2 \vee A_3)$. This process is repeated until $E^- = \varnothing$ and $E^A = \varnothing$. Therefore, a final Boolean function for this problem could be as follows (recall that the algorithm is a randomized one and thus it does not return a deterministic solution):

$$(A_2 \vee A_3) \wedge (\bar{A}_1 \vee \bar{A}_3 \vee A_4) \wedge (\bar{A}_1 \vee \bar{A}_3).$$

## 4.4 Some Computational Results

A number of computer experiments were conducted on an IBM 3090-600S mainframe computer running the VM/XA operating system, in order to investigate the effectiveness of the RA1 and RA2 heuristics on different types of problems. Some interesting results were obtained and are discussed in the following paragraphs.

The previous two heuristics RA1 and RA2 were tested on a number of different experiments. The first type of experiments used the well-known Wisconsin breast cancer database (donated by Professor Mangasarian from the University of Wisconsin, Madison, and now it can be found in the University of California at Irvine Repository of Learning Databases and Domain Theories) [Murphy and Aha, 1994]). This database contained (at the time it was obtained) 421 examples, of which 224 correspond to benign (or positive examples in our setting) and 197 to malignant cases (or negative examples in our setting). The original data were defined on nine discrete variables, each variable assuming values from the integer set [1, 10]. These data were converted into their equivalent binary data. That is, each variable was converted into four binary variables and thus the transformed database was defined on 36 (= 4 × 9) binary variables.

In addition, the RA1 heuristic was compared with the branch-and-bound method described in Chapter 3 of this book by first generating a challenging set of large random test problems. The first large random data set contained 18,120 examples defined on 15 attributes with a varied ratio of positive to negative examples. Note that this set was almost 50 times larger than the size of the previous breast cancer data set. A second large data set contained 3,750 examples defined on 14 attributes.

The measures of performance considered in these tests were of three types: (i) the accuracy of the derived system (Boolean function), (ii) the number of clauses (CNF disjunctions) of the derived system, and (iii) the CPU time required to derive a solution.

The method of determining the accuracy of a proposed system (i.e., Boolean function) was defined in two different ways. When the Wisconsin breast cancer database was used, the accuracy of a solution was defined by comparing the way an inferred system (which was derived when a part of the available data was used as the training set) and the system derived when the entire data base is used, classified a random collection of 10,000 examples. Note that this approach is similar to the testing procedures used in [Kamath, *et al.*, 1992], [Triantaphyllou, *et al.*, 1994], [Triantaphyllou, 1994], and [Triantaphyllou and Soyster, 1995b] and also in previous chapters.

For the other cases the testing procedure was different. First a collection of random examples was formed. Next, a Boolean function was formed randomly and the previous examples were classified according to that function as being either positive or negative examples. That function played the role of the *oracle* or *hidden system*. Then, the goal of the inference algorithms was to infer a close approximation of that *hidden* function. Therefore, in this case the notion of accuracy was defined as the percentage of the times the proposed and the *hidden* system agreed in classifying a random collection of 10,000 examples.

Moreover, for testing the two heuristics on the breast cancer data, two categories of Boolean functions were derived. The first category of Boolean functions used the benign set as the $E^+$ set and the malignant set as the $E^-$ set. This category of Boolean systems was denoted as system $S1$. The second category of systems treated the benign set as the $E^-$ set and the malignant set as the $E^+$ set. This category of Boolean systems was denoted as system $S2$. The purpose of formulating two categories of systems (i.e., $S1$ and $S2$) was to study the effect of the number of examples (recall that the benign and the malignant observations were 224 and 197, respectively) on the accuracies and the number of clauses in the derived systems.

Since it was required that the number of clauses be kept at a very small level, the whole clause forming process was repeated a certain number of times (defined as the value of the parameter *ITRS* in Figures 4.1 and 4.2). The value of *ITRS* equal to 150 was determined after a brief pilot study. The results are tabulated below.

| | Number of clauses | |
|---|---|---|
| **ITRS value** | **in System $S1$** | **in System $S2$** |
| 50 | 14.40 | 23.52 |
| 100 | 12.20 | 20.69 |
| 150 | 8.86 | 19.92 |
| 200 | 8.84 | 20.12 |
| 500 | 8.83 | 19.79 |
| 1,000 | 8.82 | 19.78 |

These results suggest that higher values of *ITRS* did not generate much fewer clauses (although the CPU requirement is higher now). Thus, the value of *ITRS* equal to 150 is a reasonable one. Obviously, this empirical value cannot be generalized since for different data a different *ITRS* value may be more appropriate. Therefore, we suggest that a pilot study to be undertaken before an *ITRS* value is decided. Finally, please recall that the running time of the heuristic is directly proportional to the value of *ITRS*. This is indicated in the complexities of the RA1 and RA2 heuristics as seen in Figures 4.1 and 4.2, respectively, and also stated in Theorems 4.2 and 4.4, respectively.

For the randomization of the heuristics, a candidate list of a few attributes was formed among which the representative attribute was randomly chosen. The attributes which were chosen to be in the candidate list were those which had a *POS/NEG* value within a certain percentage, say $\alpha\%$, of the maximum *POS/NEG* value in the candidate list. This could assure that the attributes which are finally chosen are in a near neighborhood of the attribute with the maximum value of the *POS/NEG* ratio in the candidate list. A good value of $\alpha\%$ seemed to be equal to 75% as it resulted in the highest accuracy in some exploratory computational experiments. These experiments are described in more detail in the following sections.

### 4.4.1 Results for the RA1 Algorithm on the Wisconsin Cancer Data

The results for Problem 1 (i.e., inference of a Boolean function with complete data) are presented in Table 4.1. The number of replications per case was equal to 50. The numbers in the parentheses indicate the standard deviations of the various observations. Individual accuracies for the benign and the malignant tumors are also presented. For instance, $B(S1)$, $M(S1)$, $B(S2)$, and $M(S2)$ represent benign and malignant accuracies for systems $S1$ and $S2$, respectively.

Figure 4.3 shows the relationship of accuracy versus the percentage of the data used. With approximately 10% of the data used as the training data, accuracy of approximately 88% was achieved with system $S1$ while a higher accuracy of 92% was achieved with system $S2$. A peak accuracy of approximately 92% was obtained with system $S1$ while a peak of 94% was obtained with system $S2$. Figure 4.4 shows the number of the derived clauses. The maximum number of clauses is 9 for system $S1$ and 20 for system $S2$.

On the average, the time required to infer the 9 clauses was *300 CPU seconds*. Resende and Feo [1995] had reported that their SAT approach took more than

**Table 4.1.** Numerical Results of Using the RA1 Heuristic on the Wisconsin Breast Cancer Database.

| % of Data | No. of Clauses (S1) | No. of Clauses (S2) | Accur. (S1) | Accur. B(S1) | Accur. M(S1) | Accur. (S2) | Accur. B(S2) | Accur. M(S2) |
|---|---|---|---|---|---|---|---|---|
| 5.0 | 1.18 | 2.28 | 0.825 | 0.810 | 0.843 | 0.906 | 0.941 | 0.867 |
| | (0.38) | (0.49) | (0.058) | (0.107) | (0.106) | (0.068) | (0.101) | (0.108) |
| 10.0 | 1.60 | 3.02 | 0.882 | 0.879 | 0.885 | 0.921 | 0.962 | 0.875 |
| | (0.57) | (0.55) | (0.032) | (0.057) | (0.055) | (0.052) | (0.073) | (0.061) |
| 15.0 | 2.06 | 3.70 | 0.880 | 0.867 | 0.896 | 0.939 | 0.995 | 0.876 |
| | (0.57) | (0.67) | (0.027) | (0.059) | (0.035) | (0.031) | (0.02) | (0.062) |
| 20.0 | 2.42 | 4.62 | 0.893 | 0.904 | 0.882 | 0.944 | 0.996 | 0.885 |
| | (0.53) | (0.75) | (0.023) | (0.04) | (0.034) | (0.027) | (0.015) | (0.048) |
| 25.0 | 2.72 | 5.14 | 0.901 | 0.916 | 0.885 | 0.945 | 0.995 | 0.887 |
| | (0.60) | (0.89) | (0.017) | (0.030) | (0.036) | (0.020) | (0.016) | (0.041) |
| 30.0 | 3.12 | 6.00 | 0.904 | 0.915 | 0.893 | 0.945 | 0.996 | 0.887 |
| | (0.71) | (1.31) | (0.014) | (0.029) | (0.033) | (0.018) | (0.011) | (0.035) |
| 35.0 | 3.40 | 6.84 | 0.903 | 0.915 | 0.890 | 0.944 | 0.998 | 0.884 |
| | (0.72) | (1.10) | (0.017) | (0.030) | (0.036) | (0.018) | (0.008) | (0.033) |
| 40.0 | 3.74 | 7.60 | 0.904 | 0.921 | 0.885 | 0.946 | 1.000 | 0.885 |
| | (0.66) | (1.34) | (0.015) | (0.023) | (0.248) | (0.185) | (0.002) | (0.037) |
| 45.0 | 4.06 | 8.60 | 0.903 | 0.914 | 0.892 | 0.950 | 1.000 | 0.893 |
| | (0.88) | (1.71) | (0.018) | (0.028) | (0.033) | (0.016) | (0.000) | (0.032) |
| 50.0 | 4.44 | 9.70 | 0.906 | 0.917 | 0.893 | 0.949 | 1.000 | 0.892 |
| | (0.78) | (1.57) | (0.018) | (0.032) | (0.037) | (0.017) | (0.001) | (0.034) |
| 55.0 | 4.90 | 10.94 | 0.902 | 0.915 | 0.887 | 0.945 | 1.000 | 0.884 |
| | (0.88) | (1.63) | (0.016) | (0.034) | (0.033) | (0.016) | (0.002) | (0.032) |
| 60.0 | 5.40 | 12.06 | 0.902 | 0.916 | 0.887 | 0.943 | 1.000 | 0.879 |
| | (0.92) | (1.54) | (0.023) | (0.037) | (0.045) | (0.018) | (0.003) | (0.038) |
| 65.0 | 5.84 | 13.46 | 0.910 | 0.918 | 0.900 | 0.945 | 1.000 | 0.881 |
| | (1.03) | (2.02) | (0.023) | (0.028) | (0.040) | (0.020) | (0.000) | (0.040) |
| 70.0 | 6.24 | 14.34 | 0.911 | 0.923 | 0.899 | 0.944 | 1.000 | 0.881 |
| | (0.93) | (2.06) | (0.022) | (0.038) | (0.041) | (0.021) | (0.002) | (0.043) |
| 75.0 | 6.92 | 15.26 | 0.910 | 0.919 | 0.900 | 0.938 | 1.000 | 0.868 |
| | (1.07) | (1.71) | (0.026) | (0.040) | (0.036) | (0.024) | (0.000) | (0.048) |
| 80.0 | 7.68 | 16.54 | 0.913 | 0.911 | 0.916 | 0.946 | 1.000 | 0.883 |
| | (0.88) | (1.56) | (0.032) | (0.046) | (0.05) | (0.024) | (0.000) | (0.049) |
| 85.0 | 7.88 | 17.44 | 0.913 | 0.920 | 0.905 | 0.944 | 1.000 | 0.880 |
| | (0.99) | (1.73) | (0.040) | (0.052) | (0.058) | (0.029) | (0.000) | (0.061) |
| 90.0 | 8.56 | 19.22 | 0.913 | 0.915 | 0.909 | 0.938 | 1.000 | 0.862 |
| | (0.75) | (1.71) | (0.046) | (0.059) | (0.071) | (0.033) | (0.000) | (0.077) |
| 95.0 | 8.88 | 19.90 | 0.917 | 0.924 | 0.911 | 0.930 | 1.000 | 0.861 |
| | (0.86) | (1.81) | (0.073) | (0.081) | (0.103) | (0.054) | (0.000) | (0.097) |

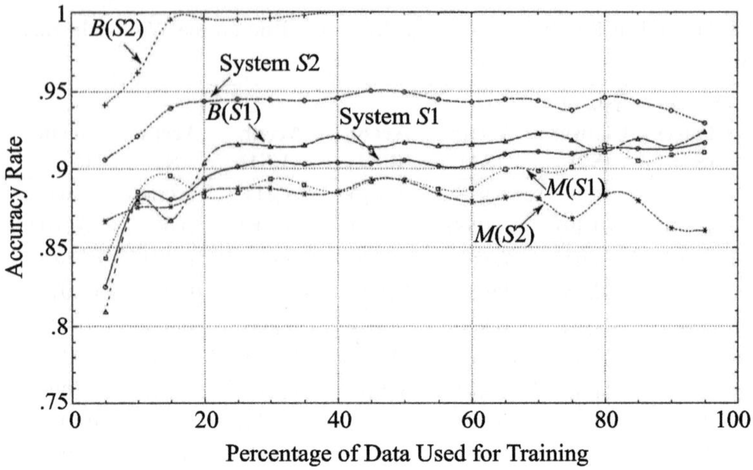


**Figure 4.3.** Accuracy Rates for Systems $S1$ and $S2$ When the Heuristic RA1 Is Used on the Wisconsin Breast Cancer Data.

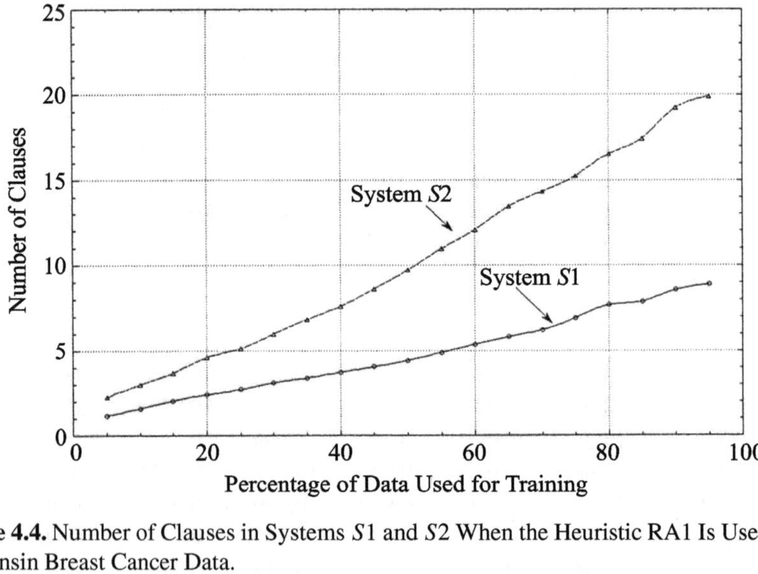


**Figure 4.4.** Number of Clauses in Systems $S1$ and $S2$ When the Heuristic RA1 Is Used on the Wisconsin Breast Cancer Data.

*27,000 CPU seconds* (on a VAX system which is 4–5 times slower than the IBM 3090-600S computer used in these tests) without being able to infer the 9 clauses from the breast cancer database. Some typical Boolean functions for systems $S1$ and $S2$ are given in Figure 4.5.

These clauses were derived from the entire set of cancer data. Recall that the cancer data had 9 discrete-valued attributes which were converted into 36 binary-valued

**System $S1$:**

| | |
|---|---|
| **Clause 1:** | $(A_1 \vee A_9 \vee A_{13} \vee A_{17} \vee A_{21} \vee A_{22} \vee A_{25} \vee A_{26} \vee A_{29} \vee A_{33} \vee A_{35})$ |
| **Clause 2:** | $(A_1 \vee A_2 \vee A_9 \vee A_{13} \vee A_{15} \vee A_{21} \vee A_{25} \vee A_{30} \vee A_{33})$ |
| **Clause 3:** | $(A_1 \vee A_3 \vee A_8 \vee A_9 \vee A_{17} \vee A_{21} \vee A_{25} \vee A_{33} \vee A_{35} \vee \bar{A}_{16})$ |
| **Clause 4:** | $(A_1 \vee A_9 \vee A_{17} \vee A_{21} \vee A_{23} \vee A_{25} \vee A_{33} \vee A_{35} \vee \bar{A}_{16})$ |
| **Clause 5:** | $(A_3 \vee A_9 \vee A_{12} \vee A_{17} \vee A_{21} \vee A_{25} \vee A_{33} \vee A_{35})$ |
| **Clause 6:** | $(A_1 \vee A_{11} \vee A_{13} \vee A_{17} \vee A_{22} \vee A_{25} \vee A_{29} \vee A_{34} \vee A_{35}$ $\vee \bar{A}_{12} \vee \bar{A}_{16} \vee \bar{A}_{20} \vee \bar{A}_{36})$ |
| **Clause 7:** | $(A_{13} \vee A_{15} \vee A_{17} \vee A_{22} \vee A_{24} \vee A_{25} \vee A_{27} \vee A_{29} \vee A_{34} \vee \bar{A}_{16} \vee \bar{A}_{36})$ |
| **Clause 8:** | $(A_4 \vee A_{23} \vee A_{35} \vee \bar{A}_2 \vee \bar{A}_{18} \vee \bar{A}_{20} \vee \bar{A}_{36})$ |
| **Clause 9:** | $(A_8 \vee A_{10} \vee A_{15} \vee A_{21} \vee A_{23} \vee \bar{A}_{16} \vee \bar{A}_{20} \vee \bar{A}_{32})$ |

**System $S2$:**

| | |
|---|---|
| **Clause 1:** | $(A_{16} \vee A_{24} \vee \bar{A}_{21})$ |
| **Clause 2:** | $(A_{24} \vee \bar{A}_{23} \vee \bar{A}_{26})$ |
| **Clause 3:** | $(A_5 \vee \bar{A}_{21} \vee \bar{A}_{26})$ |
| **Clause 4:** | $(A_{12} \vee A_{26} \vee \bar{A}_1)$ |
| **Clause 5:** | $(A_6 \vee \bar{A}_{18} \vee \bar{A}_{19})$ |
| **Clause 6:** | $(\bar{A}_9 \vee \bar{A}_{26})$ |
| **Clause 7:** | $(A_{10} \vee \bar{A}_{13})$ |
| **Clause 8:** | $(A_{30} \vee \bar{A}_3 \vee \bar{A}_{10} \vee \bar{A}_{24})$ |
| **Clause 9:** | $(A_4 \vee A_8 \vee \bar{A}_{11} \vee \bar{A}_{23})$ |
| **Clause 10:** | $(A_{15} \vee \bar{A}_7 \vee \bar{A}_{10} \vee \bar{A}_{22})$ |
| **Clause 11:** | $(A_{21} \vee \bar{A}_7 \vee \bar{A}_{14} \vee \bar{A}_{16})$ |
| **Clause 12:** | $(A_6 \vee \bar{A}_{10} \vee \bar{A}_{23})$ |
| **Clause 13:** | $(A_{12} \vee \bar{A}_{15} \vee \bar{A}_{22})$ |
| **Clause 14:** | $(A_8 \vee A_{16} \vee \bar{A}_{30})$ |
| **Clause 15:** | $(A_5 \vee A_{13} \vee \bar{A}_2 \vee \bar{A}_{11} \vee \bar{A}_{26})$ |
| **Clause 16:** | $(A_3 \vee A_5 \vee A_{15} \vee A_{27} \vee \bar{A}_6)$ |
| **Clause 17:** | $(A_5 \vee A_{13} \vee A_{18} \vee \bar{A}_6 \vee \bar{A}_{28})$ |
| **Clause 18:** | $(A_3 \vee A_{13} \vee \bar{A}_7 \vee \bar{A}_{25})$ |
| **Clause 19:** | $(A_3 \vee A_{13} \vee A_{21} \vee A_{32} \vee \bar{A}_{11} \vee \bar{A}_{20})$ |
| **Clause 20:** | $(A_3 \vee A_{15} \vee A_{21} \vee \bar{A}_7 \vee \bar{A}_{28} \vee \bar{A}_{30})$ |

**Figure 4.5.** Clauses of Systems $S1$ and $S2$ When the Entire Wisconsin Breast Cancer Data Are Used.

attributes, denoted as $A_1, \ldots, A_{36}$. System $S2$ was derived after treating the malignant examples as the $E^+$ data set and the benign examples as the $E^-$ data set. As such, the set of clauses in system $S2$ approximates the complement of system $S1$. This can be seen in Figure 4.5 where system $S1$ has 9 clauses, each clause containing on the average 10 attributes, whereas system $S2$ contains 20 clauses, each clause containing, on the average, four attributes.

**Table 4.2.** Numerical Results of Using the RA2 Heuristic on the Wisconsin Breast Cancer Database.

| % of Data | No. of Clauses (SA) | No. of Clauses (SB) | Accur. (SA) | Accur. B(SA) | Accur. M(SA) | Accur. (SB) | Accur. B(SB) | Accur. M(SB) |
|---|---|---|---|---|---|---|---|---|
| 5.0 | 1.26 | 2.08 | 0.866 | 0.882 | 0.848 | 0.869 | 0.884 | 0.921 |
| | (0.48) | (0.63) | (0.051) | (0.086) | (0.090) | (0.042) | (0.096) | (0.048) |
| 10.0 | 1.62 | 2.76 | 0.899 | 0.907 | 0.890 | 0.895 | 0.896 | 0.940 |
| | (0.49) | (0.62) | (0.031) | (0.043) | (0.065) | (0.028) | (0.055) | (0.023) |
| 15.0 | 1.88 | 3.18 | 0.916 | 0.931 | 0.899 | 0.917 | 0.897 | 0.941 |
| | (0.38) | (0.68) | (0.018) | (0.036) | (0.031) | (0.020) | (0.046) | (0.022) |
| 20.0 | 2.32 | 4.08 | 0.921 | 0.937 | 0.902 | 0.921 | 0.900 | 0.945 |
| | (0.61) | (0.84) | (0.019) | (0.030) | (0.040) | (0.022) | (0.044) | (0.023) |
| 25.0 | 2.76 | 4.84 | 0.925 | 0.941 | 0.908 | 0.934 | 0.917 | 0.953 |
| | (0.68) | (0.86) | (0.015) | (0.025) | (0.036) | (0.015) | (0.033) | (0.014) |
| 30.0 | 3.14 | 5.68 | 0.932 | 0.942 | 0.920 | 0.937 | 0.914 | 0.963 |
| | (0.75) | (0.97) | (0.013) | (0.026) | (0.024) | (0.013) | (0.028) | (0.016) |
| 35.0 | 3.60 | 5.98 | 0.940 | 0.954 | 0.924 | 0.949 | 0.936 | 0.964 |
| | (0.94) | (1.10) | (0.011) | (0.020) | (0.023) | (0.009) | (0.017) | (0.014) |
| 40.0 | 3.92 | 6.76 | 0.947 | 0.961 | 0.932 | 0.954 | 0.942 | 0.968 |
| | (0.84) | (0.93) | (0.009) | (0.015) | (0.210) | (0.009) | (0.017) | (0.014) |
| 45.0 | 4.46 | 8.14 | 0.949 | 0.958 | 0.939 | 0.954 | 0.935 | 0.977 |
| | (1.08) | (1.15) | (0.010) | (0.017) | (0.017) | (0.010) | (0.013) | (0.012) |
| 50.0 | 5.00 | 9.00 | 0.954 | 0.962 | 0.944 | 0.963 | 0.949 | 0.978 |
| | (0.98) | (1.36) | (0.009) | (0.015) | (0.015) | (0.010) | (0.020) | (0.009) |
| 55.0 | 5.38 | 9.40 | 0.959 | 0.967 | 0.950 | 0.967 | 0.952 | 0.985 |
| | (1.09) | (1.11) | (0.009) | (0.013) | (0.017) | (0.009) | (0.016) | (0.009) |
| 60.0 | 6.28 | 10.20 | 0.963 | 0.967 | 0.958 | 0.972 | 0.959 | 0.987 |
| | (1.10) | (1.25) | (0.009) | (0.014) | (0.014) | (0.008) | (0.015) | (0.008) |
| 65.0 | 6.54 | 10.78 | 0.967 | 0.973 | 0.961 | 0.977 | 0.966 | 0.990 |
| | (1.08) | (1.38) | (0.008) | (0.011) | (0.016) | (0.007) | (0.014) | (0.008) |
| 70.0 | 7.48 | 11.70 | 0.973 | 0.979 | 0.967 | 0.982 | 0.971 | 0.994 |
| | (1.30) | (1.00) | (0.007) | (0.010) | (0.013) | (0.006) | (0.011) | (0.005) |
| 75.0 | 8.24 | 12.06 | 0.978 | 0.979 | 0.976 | 0.984 | 0.973 | 0.996 |
| | (1.05) | (1.24) | (0.007) | (0.012) | (0.012) | (0.006) | (0.011) | (0.004) |
| 80.0 | 8.84 | 12.84 | 0.983 | 0.985 | 0.985 | 0.987 | 0.978 | 0.997 |
| | (1.21) | (1.17) | (0.008) | (0.011) | (0.011) | (0.006) | (0.010) | (0.004) |
| 85.0 | 9.56 | 12.34 | 0.986 | 0.988 | 0.985 | 0.992 | 0.986 | 0.998 |
| | (1.17) | (0.97) | (0.006) | (0.007) | (0.009) | (0.005) | (0.009) | (0.004) |
| 90.0 | 10.36 | 12.90 | 0.992 | 0.993 | 0.992 | 0.995 | 0.991 | 1.000 |
| | (0.97) | (0.83) | (0.004) | (0.006) | (0.007) | (0.003) | (0.006) | (0.001) |
| 95.0 | 10.98 | 12.90 | 0.996 | 0.996 | 0.996 | 0.998 | 0.996 | 1.000 |
| | (0.97) | (0.92) | (0.003) | (0.004) | (0.004) | (0.002) | (0.004) | (0.001) |

### 4.4.2 Results for the RA2 Heuristic on the Wisconsin Cancer Data with Some Missing Values

The results for Problem 2 (i.e., inference of a Boolean function with incomplete data) are presented in Table 4.2 and graphically depicted in Figures 4.6 and 4.7. Heuristic RA2 was used to solve this problem. The undecidable data were generated by "covering" (i.e., masking out) the actual values of some elements in some random examples taken from the cancer database. When the covered values were enough not to allow for the classification of that example, that example was introduced into the $E^U$ set and the covered attributes were assigned missing values.

Two systems of clauses (rules) were inferred. System *SA* was inferred with only the positive (i.e., the $E^+$) and the negative (i.e., the $E^-$) data sets. On the other hand, system *SB* was inferred from $E^+$, $E^-$ *as well as* the undecidable (i.e., the $E^U$) data set. The reason for doing this was to compare the relative benefit of including the undecidable data sets as opposed to inferring a system of clauses without the inclusion of the undecidable data.

The number of replications for each case was also equal to 50 and the numerical results are presented in Table 4.2. That number of replications produced rather acceptable confidence intervals and it was limited due to the excessive CPU time requirements. The individual benign and malignant accuracies for systems *SA* and *SB* are denoted in Table 4.2 by *B(SA)*, *M(SA)* and *B(SB)*, *M(SB)*, respectively. The graphs for the accuracies and the number of clauses derived by the RA2 algorithm with and without the inclusion of the undecidable data set $E^U$, are shown in Figures 4.6 and 4.7, respectively.

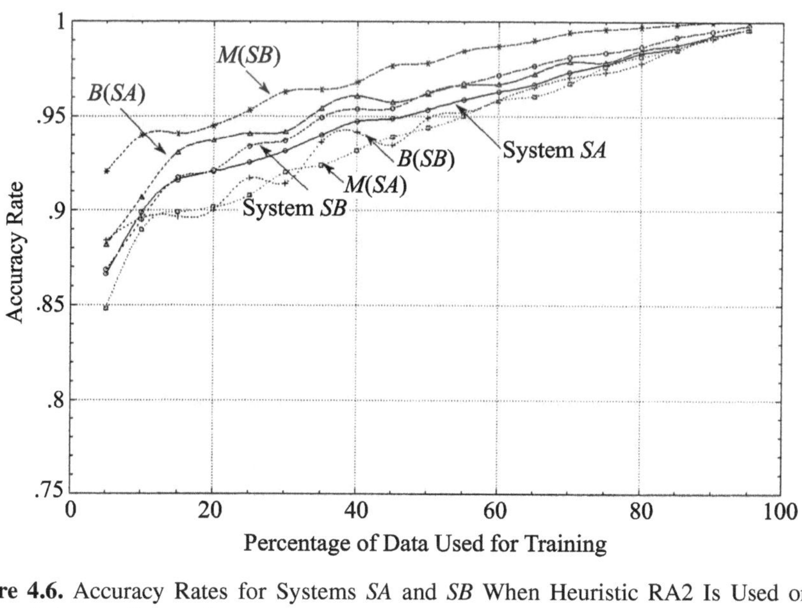


**Figure 4.6.** Accuracy Rates for Systems *SA* and *SB* When Heuristic RA2 Is Used on the Wisconsin Breast Cancer Data.

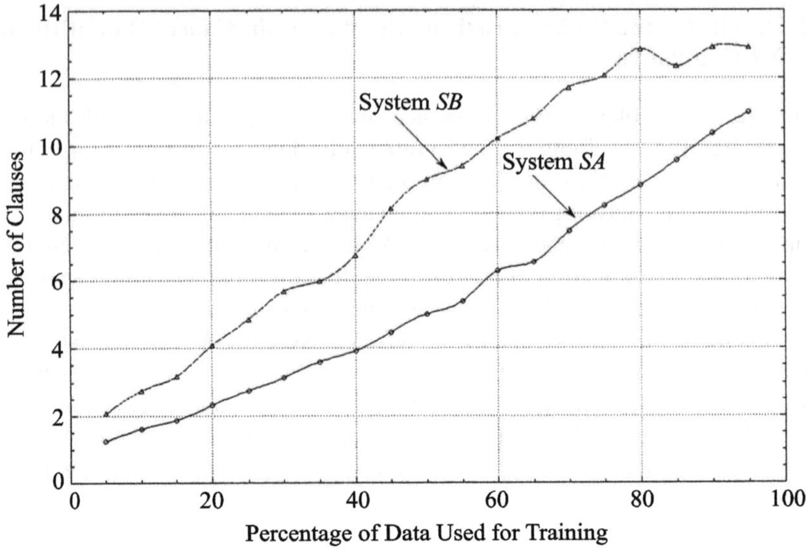


**Figure 4.7.** Number of Clauses in Systems *SA* and *SB* When Heuristic RA2 Is Used on the Wisconsin Breast Cancer Data.

As was anticipated, the accuracy obtained with the inclusion of the undecidable data set $E^U$ was *always higher* than the corresponding accuracy obtained without the inclusion of the $E^U$ data set. Therefore, it is indicated in these experiments that the use of the undecidable data set, along with the positive and negative data sets, indeed improves the quality of the inferred Boolean system. That is, the inferred system is a better approximation of the *hidden* system.

### 4.4.3 Comparison of the RA1 Algorithm and the B&B Method Using Large Random Data Sets

In order to compare the RA1 heuristic with the revised B&B method described in Chapter 3, a large data set was randomly generated and used for inferring Boolean functions. The difficulty of the problem is determined not only by the number of examples, but also by the percentage of examples used for training as compared to the total possible number of examples. For $n$ attributes, the total number of all possible distinct examples is $2^n$. The problem will be easy when there are very few examples (i.e., when $n$ is small) or when the number of the training examples approaches $2^n$.

The upper limit on the number of examples was limited by the dynamic allocation of the random access memory (RAM memory) of the IBM 3090-600S mainframe computer at the Louisiana State University (LSU) which was in use in the middle 1990s. Thus, a maximum of 18,120 examples were randomly generated for the purpose of these computational experiments when the number of attributes was set equal to 15. A maximum of 32,768 ($= 2^{15}$) distinct examples are possible when

```
DO WHILE (E⁻ ≠ ∅)
        Call Procedure RA1 to infer a single clause.
        If no examples from the E⁻ set are rejected by that clause then:
                Call the B&B method to infer a single clause.
        Update the E⁻ set.
REPEAT;
```

**Figure 4.8.** Using the RA1 Heuristic in Conjunction with the B&B Method.

$n$ is equal to 15. Therefore, this data set is a representation of one of the most difficult problems possible with $n$ equal to 15.

The RA1 algorithm terminates when all the clauses, when taken together, reject the entire set of negative examples in the $E^-$ set. A critical question is what happens if none of the clauses formed in the local search are able to reject even a single negative example. In the proof of Theorem 4.1 it was shown that a clause can always be formed which rejects exactly one negative example while accepting all positive examples. However, a disadvantage of that approach is that only a single negative example would be rejected at a time by the inclusion of such a new clause. If the number of negative examples is large and if this boundary condition is triggered often, then the system could end up with an exceptionally large number of (degenerative) clauses.

A possible alternative solution to this situation is the use of the RA1 heuristic in conjunction with the B&B method described in Chapter 3 of this book. The B&B method always guarantees to return a clause, which most of the time, rejects more than just a single negative example. However, there is also a certain trade-off to this implementation. The B&B method has an exponential time complexity. Hence, for large-size data sets, it has the potential to take large amounts of CPU time. Figure 4.8 best describes the idea of using the RA1 algorithm in conjunction with the B&B method (or for that matter, with any other single clause inference method).

The computational results are presented in Tables 4.3 and 4.4 which compare the combined RA1 heuristic and the B&B method with the stand-alone B&B method. Two sizes of random data sets were used in these tests. One set of data contained a total of 18,120 examples with a varying ratio of positive to negative examples (see also Table 4.3). The number of attributes in this data set was equal to 15.

In the other data set, the total number of examples was equal to 3,750. The number of attributes in this case was set equal to 14 (see also Table 4.4). These numbers of examples were fixed after determining the RAM memory restrictions of the IBM 3090-600S mainframe computer system at LSU.

Besides the CPU times consumed by the RA1/B&B combination and the stand-alone B&B search, the number of clauses inferred by each component method was also recorded. Moreover, the number of times the boundary condition was invoked (i.e., the number of times the B&B method was used) in the RA1/B&B combination, was recorded as well.

**Table 4.3.** Comparison of the RA1 Algorithm with the B&B Method (the total number of examples = 18,120; number of attributes = 15).

| $\|E^+\|$ | $\|E^-\|$ | Clauses by B&B in RA1/B&B | Clauses by RA1/B&B | Clauses by B&B | CPU Time RA1/B&B (in seconds) | CPU Time B&B (in seconds) |
|---|---|---|---|---|---|---|
| 264 | 17,856 | 11 | 15 | 15 | 267 | 267 |
| 690 | 17,430 | 13 | 20 | 18 | 614 | 1,265 |
| 730 | 17,390 | 16 | 23 | 26 | 2,243 | 3,302 |
| 856 | 17,264 | 20 | 27 | 23 | 5,781 | 10,983 |
| 1,739 | 16,381 | 17 | 23 | 21 | 5,266 | 6,244 |
| 1,743 | 16,377 | 36 | 45 | 46 | 3,442 | 6,016 |
| 1,773 | 16,347 | 39 | 46 | 39 | 5,150 | 10,020 |
| 2,013 | 16,107 | 24 | 26 | 25 | 2,058 | 2,000 |
| 2,298 | 15,822 | 38 | 44 | 41 | 4,777 | 4,891 |
| 2,396 | 15,724 | 23 | 24 | 31 | 2,816 | 2,583 |
| 2,400 | 15,720 | 36 | 45 | 48 | 3,719 | 4,827 |
| 2,913 | 15,207 | 35 | 40 | 45 | 4,344 | 4,532 |
| 3,090 | 15,030 | 34 | 37 | 34 | 4,889 | 4,945 |
| 3,459 | 14,661 | 38 | 40 | 32 | 12,187 | 14,547 |
| 3,574 | 14,546 | 34 | 41 | 67 | 4,980 | 9,245 |
| 3,917 | 14,203 | 46 | 53 | 52 | 10,588 | 12,232 |
| 4,781 | 13,339 | 47 | 48 | 95 | 10,243 | 19,475 |
| 5,750 | 12,370 | 29 | 30 | 29 | 7,959 | 7,944 |
| 6,503 | 11,617 | 48 | 51 | 56 | 5,316 | 9,688 |
| 6,608 | 11,512 | 34 | 37 | 52 | 3,887 | 12,632 |
| 6,989 | 11,131 | 62 | 66 | 60 | 16,719 | 17,626 |
| 9,981 | 8,139 | 42 | 44 | 43 | 12,232 | 12,146 |
| 10,554 | 7,566 | 42 | 42 | 42 | 12,681 | 12,523 |

As can be seen from the previous computational results, the combination of the RA1 heuristic with the B&B method performed significantly better than the stand-alone B&B method when one focuses on the CPU times. Figure 4.9 shows the percentage of times the B&B method was invoked when it was combined with the RA1 heuristic. On the average, the B&B method was called approximately 60% of the time in the combined RA1/B&B approach.

Figure 4.10 depicts the ratio of the number of clauses returned by the combination of the RA1/B&B method as compared to the stand-alone B&B method. As can be seen from that figure, the number of clauses returned by the combined methods is comparable to the number of clauses returned when the more greedy (and by far more CPU time demanding) B&B is used alone. Figure 4.11 depicts the absolute values of the previous numbers of clauses.

Figure 4.12 shows the ratio of the time taken by the stand-alone B&B method to the time taken by the combined RA1/B&B method. These results indicate the

**Table 4.4.** Comparison of the RA1 Algorithm with the B&B Method (the total number of examples = 3,750; number of attributes = 14).

| $\lvert E^{+}\rvert$ | $\lvert E^{-}\rvert$ | Clauses by B&B in RA1/B&B | Clauses by RA1/B&B | Clauses in B&B | CPU Time RA1/B&B (in seconds) | CPU Time B&B (in seconds) |
|---|---|---|---|---|---|---|
| 10 | 3,740 | 10 | 15 | 12 | 1 | 2 |
| 14 | 3,736 | 8 | 20 | 18 | 44 | 89 |
| 18 | 3,752 | 12 | 23 | 22 | 51 | 172 |
| 19 | 3,731 | 6 | 25 | 19 | 12 | 70 |
| 23 | 3,727 | 11 | 32 | 25 | 75 | 125 |
| 23 | 3,727 | 9 | 20 | 20 | 99 | 206 |
| 30 | 3,720 | 7 | 20 | 24 | 76 | 202 |
| 33 | 3,717 | 9 | 24 | 20 | 112 | 299 |
| 40 | 3,710 | 11 | 24 | 17 | 23 | 28 |
| 47 | 3,703 | 12 | 21 | 17 | 73 | 97 |
| 52 | 3,698 | 10 | 17 | 17 | 38 | 53 |
| 53 | 3,697 | 16 | 29 | 36 | 519 | 1,218 |
| 62 | 3,688 | 10 | 22 | 18 | 16 | 146 |
| 65 | 3,685 | 12 | 27 | 23 | 132 | 374 |
| 67 | 3,683 | 15 | 22 | 20 | 586 | 739 |
| 77 | 3,673 | 9 | 17 | 16 | 90 | 354 |
| 88 | 3,662 | 13 | 28 | 24 | 312 | 1,303 |
| 112 | 3,638 | 5 | 13 | 11 | 50 | 57 |
| 140 | 3,610 | 22 | 38 | 30 | 1,735 | 1,867 |
| 233 | 3,517 | 28 | 37 | 31 | 1,416 | 2,487 |
| 345 | 3,405 | 14 | 22 | 20 | 818 | 863 |
| 379 | 3,371 | 21 | 30 | 29 | 587 | 621 |
| 419 | 3,331 | 28 | 39 | 30 | 596 | 754 |
| 425 | 3,325 | 28 | 25 | 21 | 1,149 | 1,266 |
| 552 | 3,198 | 25 | 30 | 28 | 534 | 704 |
| 558 | 3,192 | 24 | 38 | 30 | 704 | 1,407 |
| 558 | 3,192 | 26 | 33 | 30 | 774 | 1,408 |
| 846 | 2,904 | 22 | 28 | 23 | 2,812 | 3,171 |
| 864 | 2,886 | 35 | 39 | 38 | 968 | 1,487 |
| 899 | 2,851 | 37 | 41 | 37 | 1,620 | 2,197 |
| 924 | 2,826 | 33 | 35 | 39 | 1,502 | 2,379 |
| 1,112 | 2,638 | 35 | 38 | 35 | 1,183 | 1,020 |

relative benefits of using the proposed combined approach (i.e., the RA1/B&B method) as compared to the earlier stand-alone methods (i.e., the stand-alone RA1 or the stand-alone B&B methods). The CPU time performance of the combined RA1/B&B method, when tested in terms of the previous computational experiments, was, on the average, two to three times faster when compared to the stand-alone

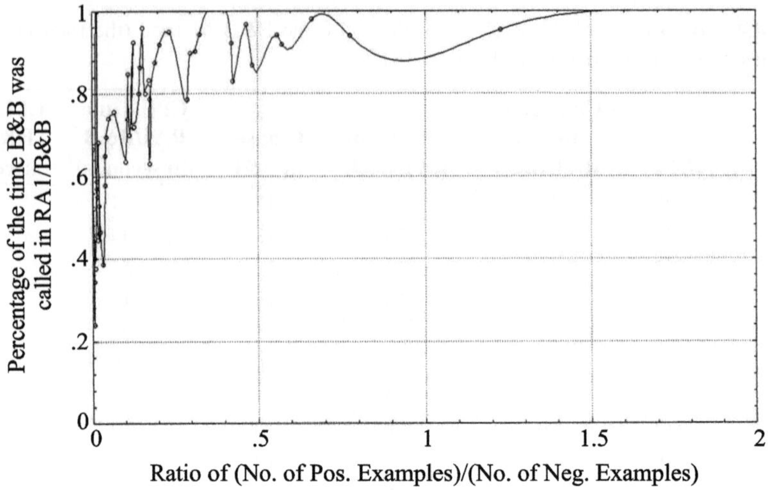


**Figure 4.9.** Percentage of the Time the B&B Was Invoked in the Combined RA1/B&B Method.

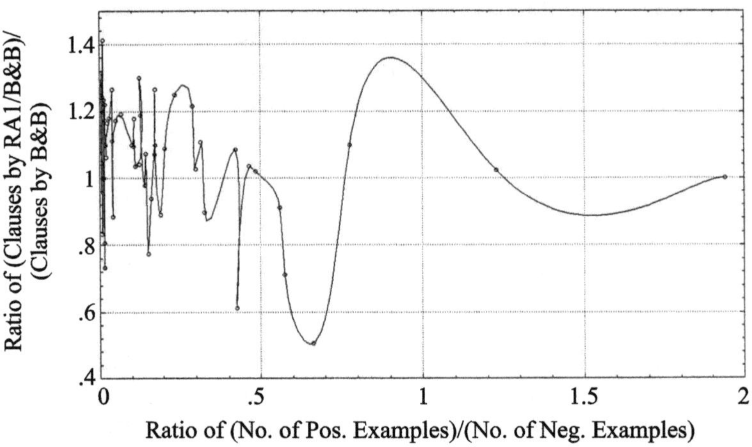


**Figure 4.10.** Ratio of the Number of Clauses by the RA1/B&B Method and the Number of Clauses by the Stand-Alone B&B Method.

B&B method. Finally, Figure 4.13 shows the actual CPU times taken by the two methods.

At present, previously obtained benchmark results which take into consideration undecidable examples are not available. Hence, computational results obtained with the RA2 algorithm were not compared with any other set of results. A logical extension of the work done so far is to develop a B&B method which would take into

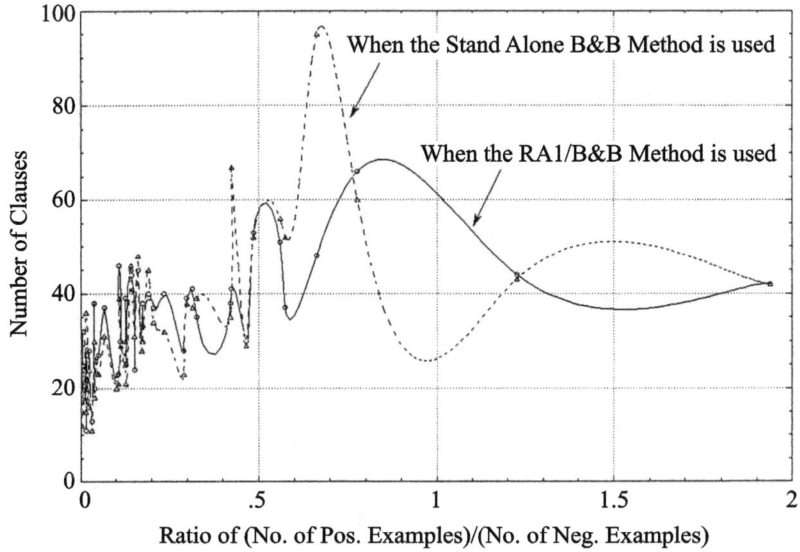


**Figure 4.11.** Number of Clauses by the Stand-Alone B&B and the RA1/B&B Method.

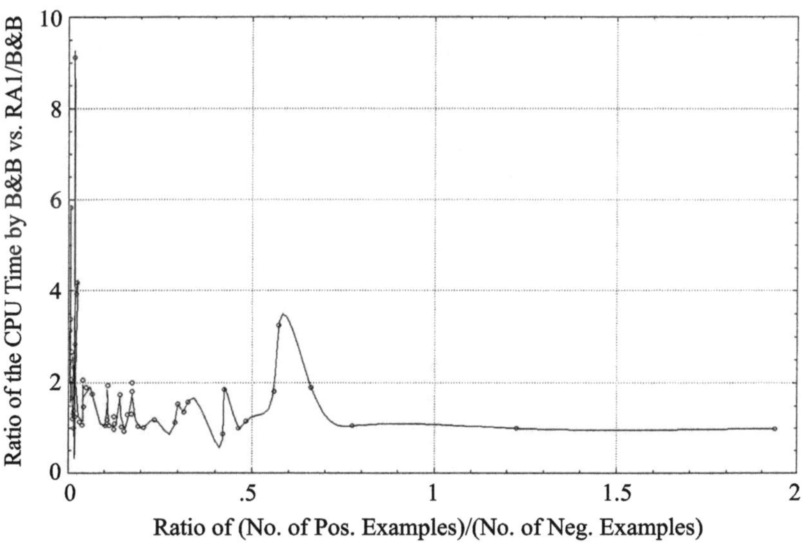


**Figure 4.12.** Ratio of the Time Used by the Stand-Alone B&B and the Time Used by the RA1/B&B Method.

consideration the undecidable examples, and compare the RA2 algorithm with this modified B&B method.

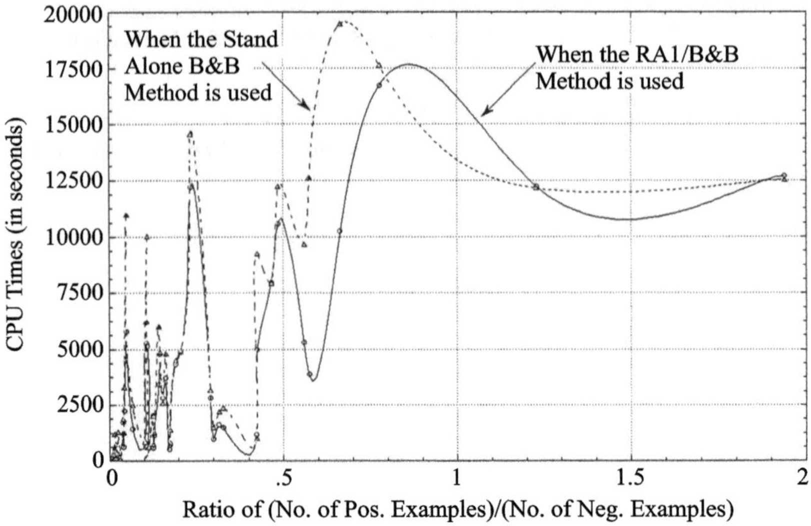


**Figure 4.13.** CPU Times by the Stand-Alone B&B and the RA1/B&B Method.

It is quite possible that the RA2 algorithm in conjunction with a modified B&B method would perform better than the present method (i.e., the stand-alone RA2 heuristic). Once this is developed, the combination of the RA2 algorithm and a modified B&B method could be tested on a large set of random examples to indicate with a high degree of certainty that the inclusion of undecidable examples indeed enhances the accuracy of the inferred systems. The above is a compelling direction for pursuing more research in this area of data mining and knowledge discovery in the future.

## 4.5 Concluding Remarks

This chapter discussed some developments in two closely related areas. The first contribution is the development of a randomized search heuristic, called RA1 (for *Randomized Algorithm 1*). This heuristic takes as input two disjoint sets of positive and negative examples (i.e., binary vectors of size $n$) and infers a Boolean function which satisfies the requirements of the input examples.

Unlike previous algorithms which were of exponential time complexity, the RA1 heuristic is of polynomial time complexity. However, it does not return small Boolean functions (in terms of the number of CNF clauses) as other more time demanding approaches require (e.g., the ones in [Kamath, *et al.*, 1992], [Triantaphyllou, *et al.*, 1994], and [Triantaphyllou, 1994] as discussed in Chapters 2 and 3).

However, computational results indicate that the RA1 heuristic returns comparably small numbers of logical clauses when it is compared with the other exponential time approaches. Moreover, as was shown in Figure 4.7 and supported by the computational results, the RA1 heuristic can be effectively combined with other methods (such as the B&B method presented in Chapter 3) and solve very large problems. This chapter also presented some test results of solving problems with more than 18,000 examples defined on 14 and also 15 binary attributes.

The second contribution described in this chapter is the development of the RA2 heuristic (for *Randomized Algorithm 2*). This algorithm can process examples in which some of the values may be missing. That is, now besides the ordinary positive and negative examples, some examples may be *undecidable* due to missing values in their descriptions (not to be confused with *unclassified* examples which simply are examples which have not been submitted for classification yet). This is the first time this kind of data been dealt with in the literature. As was anticipated, the inclusion of the undecidable data can significantly assist the search process. The above algorithms were tested on some large data sets and on the well-known breast cancer database which was originally compiled in the University of Wisconsin. As was stated earlier, an important research problem would be to develop new B&B algorithms for handling problems with undecidable examples, besides the usual positive and negative sets of examples as is the case with all the approaches so far.

The problem of extracting a small set of clauses (i.e., a Boolean function in CNF or DNF form) via logic-based approaches from classes of mutually exclusive observations is rapidly gaining a wide interest in the data mining and knowledge discovery community. This could be partly attributed to the failure of many other methods, such as neural networks, to gain the understanding and confidence of the end user (who usually does not have a computer/mathematical background). It should always be kept in mind, however, that the proposed logic-based methods should be used in deterministic environments. Some extensions into probabilistic settings are discussed later in Chapters 10 and 11. Clearly, more research in this area of high potential is required.

It should also be noted here that in all previous treatments the goal advocated in inferring a Boolean function from positive and negative training data is that of having the simplest possible function. Such a goal is in agreement with the adopted principle of maximum simplicity or Occam's razor. However, it also makes sense to pursue a different goal in inferring a Boolean function. To see the motivation, recall the three key error rates introduced in Chapter 1, namely, the false-positive, false-negative, and undecidable (or unclassifiable) error rates. Oftentime, especially in critical areas of application of data mining technologies, these three error rates may be associated with profoundly different penalty costs. As was mentioned in Chapter 1, making a diagnostic mistake of the false-positive type (i.e., for instance, recommending that a person is healthy while in reality he/she has a serious medical condition) is way more "costly" than the other way around, i.e., having a false-negative case. Therefore, one may wish to minimize (or at least significantly reduce) the following total misclassification cost function:

$$\begin{aligned}\text{Total Misclassification Cost} = {} & \text{Cost1} \times \text{false-positive rate} \\ & + \text{Cost2} \times \text{false-negative rate} \\ & + \text{Cost3} \times \text{undecidable rate}.\end{aligned} \tag{4.1}$$

The above goal is being pursued in some recent work as described in [Pham and Triantaphyllou, 2007; 2008; 2009a; 2009b]. This is done by first decomposing the sets of the training examples into smaller sets. This is achieved by using certain data homogeneity or convexity properties and then solving the smaller inference as defined by these partitions. The smaller problems attempt to find an optimal balance between data overfitting and model overgeneralization. These strategies can be applied in conjunction with various data mining approaches and not only those that are based on mathematical logic.

One may argue that minimizing the above total misclassification cost is a very important objective. However, what if the derived models constructed this way are way too complex? For instance, what if when logic methods are used and the derived Boolean functions have too long CNF or DNF representations? Then, such knowledge might be too complex to be validated by domain experts or be trustworthy by such experts. On the other hand, shorter expressions are easier to be validated and implemented in real-life applications when compared to very complex ones.

Therefore, the problem of inferring the "best" new knowledge is not a well-defined one. That is, the real goal may not be accurately captured by an objective function given as expression (4.1). Perhaps, the truly best goal is a weighted *combination* of the maximum simplicity principle and a significant reduction of the total misclassification cost given as expression (4.1). Clearly, such an important issue is right at the core of the data mining and knowledge discovery field and could be resolved by doing more research in this area. Such research could aim at specific application domains by multidisciplinary groups, which should also include domain experts.

# Chapter 5

# An Approach to Guided Learning of Boolean Functions

## 5.1 Some Background Information

In most of the previous treatments it was assumed that somehow we have available two disjoint sets of training data described by binary vectors, that is, the collections of the positive and negative examples. Then the problem was how to infer a Boolean function that "fits these data." In other words, a Boolean function in CNF or DNF form that satisfies the requirements of the positive and negative examples as described in Chapters 2 and 3. It is hoped at this point that the inferred Boolean function will accurately classify all remaining examples not included in the currently available positive and negative examples.

Next suppose that the analyst has the capability to recommend which example to consider as the next input point. Of course, this example must be one of the unclassified ones. Such an example is not included in the current positive or negative examples. It is assumed that the analyst can somehow determine the structure of the next example and that example is submitted to an oracle for its actual classification. Such an oracle may be a series of tests or the opinion of a highly paid expert. In other words, this process may involve some type of *cost*. Thus, one may wish to get only a few additional examples in such a way that the understanding of the system under consideration will improve quickly.

For any new example, after its class membership has been determined by the oracle, there are two possibilities (as before, we assume that the system is deterministic). The classification by the oracle agrees with the classification of the inferred system (pattern, Boolean function, neural network, etc.) as defined so far, or the two disagree. The most beneficial case is when one gets as a new example one which reveals a *disagreement* between the two, when they indeed are different. In the opposite case, the classification of the new example would be of no value, as it would not provide an opportunity to improve our understanding of the system under consideration.

The above can be visualized as follows. Suppose that some training data points are available and some data mining algorithm (and not only ones which infer Boolean functions) is applied on these data points. The result would be the extraction of

E. Triantaphyllou, *Data Mining and Knowledge Discovery via Logic-Based Methods*, Springer Optimization and Its Applications 43, DOI 10.1007/978-1-4419-1630-3_5,

a model which in turn can be used to accurately classify the entire population of examples.

Next suppose that the *same* data mining algorithm is applied on the *same* training data as before, but now the positive examples are treated as negative and the negative examples are treated as positive. We will call the new model the *negative model* and the original model the *positive model*. These two names are quite arbitrary but indicate the way the two models are derived. Some approaches for learning from examples may derive two *symmetric systems*. That is, each system is the complement (opposite) of the other. Then the proposed guided learning strategy is not applicable. The OCAT approach with the B&B algorithms and the fast heuristics discussed in Chapters 2, 3, and 4 do not derive symmetric systems, thus the proposed guided learning strategy is applicable.

For such approaches each model (i.e., system inferred from training examples) splits the entire population of examples into two regions (the *positive region* and the *negative region*) in different ways (i.e., not symmetric). In such cases it is possible that the two previous models, when they are considered together, may split the entire population into four regions as follows (see also Figure 5.1 which is conceptual in nature):

(1) The region which the positive model classifies as positive and the negative system as positive. That is, examples in this region are accepted by both models. This is **Region A** in Figure 5.1.
(2) The region which the positive model classifies as negative and the negative model as positive. That is, examples in this region are accepted by one (the negative model) and rejected by the other (the positive model). This is **Region B** in Figure 5.1.
(3) The region which the positive model classifies as positive and the negative model as negative. That is, examples in this region are accepted by one (the positive model) and rejected by the other (the negative model). This is **Region C** in Figure 5.1.
(4) The rest of the area which has the examples rejected by both models. This is **Region D** in Figure 5.1.

The key question now is how to select a new example in a guided learning mode in a way that can lead to fast approximation of the two "hidden" systems. From the previous discussion it follows that if an example is selected such that it is currently accepted by both models or currently rejected by both models (that is, from Region A or Region D, respectively), then when it is classified by the oracle, one of the two systems will need to be modified. For instance, if an example is selected from Region A, and the oracle declares it to be a positive one, then the "negative" model will need to be retrained by inferring a new model which satisfies the requirements of the previous training data but it should also not accept the new example. If the new example is classified by the oracle as negative, then the "positive" model needs to be determined again. An analogous strategy can be followed if a new example is selected from Region D.

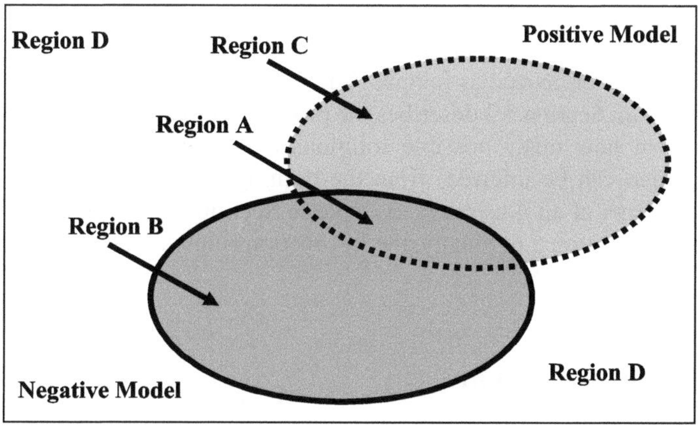


**Figure 5.1.** All Possible Classification Scenarios When the Positive and Negative Models Are Considered.

On the other hand, if an example is selected, say, from Region B, and the oracle classifies it as negative, then *neither* of the two models needs to be retrained. However, if that example from Region B is classified as positive by the oracle, then *both* of the models need to be retrained. That is, now the action at each step may be to retrain neither model or both models. However, when examples are selected from Region A or Region D (i.e., the areas with the undecidable/unclassifiable examples), then *always* one of the two models needs to be retrained. Another related research question at this point is how to select a new example from a given region. The above considerations are the foundation of the theory that is developed in later sections of this chapter.

Besides improving the performance of an existing intelligent system, a learning mechanism can assist in creating the initial knowledge base of that intelligent system. That is, it can assist in the *knowledge acquisition* phase. By asking the oracle a short sequence of key questions, we hope that the knowledge base of an intelligent system can be configured accurately and efficiently.

In the context of this chapter these questions refer to the *classification of examples*. That is, the oracle is presented with a new example, one at a time. Then, the oracle is asked to classify this example either as positive or as negative. Although the logic (also known as rules, system, clauses, Boolean function or expression) may not be explicitly known to the oracle, it is assumed that the oracle can classify new examples correctly. The *inductive inference problem* is to derive the "hidden" system (also called the *hidden logic*) from the classifications of sampled examples.

*The problem examined in this chapter is how to determine new examples.* It is assumed that two initial sets of positive and negative examples are given. Since these initial sets are a small sample of all the possibilities, new examples may be required.

An effective sequence of new examples should be able to lead to the accurate inference of the " hidden logic" by considering relatively few new examples.

This chapter is organized as follows. The following section describes the guided learning problem. Section 5.3 describes the proposed approach. Section 5.4 analyzes the problem of how many possible solutions exist (i.e., the number of Boolean functions which can be inferred) from the training data. The proposed method is described in terms of an illustrative example in Section 5.5. Some empirical results are analyzed in Section 5.6. Finally, the last section summarizes the main points and conclusions of this chapter.

## 5.2 Problem Description

Suppose that there exists a "hidden logic." In other words, there is a system that we would like to infer from collections of positive and negative examples. Although we *cannot explicitly* identify the structure of the "hidden system," it is assumed that it is possible to correctly classify any new examples according to this "hidden" system. This can occur, for instance, by interviewing an oracle.

To help fix ideas, suppose that the following represents the "hidden logic":

$$(\bar{A}_1 \vee \bar{A}_4 \vee A_6)(\bar{A}_2 \vee A_8) \wedge (A_2).$$

This system is considered to be *unknown* to the user. By *user* we mean here the person (or another computer system) which wants to infer the "hidden logic" from collections of positive and negative examples. Next, let the following sets $E^+$ and $E^-$ represent two collections of positive and negative examples, respectively. These are the examples which an oracle has already classified according to the "hidden logic" which remains unknown to the analyst.

$$E^+ = \begin{bmatrix} 1 & 1 & 1 & 0 & 1 & 0 & 1 & 1 \\ 0 & 1 & 0 & 0 & 0 & 1 & 0 & 1 \\ 0 & 1 & 1 & 0 & 0 & 1 & 0 & 1 \end{bmatrix} \text{ and}$$

$$E^- = \begin{bmatrix} 1 & 0 & 1 & 1 & 0 & 0 & 1 & 0 \\ 0 & 1 & 1 & 1 & 1 & 0 & 0 & 0 \\ 1 & 1 & 1 & 1 & 0 & 1 & 1 & 0 \\ 0 & 0 & 1 & 1 & 0 & 0 & 1 & 1 \\ 0 & 0 & 1 & 0 & 1 & 1 & 0 & 0 \\ 1 & 0 & 1 & 0 & 0 & 0 & 1 & 0 \\ 0 & 0 & 1 & 0 & 0 & 1 & 0 & 0 \end{bmatrix}.$$

Given the above examples, we want to determine a set of clauses (i.e., a Boolean function) which correctly classify the previous examples.

Next, we apply the OCAT approach (see also Chapters 2 and 3) with the RA1 heuristic as described in Chapter 4. When the OCAT approach is applied on the previous $E^+$ and $E^-$ sets of examples, the following CNF system is derived (we call it system $S_{\text{SAMPLE}}$ to emphasize that it has been derived from sampled data):

$$(\bar{A}_3 \vee A_8) \wedge (A_2). \tag{5.1}$$

The system proposed above may or may not be a good approximation of the "hidden logic." Recall that any sampling process is subject to random variation.

Now suppose that the user can supply the oracle with additional examples for correct classification. Then, the *main problem* examined in this chapter is *how to generate the next example.* One obvious approach is to generate the next example *randomly*. However, this may result in generating many examples and still not achieving a good approximation of the unknown system. It is obviously desirable to consider a sequence of new examples which can lead to a good approximation of the unknown system as quickly as possible.

When a new example is considered, it is given to the oracle for the correct classification. Two situations can occur. First, the current Boolean function (which is attempting to represent the unknown "hidden logic") classifies the new example in a manner *identical* with the oracle (which always correctly classifies each example). In the second case, the new example is classified in the *opposite* way by the oracle and the current Boolean function. If the current Boolean function is not yet a good approximation of the "hidden logic," then the last case is the most desirable scenario. This is true, because in this case one can reexecute a learning from examples algorithm (for instance, the OCAT approach) again and, hopefully, derive a closer approximation of the unknown system.

If the current version of the Boolean function is not an accurate approximation of the "hidden logic" and one generates new examples which fail to reveal any contradictions, then additional costs are incurred in classifying new examples, but *no improvement* is gained. Clearly, it is desirable to use a strategy for determining the next example, such that any possible contradiction between the current version of the Boolean function and the "hidden logic" will *surface early in the interviewing process*. The next section presents the development of such a strategy.

## 5.3 The Proposed Approach

Consider some sets of positive and negative examples, denoted as $E^+$ and $E^-$, respectively, defined on $n$ attributes. Let $S_{\text{SAMPLE}}$ denote a Boolean function which correctly classifies the sample data, i.e., the examples in $E^+$ are classified as positive and the examples in $E^-$ are classified as negative. When the proposed guided learning strategy is applied, the derived system will be denoted as $S_{\text{GUIDED}}$. Also, denote as $S_{\text{HIDDEN}}$ the "hidden logic" Boolean function and $\bar{S}_{\text{HIDDEN}}$ the complement (i.e., the negation) of $S_{\text{HIDDEN}}$. Hence, if $S_{\text{HIDDEN}}$ is

$$(\bar{A}_2 \vee A_3) \wedge (A_1 \vee A_2),$$

then $\bar{S}_{\text{HIDDEN}}$ is

$$\overline{(\bar{A}_2 \vee A_3) \wedge (A_1 \vee A_2)}.$$

Our objective is to sequentially modify and improve $S_{\text{GUIDED}}$ so that

$$S_{\text{GUIDED}} \rightarrow S_{\text{HIDDEN}},$$

when additional examples are generated and included either in $E^+$ or in $E^-$. If the sequence of distinct examples generated so far is denoted as $v_1, v_2, v_3, \ldots, v_k$, then at least when $k = 2^n$, one must, by definition (since all possible examples have been generated), obtain

$$S_{\text{GUIDED}} = S_{\text{HIDDEN}}.$$

The objective of our selection strategy is to choose a sequence of examples so that

$$S_{\text{GUIDED}} \approx S_{\text{HIDDEN}},$$

even when $k$ is rather small (maybe only a tiny fraction of $2^n$). We view the problem from a local perspective only. *In particular, if k examples have already been generated, then what should be the $k + 1^{\text{st}}$ example?*

The method by which we propose to select the $k + 1^{st}$ example is based on the observation that for any example $v$, either $S_{\text{HIDDEN}}$ or $\bar{S}_{\text{HIDDEN}}$ must classify the example as positive, but *not* both. Denote $S_{\text{HIDDEN}}(v) = 1$ if the Boolean function $S_{\text{HIDDEN}}$ classifies the example $v$ as positive, and $S_{\text{HIDDEN}}(v) = 0$ if it classifies it as negative. Then, for any example $v$ the following relation is always true:

$$S_{\text{HIDDEN}}(v) + \bar{S}_{\text{HIDDEN}}(v) = 1.$$

Next, consider a Boolean function, $S_{\text{GUIDED}}$, determined by sampling $k$ examples and generating a set of clauses which correctly classify the $k$ examples. $S_{\text{GUIDED}}$ is an approximation to $S_{\text{HIDDEN}}$.

The OCAT approach described in earlier chapters is one strategy for determining $S_{\text{GUIDED}}$. However, **any** learning algorithm such as OCAT could also be applied to a type of *dual problem*, i.e., a problem in which the positive examples are treated as negative and vice versa. Let $S_{\text{R-GUIDED}}$ be a Boolean function generated when the $k$ examples are assigned *reverse* truth values. That is, the positive examples are treated as negative and the negative examples as positive. In this manner, $S_{\text{R-GUIDED}}$ would be an approximation to $\bar{S}_{\text{HIDDEN}}$. If, indeed, $k = 2^n$, then the following is true:

$$S_{\text{GUIDED}}(v) + S_{\text{R-GUIDED}}(v) = 1, \tag{5.2}$$

for all examples $v$.

However, in general, for $k < 2^n$ one should expect that some examples will exist for which the sum in (5.2) will be 0 or 2, i.e., some example $v$ will be classified either as negative (sum is equal to 0) or as positive (sum is equal to 2) by *both* Boolean functions. Such an example is the key to our approach. The existence of such an example means that exactly one of our two example generated Boolean functions (i.e., $S_{\text{GUIDED}}$ and $S_{\text{R-GUIDED}}$) is in error. Either $S_{\text{GUIDED}}$ or $S_{\text{R-GUIDED}}$ must be modified to *correctly* classify the new example plus all the previous training examples. These observations are summarized in the following theorem [Triantaphyllou and Soyster, 1996]:

**Theorem 5.1.** *Suppose that there exists an example* $v \in \{0, 1\}^n$ *such that*

$$S_{\text{GUIDED}}(v) + S_{\text{R-GUIDED}}(v) = 0 \quad or: \tag{5.3a}$$

$$S_{\text{GUIDED}}(v) + S_{\text{R-GUIDED}}(v) = 2. \tag{5.3b}$$

*Furthermore, suppose that the example v is classified by the oracle as either positive or negative. Then, one and only one of the following situations is true:*

1) *If (5.3a) holds and v is a positive example, then system* $S_{\text{GUIDED}}$ *is not valid.*
2) *If (5.3a) holds and v is a negative example, then system* $S_{\text{R-GUIDED}}$ *is not valid.*
3) *If (5.3b) holds and v is a positive example, then system* $S_{\text{R-GUIDED}}$ *is not valid.*
4) *If (5.3b) holds and v is a negative example, then system* $S_{\text{GUIDED}}$ *is not valid.*

Therefore, the overall strategy, starting with two Boolean functions, is to attempt to generate a sequence of new examples $v_{k+1}, v_{k+2}, v_{k+3}, \ldots, v_m$, where each example is appropriately classified, as positive or negative, by the oracle. Each additional example should have the property that it invalidates either $S_{\text{GUIDED}}$ or $S_{\text{R-GUIDED}}$, i.e., one of the two Boolean functions must be modified. In doing so, it is expected that $S_{\text{GUIDED}}$ and $S_{\text{R-GUIDED}}$ become more closely aligned with $S_{\text{HIDDEN}}$ and $\bar{S}_{\text{HIDDEN}}$, respectively.

How does one find an example that invalidates either $S_{\text{GUIDED}}$ or $S_{\text{R-GUIDED}}$? Conceptually it is quite simple. One strategy is to formulate and solve at most two *clause satisfiability problems*. The satisfiability (or SAT) problem is NP-complete and can be defined as follows (see, for instance, [Hooker, 1988a]):

*Consider the m CNF clauses* $C_1, C_2, C_3, \ldots, C_m$ *involving the n attributes* $A_1, A_2, A_3, \ldots, A_n$, *and the Boolean expression* $C_1 \wedge C_2 \wedge C_3 \wedge \cdots \wedge C_m$. *The expression is* **satisfiable** *if there exists an assignment of truth values which makes the Boolean expression true.*

The clause satisfiability problem has been examined with noticeable success [Hooker, 1988a; 1988b]. A related development reported in [Kamath, *et al.*, 1992] uses an interior point algorithm developed by Karmakar and his associates in [Karmakar, *et al.*, 1991] with considerable success. Also, some problem preprocessing techniques can be found in [Cavalier, Pardalos, and Soyster, 1990].

The two satisfiability problems of interest in this chapter are as follows: Determine an example $v$ which results in truth value for

$$\bar{S}_{\text{GUIDED}}(v) \wedge \bar{S}_{\text{R-GUIDED}}(v) \quad \text{or:} \tag{5.4}$$

$$S_{\text{GUIDED}}(v) \wedge S_{\text{R-GUIDED}}(v). \tag{5.5}$$

If relation (5.4) is true (i.e., it is satisfied), then the example $v$ is evaluated as negative by both systems (Boolean functions). Similarly, if relation (5.5) is true, then $v$

is evaluated as positive by both systems. Observe that relations (5.4) and (5.5) are *equivalent* to relations (5.3a) and (5.3b), respectively.

If an example is found which satisfies either (5.4) or (5.5), then one of the two Boolean functions is modified and the same process is repeated. To illustrate this, suppose that a new example $v$ satisfies (5.5) and the oracle classifies the example $v$ as positive (i.e., it evaluates it to true value). In this case $S_{\text{R-GUIDED}}$ does not classify $v$ correctly and thus it must be updated. If the oracle classifies $v$ as negative (i.e., it evaluates it to false value), then the example invalidates $S_{\text{GUIDED}}$ and thus now $S_{\text{GUIDED}}$ must be updated. If the example $v$ satisfies (5.4), a similar analysis is applicable. (Note that if we find an example $v$ which satisfies (5.4), then there is no need to also search for an example which satisfies (5.5).) If no example can be found to satisfy (5.4), then we search to find an example which satisfies (5.5).

But, suppose that there are no examples which satisfy (5.4) or (5.5). Does this mean that

$$S_{\text{GUIDED}} \equiv S_{\text{HIDDEN}}?$$

Unfortunately, *the answer is no*. As a trivial demonstration of why this is true, consider the case in which $S_{\text{HIDDEN}} = (A_1 \vee A_2)$. Next, consider the input examples to be defined as follows: $E^+ = [10]$ and $E^- = [00]$. Then, OCAT returns $S_{\text{GUIDED}} = A_1$, and $S_{\text{R-GUIDED}} = \bar{A}_1$. Clearly, although neither relation (5.4) nor (5.5) is satisfied in this case $S_{\text{GUIDED}}$ is different than $S_{\text{HIDDEN}}$.

In cases like this we revert to a *random* search process. Examples are randomly generated, say $v_{k+1}, v_{k+2}, v_{k+3}, \ldots, v_m$. The oracle appropriately classifies $v_{k+1}$ (as positive or as negative) and one evaluates

$$S_{\text{GUIDED}}(v_{k+1}) \tag{5.6}$$

and

$$S_{\text{R-GUIDED}}(v_{k+1}) \tag{5.7}$$

for consistency. That is, if $v_{k+1}$ is positive, then (5.6) should be true and (5.7) false, and if $v_{k+1}$ is negative, then the converse must be true. This raises the question of a termination criterion. How long should one generate new examples?

Clearly, there are two factors that one needs to take under consideration. One is the *cost* of generating and classifying new examples. The second factor is the *desired accuracy*. In general, one expects that the more examples one uses to infer a set of clauses, the more accurate the inferred system would be.

Therefore, it is recommended that if no inconsistency is determined for some large value of $m$ (which value depends on the cost of classifying new examples), the process is terminated and $S_{\text{GUIDED}}$ is our approximation to the "hidden logic." The proposed strategy is depicted in Figure 5.2, and is demonstrated by means of an illustrative example in Section 5.5. In Figure 5.2 a very simple termination criterion is used by introducing an upper limit (i.e., the value of MAX) on the number of the new examples.

A confidence interval on the probability that the derived system $S_{\text{GUIDED}}$ will disagree with the "hidden logic" can be found as follows: Suppose that $S_{\text{GUIDED}}$ has just

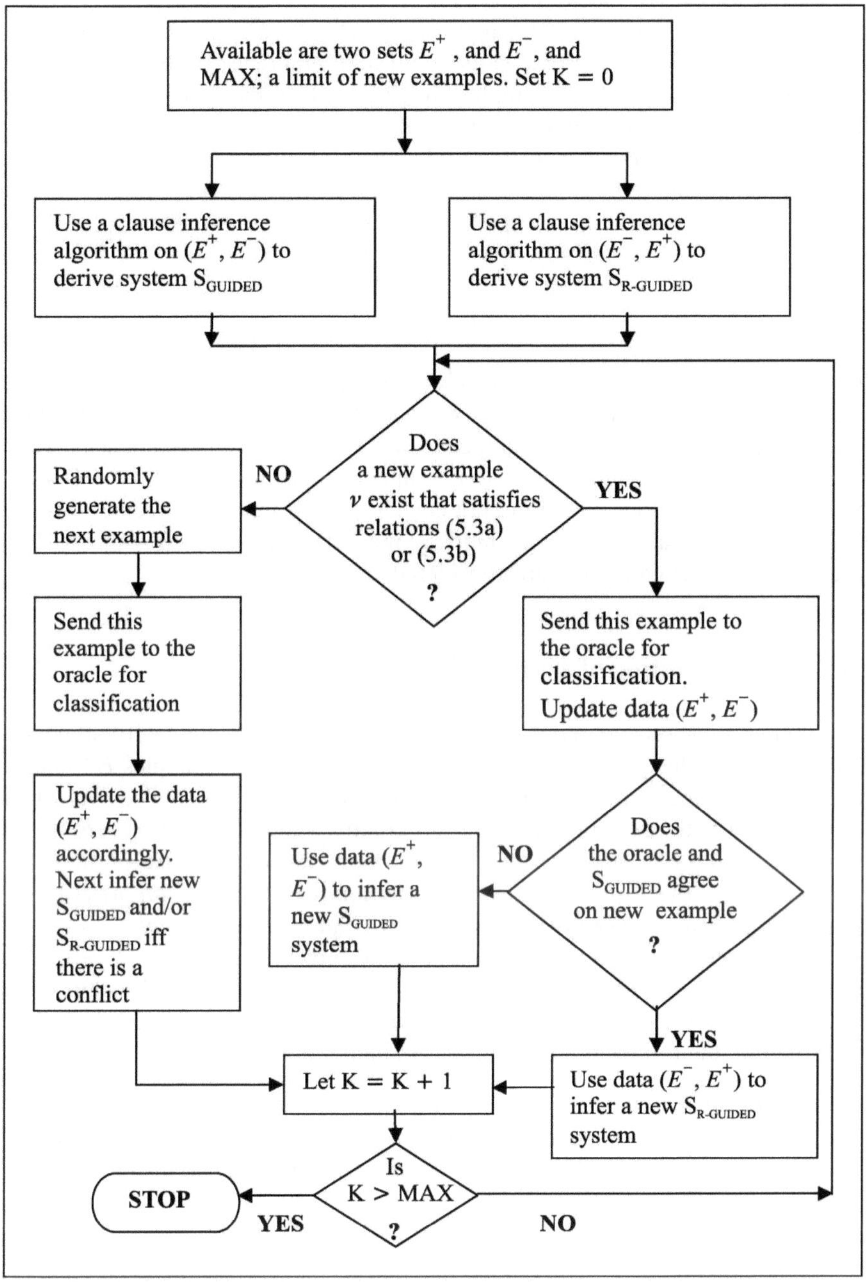


**Figure 5.2.** Flowchart of the Proposed Strategy for Guided Learning.

been redefined when the $i$-th example was presented (that is, $i = |E^+| + |E^-|$). Let the number of disagreements between $S_{GUIDED}$ and $S_{HIDDEN}$ be $r$. Apparently, these disagreements are among the remaining $2^n - i$ (where $n$ is the number of attributes) unclassified examples. A reasonable model of total ignorance of the outcomes (i.e., the way the remaining $2^n - i$ examples are classified) of the remaining steps is that all permutations of these outcomes are equally likely.

The probability, denoted as $Q_r(x)$, of having $x$ or more additional steps without a disagreement is then

$$\frac{\binom{2^t - i - r}{x}}{\binom{2^- i}{x}}.$$

Note that this probability, besides the values of $x$ and $r$, also depends on $i$.

Given $x$, then find the greatest $r$ such that $Q_r(x) \geq \alpha$ (where $\alpha$ might be 0.05 to correspond to a 5% level of significance or a 95% level of confidence). This is (under the random permutation model) an upper confidence bound on $r$. One may observe that the previous model did not consider any pattern of agreements in the $i$ examples already classified by the "hidden logic." Therefore, it is possible to derive tighter bounds, when all issues are considered.

## 5.4 On the Number of Candidate Solutions

An important issue related to this problem is to determine the number of all possible *distinct* systems which satisfy the requirements of two sets of examples $E^+$ and $E^-$. Two systems, defined on $n$ attributes, are called distinct if they are not equivalent, that is, if they classify the examples in $\{0, 1\}^n$ differently. Suppose that $|E^+|+|E^-| < 2^n$. Then, *the number of distinct systems which satisfy the requirements of the current positive and negative examples is equal to the number of all possible ways that the remaining examples can be divided into positive and negative examples.*

Let $L$ denote the number of the remaining examples. That is, $L = 2^n - (|E^+| + |E^-|)$. Since each row of the truth table for the $L$ unclassified examples can be independently a positive or a negative example, the answer to the previous question is $2^L$. Therefore, the following Lemma 5.1 [Triantaphyllou and Soyster, 1996] is true:

**Lemma 5.1.** *Suppose that $E^+$ and $E^-$ are the sets with the positive and negative examples, respectively. Then $K$, the number of distinct systems which satisfy the requirements of these examples, is given by the following formula:*

$$K = 2^L, \text{ where } L = 2^n - (|E^+| + |E^-|).$$

The above number $K$ is extremely large even for very small values of $n$. The OCAT approach tries to determine a rather compact system among this extraordinarily large number of possible solutions. The value of $K$ is the size of the hypothesis space and is used in the next section to quantify the *space complexity* of a learning algorithm. In the next section these ideas are further demonstrated via an illustrative example.

## 5.5 An Illustrative Example

Consider the two collections of positive and negative examples which were given in Section 5.2. There are 3 positive and 7 negative examples and the number of attributes is equal to 8. Recall that it was assumed that the system $S_{\text{HIDDEN}}$ (i.e., the "*hidden logic*") is as follows:

$$(\bar{A}_1 \vee \bar{A}_4 \vee A_6) \wedge (\bar{A}_2 \vee A_8) \wedge (A_2).$$

In this illustrative example, we use the OCAT approach in order to derive a CNF system from the given examples. When the OCAT approach is applied on the $(E^+, E^-)$ sets of examples, the following system (Boolean function in CNF), $S_{\text{GUIDED}}$, is derived:

$$(\bar{A}_3 \vee A_8) \wedge (A_2).$$

From Lemma 5.1 it follows that the total number of systems, denoted as $K$, which satisfy the requirements of these examples (and hence, are candidates to be the "hidden logic") is $2^{246}$. This happens although this is a trivial size Boolean function inference problem. The value 246 in the previous expression is derived from $2^8 - (3 + 7) = 246$.

To get a feeling of this incredible number consider the following facts. The value of $2^{10}$ is just greater than one thousand, the value of $2^{20}$ is just greater than one million, the value of $2^{50}$ is just greater than one quadrillion (i.e., more than one thousand trillions), and the value of $2^{200}$ is higher than the number of all the atoms in the matter of all the stars and planets which are visible during a clear night sky! Simply put, the value of the space of all candidate solutions (i.e., the size of the hypothesis space), which is equal to $2^{246}$ for this problem, cannot be expressed in a way that a human mind can comprehend. This is true despite the fact that the size of this illustrative example is trivial. For real-life problems the above situation becomes definitely impossible even to attempt to comprehend in terms of the magnitude of the hypothesis space.

Next, consider the system which is derived when the positive examples are treated as negative and the negative examples as positive. When the OCAT approach is applied on the $(E^-, E^+)$ data (i.e., the roles of the positive and negative examples are now reversed), then the following system (Boolean function in CNF), $S_{\text{R-GUIDED}}$, is derived:

$$(\bar{A}_2 \vee \bar{A}_8).$$

The next issue to investigate is to see whether there is an example which satisfies relation (5.3a) or (5.3b). It can be observed that the new example (i.e., which is still unclassified) (0 1 0 1 1 0 1 0) is classified as *positive by both systems*. That is, this example makes (5.3b) true. This example can be determined by finding a feasible solution of the clause satisfiability problem formed when the two systems $S_{\text{GUIDED}}$ and $S_{\text{R-GUIDED}}$ are taken together. That is, the Boolean function of the satisfiability problem is

$S_{\text{SAMPLE}} \wedge S_{\text{R-SAMPLE}}$, or:

$$((\bar{A}_3 \vee A_8) \wedge (A_2)) \wedge (\bar{A}_2 \vee \bar{A}_8) = (\bar{A}_3 \vee A_8) \wedge (A_2) \wedge (\bar{A}_2 \vee \bar{A}_8).$$

Following Theorem 5.1, either system $S_{\text{GUIDED}}$ or system $S_{\text{R-GUIDED}}$ will be revised and updated. Now suppose that the oracle classifies the new example as a *negative* one. Hence, it follows that system $S_{\text{GUIDED}}$ is invalid.

Next, this new example is added to the current collection of the negative examples and the OCAT approach is reapplied on the updated input data. The *new (i.e., updated) version* of the system $S_{\text{GUIDED}}$ is as follows:

$$(A_8) \wedge (A_2).$$

Since the new example did not reveal any inaccuracies for system $S_{\text{R-GUIDED}}$, this system needs no modification at this time. One may notice that the new version of the system $S_{\text{GUIDED}}$ and the current version of the system $S_{\text{R-GUIDED}}$ classify all examples in $\{0, 1\}^8$ in exactly the opposite manner (i.e., they are the complement, or the negation, of each other). Therefore, no new examples can be determined as above. As indicated earlier, this does not necessarily imply that the current version of the system $S_{\text{GUIDED}}$ is equivalent to the " hidden logic" (i.e., system $S_{\text{HIDDEN}}$). In situations like the above, it is proposed that the next example be generated *randomly.* Figure 5.2 summarizes the main steps of the proposed guided learning strategy.

At this point it is interesting to make some additional observations. In this illustrative example the structure of the "hidden logic" is known. Therefore, it is possible to estimate how closely the proposed system (i.e., system $S_{\text{GUIDED}}$) approximates the "hidden logic" $S_{\text{HIDDEN}}$. To accomplish this task in this illustrative example, all the remaining 246 ($= 2^8 - (3+7)$) unclassified examples have been evaluated by $S_{\text{GUIDED}}$ and $S_{\text{HIDDEN}}$ and compared. The 246 examples, as evaluated by $S_{\text{GUIDED}}$ and $S_{\text{HIDDEN}}$, agree 84% of the time. Hence, if a new example is chosen at random, there is 84% chance that $S_{\text{GUIDED}}$ will correctly classify it. Furthermore, when the *updated version* of the system $S_{\text{GUIDED}}$ is compared with the " hidden logic" system $S_{\text{HIDDEN}}$ they agree 97% of the time.

It should also be stated here that when a new example is considered and system $S_{\text{GUIDED}}$ is updated, the new system is *not always* a closer approximation of the "hidden logic." It is sometimes possible that the updated system is a *worse* approximation of the "hidden logic." Hence, convergence in terms of this measure of performance is not necessarily monotonically increasing.

The size of the hypothesis space is influential in determining the *sample complexity* of a learning algorithm, that is, the number of examples needed to accurately approximate a target concept. The presence of *bias* in the selection of a hypothesis from the hypothesis space can be beneficial in reducing the sample complexity of a learning algorithm [Mitchell, 1980], [Natarajan, 1989]. Usually the amount of bias in the hypothesis space $H$ is measured in terms of the *Vapnik–Chernovenkis dimension*, denoted as $VC\text{dim}(H)$ [Vapnik, 1982], [Haussler, 1988]. A well-known theoretical result regarding the $VC\text{dim}(H)$ is due to [Vapnik, 1982] and states that the sample complexity is at most equal to (note that $\varepsilon$ and $\delta$ are as defined in Section 5.1)

$$\frac{1}{\varepsilon(1-\sqrt{\varepsilon})}\left(2VC\dim(H)l\frac{6}{\varepsilon}n+\ln\frac{2}{\delta}\right).$$

This is better than some other bounds given in [Blumer, *et al.*, 1989]. However, the previous bound is still an overestimate [Haussler and Warmuth, 1993].

The proposed strategy makes no assumption regarding the hypothesis space. In the computational experiments reported in this chapter, the OCAT approach was used to infer a Boolean function from positive and negative examples. The OCAT approach has a tendency to return CNF (or DNF) functions with very few clauses (i.e., disjunctions or conjunctions, respectively).

Therefore, when the OCAT approach is combined with the proposed guided learning strategy, only then the hypothesis space is biased in favor of functions with small representations. That is, the proposed guided learning strategy makes no assumption regarding the hypothesis space. However, the behavior of the proposed strategy can be influenced by the nature of the algorithm used to infer the Boolean function. Why inferring functions with a few terms is desirable, was best explained in Section 1.3.2.

The OCAT and SAT approaches are NP-complete (for more details see Chapters 2 and 3). This chapter *does not* deal with the problem of inferring a Boolean function from collections of examples (this was the central topic of earlier chapters). Instead, it examines the problem of what should be the next example to be considered if one wishes to correctly infer a "hidden logic" by using a short sequence of new examples.

In this chapter we do not limit the form of the target Boolean function. The OCAT and SAT approaches used to infer the function of the proposed guided learning strategy, do not restrict themselves in deriving $k$-CNF or $k$-DNF functions. If the restriction to derive a $k$-CNF or $k$-DNF function is imposed, then the developments are exactly the same as with the unrestricted case.

The present guided learning problem mentions the issue whether one can infer a Boolean function when a new example is considered. Obviously, a brute force way is to solve the function inference problem from the beginning. A more efficient way is to try to devise an *incremental learning* approach in inferring a function when only one new example is introduced and the OCAT or the SAT approaches are used. For instance, this is the case in [Utgoff, 1988] where an incremental approach to the original *ID3* algorithm [Quinlan, 1986] is described. Although this is an interesting issue, it is beyond the scope of this chapter. Instead, this is the main topic of the next chapter.

## 5.6 Some Computational Results

A number of computer experiments were conducted and reported in [Triantaphyllou and Soyster, 1996] in order to investigate the effectiveness of the proposed strategy compared with random input learning (that is, when new examples are generated randomly). These experiments were approached in a manner similar to the illustrative example of the previous section. At first, a "hidden logic" was generated. The "hidden logic" systems considered in this chapter are based on the systems described in [Kamath, *et al.*, 1992], [Triantaphyllou, 1994] and also in Table 3.3 in Chapter 3. The only difference is that now the logical OR and AND operators are interchanged. In this way we deal with CNF systems instead of the DNF systems in [Kamath, *et al.*, 1992].

The systems with IDs (see also Tables 5.1a and 5.1b) 8A, 8B, 8C, 8D, and 8E are defined on 8 attributes. Similarly, systems 16A, 16B, 16C, 16D, and 16E are defined on 16 attributes. Finally, systems 32A, 32C, and 32D are defined on 32 attributes. Systems 32B and 32E (described in [Kamath, *et al.*, 1992]) were not considered due to excessive CPU requirements in obtaining computational results.

Each system was tested on 20 sequences of examples. For each such sequence initially ten examples were randomly generated and classified as either positive or negative by the oracle (i.e., the "hidden logic"). Each example was a vector with $n$ elements (where $n$ is the number of attributes). Each element was either 0 or 1 with probability 0.50. After an example was generated this way, it was classified by the "hidden logic" as either positive or negative.

Next, the OCAT algorithm was implemented to generate an initial version of $S_{\text{GUIDED}}$. What followed was the iterative generation of additional examples by the two different methods; GUIDED and RANDOM. Let $S_{\text{RANDOM}}$ be the Boolean function generated from the initial examples and the sequence of additional examples which were generated randomly (and $S_{\text{GUIDED}}$ is the Boolean function generated from the GUIDED input).

**Table 5.1a.** Some Computational Results Under the Random Strategy.

| System ID | Number of Examples Under RANDOM | | | |
|---|---|---|---|---|
| | MIN | Average | MAX | St. Dev. |
| 8A | 29 | 59.55 | 104 | 18.85 |
| 8B | 18 | 89.30 | 194 | 53.60 |
| 8C | 19 | 62.50 | 125 | 31.69 |
| 8D | 23 | 48.40 | 114 | 23.95 |
| 8E | 10 | 12.90 | 19 | 3.04 |
| 16A | 85 | 167.70 | 305 | 48.90 |
| 16B | 57 | 90.00 | 201 | 32.86 |
| 16C | 74 | 132.20 | 274 | 49.20 |
| 16D | 81 | 134.85 | 202 | 35.55 |
| 16E | 90 | 165.70 | 286 | 46.70 |
| 32A | 53 | 105.85 | 158 | 30.88 |
| 32C | 122 | 339.10 | 510 | 130.10 |
| 32D | 57 | 122.70 | 228 | 43.09 |

**Table 5.1b.** Some Computational Results Under the Guided Strategy.

| System ID | Number of Examples Under GUIDED | | | |
|---|---|---|---|---|
| | MIN | Average | MAX | St. Dev. |
| 8A | 18 | 34.85 | 70 | 11.02 |
| 8B | 15 | 50.90 | 153 | 37.09 |
| 8C | 20 | 35.60 | 65 | 12.46 |
| 8D | 18 | 42.80 | 207 | 40.95 |
| 8E | 10 | 12.30 | 24 | 3.26 |
| 16A | 36 | 67.95 | 125 | 22.19 |
| 16B | 45 | 68.20 | 86 | 11.57 |
| 16C | 53 | 83.45 | 114 | 17.24 |
| 16D | 48 | 88.85 | 171 | 32.94 |
| 16E | 84 | 118.90 | 176 | 26.80 |
| 32A | 48 | 77.40 | 146 | 25.81 |
| 32C | 93 | 115.65 | 151 | 17.35 |
| 32D | 60 | 95.55 | 142 | 25.91 |

In general, random examples in these experiments were generated as described above. In this way, after a sufficient number of random examples were classified by the "hidden logic," the collection of the negative examples would be a representative sample of the total population of the negative examples of the "hidden logic." The same issue is also true regarding the positive examples. Therefore, the computational experiments made no assumption regarding the distribution of the examples and the proposed guided learning strategy applies to any arbitrary distribution of the examples. After each pair of new examples was generated (one from GUIDED and one from RANDOM), the updated $S_{\mathrm{GUIDED}}$ and $S_{\mathrm{RANDOM}}$ systems were tested for convergence to $S_{\mathrm{HIDDEN}}$.

Convergence of $S_{\mathrm{GUIDED}}$ (or $S_{\mathrm{RANDOM}}$) was assumed to have occurred if 10,000 randomly generated examples were classified correctly by $S_{\mathrm{GUIDED}}$ (or $S_{\mathrm{RANDOM}}$). This is admittedly an approximation, but our real interest is comparing the relative speed of convergence of $S_{\mathrm{GUIDED}}$ and $S_{\mathrm{RANDOM}}$ with $S_{\mathrm{HIDDEN}}$. The comparison is simply how many additional examples are needed to correctly classify a random set of 10,000 observations.

Overall, the GUIDED strategy required on the average about 42% fewer examples than the RANDOM strategy for the entire set of 13 problems. The results for the individual problems are provided in Tables 5.1a and 5.1b. Note, for instance, that for the problem with ID equal to 8A, the GUIDED strategy required on the average 34.85 examples (see also Table 5.1b) to converge to a system equivalent to system 8A (used as the "hidden logic"). The same number under the RANDOM strategy is 59.55 (see also Table 5.1a). Tables 5.1a and 5.1b also present the MIN, MAX, and standard deviations of the gathered observations. Figures 5.3a to 5.3d depict analytical results of the performance of the two strategies for four of the previous systems (selected randomly). These plots are in agreement with the summary

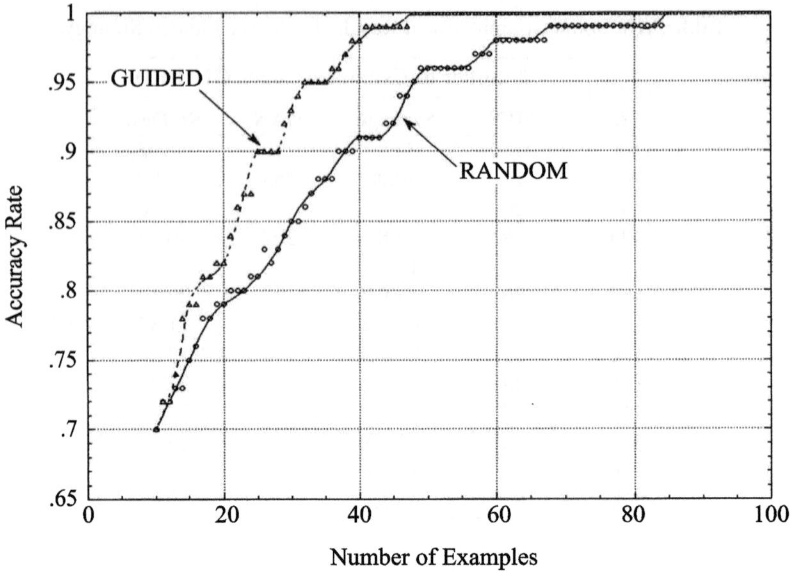


**Figure 5.3a.** Results When “Hidden Logic” Is System 8A.

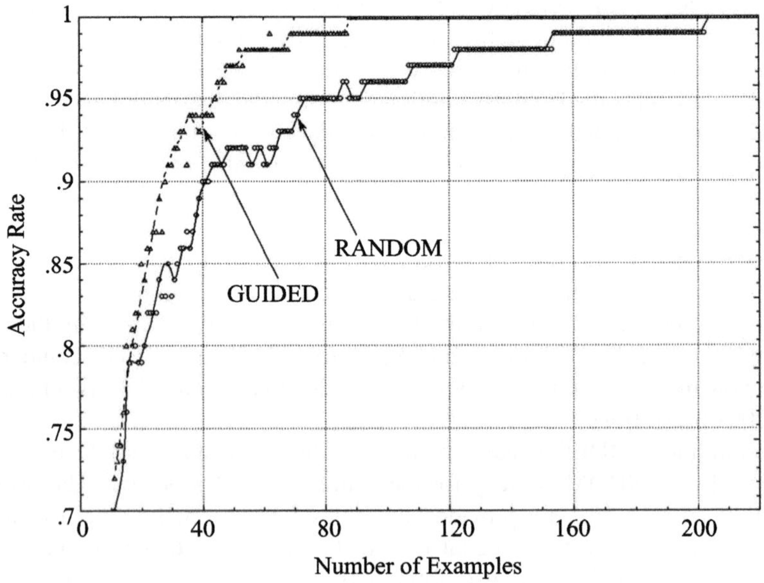


**Figure 5.3b.** Results When “Hidden Logic” Is System 16A.

results presented in Tables 5.1a and 5.1b. It is also evident that strategy GUIDED outperformed, in almost all cases, the RANDOM strategy.

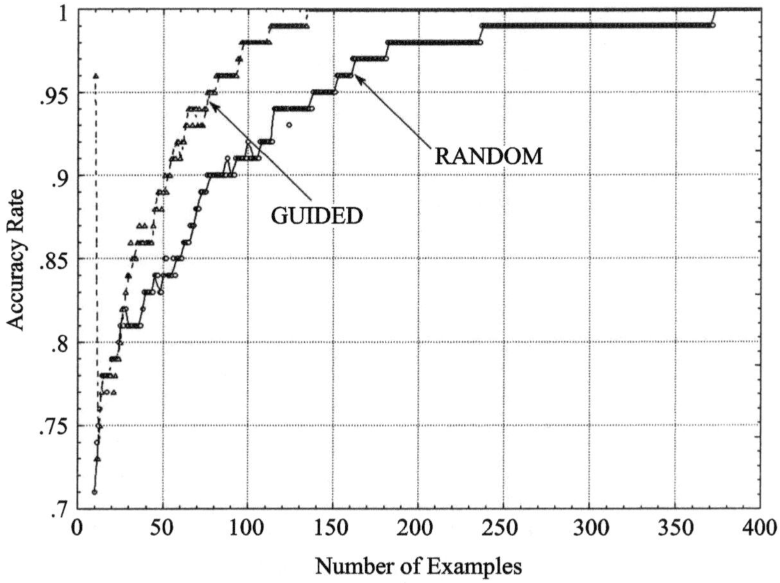


**Figure 5.3c.** Results When "Hidden Logic" Is System 32C.

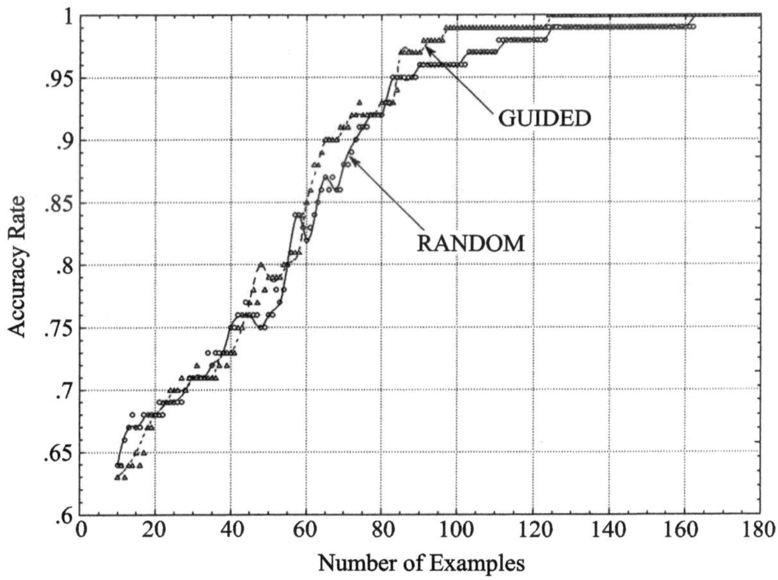


**Figure 5.3d.** Results When "Hidden Logic" Is System 32D.

Figure 5.4 depicts what occurs as a single sequence of examples was generated for system 8B. This is a rather representative case. Note how $S_{\text{GUIDED}}$ uniformly

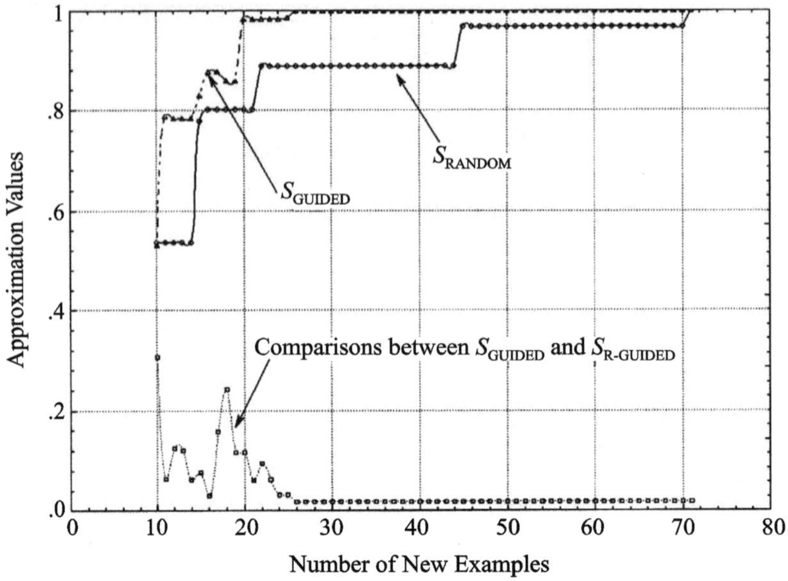


**Figure 5.4.** Comparisons between systems $S_{\text{RANDOM}}$, $S_{\text{GUIDED}}$, and $S_{\text{R-GUIDED}}$ when new examples are considered (system $S_{\text{HIDDEN}}$ is $(\bar{A}_1 \vee \bar{A}_4 \vee A_6) \wedge (\bar{A}_2 \vee A_8) \wedge (A_2)$).

dominates $S_{\text{RANDOM}}$. The bottom curve indicates how $S_{\text{GUIDED}}$ and $S_{\text{R-GUIDED}}$ become complements of each other. As can be seen from this figure, both strategies start from the same point (recall that initially there are 10 random examples). However, with the guided input strategy the inferred system reaches 100% accuracy much sooner (i.e., just after 16 new examples, while with random input it takes 61 new examples).

Given a set of data, the derived system $S_{\text{GUIDED}}$ and a new example, the new proposed system may or may not be more accurate than its previous version. This depends on two factors: (1) on which approach was used to infer the proposed system and (2) on the particular data. This is best described in the following illustrative example.

Suppose that the target function is $(\bar{A}_1 \vee \bar{A}_4 \vee A_6) \wedge (\bar{A}_2 \vee A_8) \wedge (A_2)$ (i.e., it is system 8B). Let the two sets of positive and negative examples be as follows:

$$E^+ = \begin{bmatrix} 0 & 1 & 1 & 1 & 1 & 1 & 0 & 1 \\ 1 & 1 & 1 & 1 & 0 & 1 & 0 & 1 \\ 1 & 1 & 1 & 0 & 0 & 0 & 1 & 1 \\ 0 & 1 & 0 & 0 & 0 & 0 & 1 & 1 \\ 0 & 1 & 0 & 0 & 1 & 1 & 0 & 1 \end{bmatrix} \quad \text{and}$$

$$E^- = \begin{bmatrix} 0 & 0 & 0 & 1 & 1 & 0 & 1 & 0 \\ 0 & 1 & 1 & 1 & 1 & 1 & 1 & 0 \\ 1 & 0 & 0 & 1 & 1 & 0 & 0 & 1 \\ 0 & 0 & 1 & 0 & 0 & 0 & 1 & 1 \\ 1 & 1 & 0 & 0 & 1 & 1 & 0 & 0 \\ 1 & 0 & 0 & 1 & 1 & 1 & 0 & 1 \\ 0 & 0 & 1 & 0 & 1 & 0 & 0 & 0 \\ 1 & 0 & 0 & 0 & 1 & 0 & 1 & 1 \\ 1 & 0 & 0 & 0 & 0 & 0 & 0 & 0 \\ 1 & 0 & 1 & 0 & 0 & 1 & 0 & 0 \\ 0 & 0 & 1 & 0 & 0 & 0 & 0 & 0 \\ 1 & 0 & 1 & 0 & 0 & 1 & 0 & 1 \\ 0 & 1 & 1 & 1 & 0 & 1 & 1 & 0 \\ 1 & 0 & 1 & 1 & 1 & 1 & 0 & 1 \\ 1 & 1 & 0 & 1 & 0 & 0 & 0 & 1 \end{bmatrix}.$$

When the OCAT approach is used on the previous data, then $S_{\text{GUIDED}}$ is

$$(A_3 \vee A_6 \vee A_7) \wedge (A_8) \wedge (A_2)$$

and $S_{\text{R-GUIDED}}$ is $(A_1 \vee \bar{A}_2 \vee \bar{A}_8) \wedge (A_5 \vee \bar{A}_2 \vee \bar{A}_3 \vee \bar{A}_8)$.

When system $S_{\text{GUIDED}}$ is compared with the "hidden logic" system, then the derived accuracy is 0.95. Let (1 1 0 1 0 0 1 1) be the next example to consider. It can be observed that this example is classified identically (e.g., as positive) by both of the previous two systems. Next, this example is also classified by the "hidden logic," and a contradiction between the proposed system and the "hidden logic" is revealed. Therefore, the set of negative examples is augmented by that example, and the OCAT approach is used again to derive the new version of system $S_{\text{GUIDED}}$. The updated system is

$$(\bar{A}_4 \vee \bar{A}_7) \wedge (A_3 \vee \bar{A}_1) \wedge (A_2).$$

The corresponding accuracy rate is now equal to 0.77, which is *smaller* than the accuracy of the previously proposed system (although more examples are now used as input). A similar situation is also depicted in Figure 5.4. In this figure the curve which corresponds to the guided strategy illustrates this phenomenon. Note that when the number of examples is 19, there is a *valley* in this curve.

It should be stated here that in these experiments no satisfiability formulation was used in the GUIDED strategy in order to accomplish the task of finding the next example. Instead, a high number (i.e., 10,000) of randomly generated examples were used to find an example which would be classified identically by the two systems $S_{\text{GUIDED}}$ and $S_{\text{R-GUIDED}}$. This was done for the sake of simplicity in order to keep the CPU requirements low.

As can be seen from the results in Tables 5.1a and 5.1b, the guided input strategy was superior to random input strategy almost all the time. Only for the cases of systems 8E and 32E this was not the case. As was anticipated, most of the time, the guided input learning strategy required considerably less examples in order to correctly infer a "hidden logic."

Professor Mangasarian from the University of Wisconsin and his associates have extensively used in their pattern recognition studies (see, for instance, [Mangasarian, *et al.*, 1991]) a database with observations regarding nine cytological characteristics of breast tumors. Information for each tumor is derived by analyzing biological material extracted by using fine needle aspirates (FNAs). Each tumor was also classified as benign or malignant. At the time we performed our experiments were 421 cases, of which 224 were benign while the remaining 197 were malignant. These data were also available at that time to the general public by anonymous "*ftp*" or regular Web downloading from the Machine Learning Database Repository at the University of California at Irvine, Department of Computer Science.

We transferred the data into the equivalent binary data and performed a series of computational experiments as follows. At first a 10% random collection of the original data was considered. Next, we generated the next example by using random input and also by using guided input (in two independent scenarios as before). That is, we applied an experimental procedure as with the previous experiments. However, now there is no "hidden logic" available, and thus we compared the accuracy of the derived systems in terms of how well they classified the *remaining* data (which were used as the testing data).

For each experiment we used 50 random replications of it. The results of these tests are summarized in Table 5.2 and are also depicted in Figures 5.5a and 5.5b. In these results both the number of derived rules (i.e., clauses in CNF) and the accuracy rates were recorded. As can be easily verified from these results, once again the proposed guided strategy significantly outperformed the random input strategy. In these experiments as Boolean function inference algorithm we used the randomized heuristic RA1 described in Chapter 4.

As a final comment it should be stated here that the accuracy rates in Figure 5.5b did not necessarily reach the 100% value as the percent of the training data increased, because we were not comparing the inferred systems with a "hidden logic" but with the way they classified the *remaining* available data.

An interesting issue is to try to determine a way to conclude whether $S_{\text{GUIDED}}$ is a close approximation to $S_{\text{HIDDEN}}$. Intuitively, one expects that the closer the two systems $S_{\text{GUIDED}}$ and $S_{\text{HIDDEN}}$ become, the more apart the two systems $S_{\text{GUIDED}}$ and $S_{\text{R-GUIDED}}$ should become. One measure of the "closeness" of the two systems $S_{\text{GUIDED}}$ and $S_{\text{HIDDEN}}$ is determined by the percentage of 10,000 randomly generated examples which are classified identically by both systems.

This situation can be seen in Figure 5.4. One may observe that in Figure 5.4 the systems $S_{\text{GUIDED}}$ and $S_{\text{HIDDEN}}$ become very close to each other (the top curve converges to value 1.00) when the bottom curve (which indicates the closeness of the systems $S_{\text{GUIDED}}$ and $S_{\text{R-GUIDED}}$) approaches the value 0. This is a rather representative case and other experiments demonstrated similar behavior. This observation suggests the following *empirical test for system validation.* Suppose that one has generated a number of examples and has specified the two systems $S_{\text{GUIDED}}$ and $S_{\text{R-GUIDED}}$. Then, by using a sample of 10,000 randomly generated examples,

**Table 5.2.** Computational Results When the Wisconsin Breast Cancer Data Are Used.

| | With Random Input | | With Guided Input | |
|---|---|---|---|---|
| % of Data Used for Training | No. of Clauses | Accuracy Rate | No. of Clauses | Accuracy Rate |
| 10 | 1.60 | 0.88 | 1.63 | 0.87 |
| 15 | 2.06 | 0.88 | 2.26 | 0.91 |
| 20 | 2.42 | 0.89 | 3.09 | 0.92 |
| 25 | 2.72 | 0.90 | 3.91 | 0.94 |
| 30 | 3.12 | 0.90 | 4.43 | 0.95 |
| 35 | 3.40 | 0.90 | 5.29 | 0.97 |
| 40 | 3.74 | 0.90 | 5.94 | 0.97 |
| 45 | 4.06 | 0.90 | 6.77 | 0.98 |
| 50 | 4.44 | 0.90 | 7.23 | 0.98 |
| 55 | 4.90 | 0.90 | 7.74 | 0.99 |
| 60 | 5.40 | 0.90 | 8.11 | 0.99 |
| 65 | 5.84 | 0.91 | 8.11 | 0.99 |
| 70 | 6.24 | 0.91 | 8.14 | 0.99 |
| 75 | 6.92 | 0.91 | 8.17 | 0.99 |
| 80 | 7.68 | 0.91 | 8.34 | 0.99 |
| 85 | 7.88 | 0.91 | 8.60 | 0.99 |
| 90 | 8.56 | 0.91 | 8.77 | 0.99 |
| 95 | 8.88 | 0.92 | 8.83 | 0.99 |

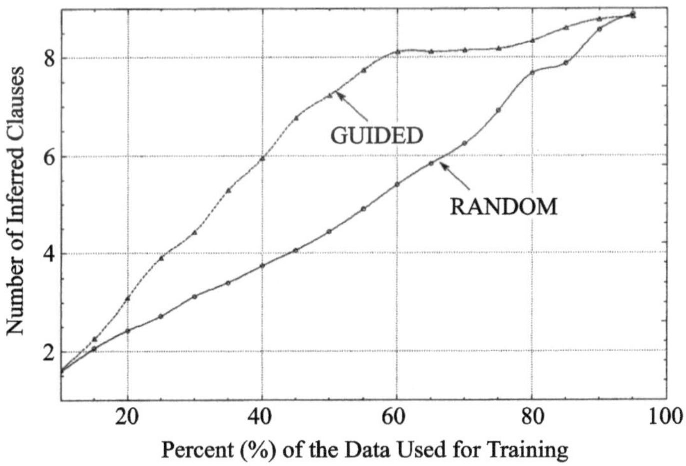


**Figure 5.5a.** Results When the Breast Cancer Data Are Used. The Focus Is on the Number of Clauses.

the two systems $S_{\text{GUIDED}}$ and $S_{\text{R-GUIDED}}$ can be compared on how often they agree in classifying these examples. If they agree very little (i.e., the approximation rate

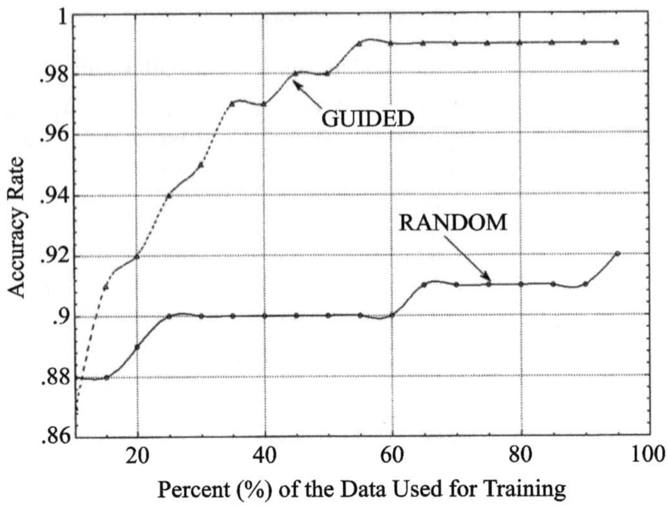


**Figure 5.5b.** Results When the Breast Cancer Data Are Used. The Focus Is on the Accuracy Rates.

is very low), then it is very likely that the system $S_{\text{GUIDED}}$ is a *good approximation* of the corresponding "hidden logic." However, if the approximation rate is very high, then it is rather unlikely that the system $S_{\text{GUIDED}}$ is an accurate approximation of the "hidden logic." Finally, it should be stated here that some additional computational experiments on this strategy are described in Section 13.7. In that study the training and testing examples are defined after the analysis of text documents. Similarly with the results of this chapter, those results also support the potential of this strategy for guided learning.

## 5.7 Concluding Remarks

This chapter discussed the development of a strategy for guided learning of Boolean functions from examples. The proposed method is based on the comparison of two Boolean functions. The first function is the one derived from the positive and negative examples. The second function is derived by treating the original positive examples as negative and the original negative examples as positive. If it is possible to find a new example which is classified identically by both systems, then this example is considered next in the learning process. If no such example can be found, then the next example is generated randomly.

The computational results in this chapter suggest that most of the time it is possible to find an example which is classified identically by both systems. In this way, the new example reveals that one of the two systems is inaccurate. Furthermore, the same computational results demonstrate that, on the average, the proposed strategy is significantly more effective than random learning. That is, most of the time in our computational experiments it was possible to infer a "hidden logic" much faster when the new examples were generated according to the proposed guided learning strategy, than when the examples were generated randomly.

An interesting issue for future research would be to expand the proposed methodology to more general situations. The present chapter focused on the case of dealing with Boolean functions. There is no reason why the proposed methodology cannot be expanded to cover more general cases. Another direction for future research in this area is to develop strategies which can select new examples from Region B or Region C as defined in Figure 5.1. In this way there is a possibility to update *both* inferred systems at a given iteration simultaneously.

When a new example is selected from Region A or D (see also Figure 5.1), then an interesting idea is to select examples which are "deeply" inside these regions. That is, to select the most representative ones. Of course, it is not obvious how can one define this concept of "depth." However, when that happens, it is reasonable to assume that the system in error with the oracle will be revised in a major manner as the revealing example is well inside the disputed region. Hopefully, such a strategy would cause more significant improvements at each iteration than the current strategy which finds an example that just invalidates one of the two systems.

From the previous discussions it follows that the proposed approach can be applied when it is possible to derive the two systems $S_{\text{GUIDED}}$ and $S_{\text{R-GUIDED}}$. Then, by determining as the next example an example which is classified identically by the two systems, the convergence of the proposed hypothesis to the target concept could be expedited. Clearly, more work is needed in this critical area of data mining and knowledge discovery from databases.

# Chapter 6

# An Incremental Learning Algorithm for Inferring Boolean Functions

## 6.1 Some Background Information

The previous chapter studied the guided learning problem. In that setting, the analyst has the option to select which unclassified example to send to the oracle for classification and use that information to improve the understanding of the system under consideration. When the new example would unveil the need for an update, one had to use all the existing training examples, plus the newly classified example, to infer a new (and hopefully more accurate) pattern in the form of a Boolean function or other data mining model.

This chapter studies a very closely related problem to the guided learning problem examined in the previous chapter. This is the problem of inferring a Boolean function in an *incremental* way. The developments presented in this chapter are based on the results presented in [Nieto Sanchez, Triantaphyllou, *et al.*, 2002]. According to this approach, instead of running an algorithm for inferring a Boolean function from the beginning, now the goal is to try to modify the current Boolean function in a way that satisfies the requirements imposed by the existing training data, plus the newly classified example.

Thus, Chapter 6 introduces a new incremental learning from examples (ILE) algorithm for the inference of a Boolean function from examples. The derived functions are in disjunctive or conjunctive normal form (DNF or CNF, respectively) and emphasis is given on having as few DNF or CNF clauses as possible.

In this chapter the new algorithm is combined with an existing algorithm for nonincremental learning from examples (NILE) of Boolean functions from two collections of examples. However, the proposed incremental approach can be combined with *any* nonincremental (NILE) approach for deriving a Boolean function from examples. In this study the NILE algorithm used with the proposed ILE approach is the OCAT approach (see also Chapters 2 and 3). Thus, in this chapter the new approach will be called IOCAT (for *Incremental* OCAT).

In order to assess the comparative value of the new approach versus the old one (i.e., with the nonincremental OCAT approach), we used examples derived by analyzing almost 3,000 text documents from the TIPSTER collection of documents

E. Triantaphyllou, *Data Mining and Knowledge Discovery via Logic-Based Methods*,
Springer Optimization and Its Applications 43, DOI 10.1007/978-1-4419-1630-3_6,

[Harman, 1995], [Voorhees, 1998]. For this purpose we used the document surrogate concept as introduced by Salton [1989] in order to represent text documents as binary vectors. The TIPSTER collection is often used to evaluate information retrieval systems and machine learning algorithms.

As classes for the training examples, we used documents from four document categories. These were documents related to the Department of Energy (DOE), the Wall Street Journal (WSJ), the Associated Press (AP), and technical documents from the ZIPFF collection. In addition, in order to define two disjoint classes, the following three class-pairs were formed: (DOE vs. ZIPFF), (AP vs. DOE), and (WSJ vs. ZIPFF). The various algorithms were compared in terms of three measures of performance as follows: (i) the CPU time requirements, (ii) the accuracy of the derived Boolean functions, and (iii) the number of clauses in the derived Boolean functions.

This chapter is organized as follows. Section 6.2 presents a formal description of the data mining problem studied in this chapter. Section 6.3 briefly reviews the main parts of the related literature. Section 6.4 describes the proposed IOCAT (i.e., the incremental OCAT) algorithm. Sections 6.5 and 6.6 present and discuss the results of a computational study. The chapter ends with a conclusions section.

## 6.2 Problem Description

As before, suppose that some collections of examples from two disjoint classes are somehow made available to a computerized classification system. Each example is a binary vector defined on $n$ attributes. Each example comes with a class membership designation. Furthermore, this setting is assumed to be deterministic and no errors are considered. These two collections form the training examples.

As was also stated earlier, the task of a classification system is to analyze the information embedded in the training examples and infer a model that best captures the behavior of the hidden system. That is, we assume that there is a system that can classify these, and also more, examples. Thus, a main challenge is to use the available training examples to infer a Boolean function that in turn can be used to accurately classify new (and thus unclassified) examples.

For a given set of examples, the learned (inferred) Boolean function may not be an accurate representation of the hidden system. This is especially true if the sizes of the two collections of the training examples are limited or they are not representative of the entire population of examples. The very next example may negate the current Boolean function, and thus it can initiate a revision of this function. This is the basis for the guided learning problem studied in the previous chapter.

A fundamental problem closely associated with guided learning is how to best modify an existing Boolean function when the classification of a new example reveals that the current Boolean function is inaccurate. One approach (the brute force approach) is to reconstruct the entire function *from the beginning* by using the entire sets of the training examples augmented with the new example. An alternative approach might be to repair only a few clauses of the existing Boolean function in a way that the modified function correctly classifies all the available training examples (i.e., the

$$E^+ = \begin{bmatrix} 1 & 0 & 1 & 0 \\ 0 & 0 & 0 & 1 \\ 1 & 1 & 1 & 1 \\ 0 & 0 & 0 & 0 \\ 1 & 0 & 0 & 0 \\ 1 & 1 & 1 & 0 \end{bmatrix}, \quad \text{and} \quad E^- = \begin{bmatrix} 0 & 1 & 0 & 0 \\ 1 & 1 & 0 & 0 \\ 0 & 0 & 1 & 1 \\ 1 & 0 & 0 & 1 \end{bmatrix}$$

$$F = (A_3 \vee \bar{A}_2) \wedge (\bar{A}_4 \vee A_2 \vee \bar{A}_1)(A_1 \vee \bar{A}_3)$$

**Figure 6.1.** A Sample Training Set of Six Positive Examples and a Set of Four Negative Examples and a Boolean Function Implied by These Data.

old training examples plus the new one that revealed the need for a change). This is exactly the main problem of interest in this chapter.

## 6.3 Some Related Developments

Figure 6.1 shows two mutually exclusive sets of binary examples. The first set, denoted as $E^+$, represents the first class of training example and it is called the set with the positive examples. Similarly, the second set, denoted as $E^-$, represents the set with the negative training examples. All these examples are defined by the presence (i.e., "1" value) or absence (i.e., "0" value) of four attributes $A_i$ (for $i = 1, 2, 3, 4$). A Boolean function, denoted as $F$, that satisfies the requirements of these examples is also provided in Figure 6.1.

By definition, the "hidden system" (unknown Boolean function which we try to infer) accepts each positive example while it rejects each negative one. Consequently, the inferred Boolean function should also evaluate each positive example as true and each negative example as false. Later, such examples are used to evaluate the performance of the proposed approach on some large-scale simulated problems. In the computational studies that follow, these examples were defined by properly analyzing text documents. A text document can be considered as text defined over a finite set of *keywords*. Then, the presence or absence of a keyword can be indicated with the 1/0 value of a binary attribute. The binary vectors formed this way are called *document surrogates* or just *surrogates* in the text analysis literature (see, for instance, [Salton, 1968], [Salton, 1989], [Cleveland and Cleveland, 1983], and [Meadow, 1992]). In our tests, we used such examples that were defined on 800 binary attributes and were extracted by analyzing a total of almost 3,000 text documents (examples).

In general, let $\{A_1, A_2, A_3, \ldots, A_n\}$ represent a set of $n$ Boolean attributes. Also, let $v$ be a binary vector that is defined on these $n$ attributes. Furthermore, let $F$ be a Boolean function that evaluates to either 0 or 1 depending on the combination of the values of the attributes in vector $v$. That is, $F(v) = 1$ or $F(v) = 0$, depending on whether the vector is a positive or negative example, respectively.

The OCAT approach will be used extensively here as a demonstration data mining approach. However, *any other approach* which can infer a Boolean function from two collections of disjoint examples is applicable as well. Recall that the logic of the OCAT approach is given in Figure 2.1.

As was described there, the main step in the OCAT approach is Step 2 in Figure 2.1. Some algorithms for dealing with the problem in Step 2 are given in Chapters 2, 3, and 4. These include two branch-and-bound (B&B) approaches and some fast heuristics. A heuristic which will also be used extensively in this chapter is the RA1 one and it was described in Section 4.2 and summarized in Figure 4.1. The proposed IOCAT approach uses this heuristic.

Recall that the logic of that heuristic is based on the two functions termed $POS(a_i)$ and $NEG(a_i)$. Function $POS(a_i)$ returns the number of positive examples accepted by the current clause (in CNF) under construction if the attribute $a_i$ (where $a_i$ is either $A_i$ or $\bar{A}_i$) is included in the clause under construction. A similar interpretation applies for the $NEG(a_i)$ function with regard to the negative examples. If $NEG(a_i) = 0$ in Step 2 in Figure 4.1, then the attribute $a_j$ is given a very high priority for inclusion in the clause being formed. That heuristic was also combined with some randomization techniques and with the revised B&B approach presented in Chapter 3.

The main step of the OCAT approach (i.e., Step 2 in Figure 2.1) involves the inference of a Boolean function from two collections of examples. In a guided learning mode, when new examples are presented one at a time, if the Boolean function is inferred from the beginning each time, this is known as *nonincremental learning from examples* (or NILE). This may be computationally expensive when one already has a Boolean function which satisfies the requirements of all the available examples but the very last one. Some extensive surveys of NILE learning can be found, for instance, in [Hunt, *et al.*, 1966], [Michalski and Larson, 1978], [Michalski, 1985], [Reine and Michalski, 1986], [Schlimmer, 1987], [Schlimmer and Fisher, 1986], and [Utgoff, 1989; 1998].

On the other hand, *incremental learning from examples* (or ILE) may be an attractive strategy to modify an existing function when it misclassifies a newly introduced training example. Among the first contributions in ILE is the *Concept Learning System* (CLS) [Hunt, *et al.*, 1966]. In CLS prior observations were selected at random and were replaced with new examples in order to reconstruct the new knowledge. The CLS approach was soon abandoned because the learning rates were slow. In [Michalski and Larson, 1978], the *AQ system* [Michalski, 1973] was adapted to learn incrementally by limiting the number of examples needed to reconstruct the faulty knowledge, which was expressed in the DNF form. The AQ system repaired this knowledge by using a Euclidean distance measure to identify new examples that were good concept representatives. Its goal was to reconstruct only those portions of the knowledge (i.e., a set of clauses that describe an individual concept) that caused the misclassification.

Later, Reine and Michalski [1986] extended the AQ system into the *GEM system* which repairs only individual clauses of a DNF expression. In the GEM system only the faulty conjunctive clauses were submitted to a generalization procedure along with the observations it currently covered and those that triggered the classification inconsistency. The results of this system suggested that: (i) ILE methods may yield more complex concept descriptions than NILE methods and (ii) knowledge updates might be less expensive using ILE methods than with the NILE methods.

Next, suppose that two sets of training examples have, somehow, become available. As before, one set will be the "positive" examples and the other the "negative" examples (these names are assigned arbitrarily). Then a function inference algorithm is applied on these training examples and a single Boolean function is inferred. This Boolean function is derived in an attempt to infer the "hidden" system that classified these training examples. Since this function accepts all the positive examples while it rejects all the negative ones, we will call it "the set with the positive rules" or just the "positive rules" (since the clauses of a Boolean function in CNF or DNF can also be viewed as a set of rules). For convenience, this function will be denoted as $R^+$. Next, one can use the same Boolean function inference algorithm to construct the "negative rules" (to be denoted as $R^-$) by simply switching the roles of the training examples, that is, by treating the initial negative examples as the positive examples and vice-versa. Obviously, the negative rules (negative Boolean function) will reject all the positive examples while they will accept all the negative ones. Some methods may create symmetric systems (i.e., two systems which are complements of each other). This is not the case with the OCAT approach.

These two functions (i.e., the positive and the negative rules denoted as $R^+$ and $R^-$, respectively) can play a pivotal role in an incremental learning setting. This idea was first applied on the guided learning strategy described in Chapter 5. In that strategy, the OCAT approach was used to infer the previous two Boolean functions in an incremental learning environment. That is, it was assumed that the analyst had control on determining the composition of the next example to be sent for classification to the oracle and then to be included in the training examples. In the strategy described in Chapter 5 the selected next example was one that was classified (before sending it to the oracle for the actual class classification) *identically* by both functions. In this way, it was secured that after the actual class membership was determined, then one of the two functions will be modified and hopefully its classification accuracy would be improved. The empirical results reported in the previous chapter strongly suggested that this guided learning strategy is superior to just randomly selecting the next example for inclusion in the two training sets.

The above issues are best formalized as follows. Suppose that the oracle classifies the new example. If the oracle classifies this new example as a positive one, then it will be denoted as $e^+$. Otherwise (i.e., if it is classified as a negative one), it will be denoted as $e^-$. When the new example is fed to the two Boolean functions $R^+$ and $R^-$, then one and only one of the following three scenarios is possible.

1. It has been classified *correctly* if and only if:
   (a) $R^+(e^+) = 1$ and $R^-(e^+) = 0$; or:
   (b) $R^+(e^-) = 0$ and $R^-(e^-) = 1$.

2. It has been classified *incorrectly* if and only if:
   (c) $R^{+}(e^{+}) = 0$ and $R^{-}(e^{+}) = 1$; or:
   (d) $R^{+}(e^{-}) = 1$ and $R^{-}(e^{-}) = 0$.
3. The new example triggers an *undecided* situation if and only if:
   (c) $R^{+}(e^{+}) = 1$ and $R^{-}(e^{+}) = 1$; or:
   (d) $R^{+}(e^{-}) = 1$ and $R^{-}(e^{-}) = 1$; or:
   (e) $R^{+}(e^{+}) = 0$ and $R^{-}(e^{+}) = 0$; or:
   (f) $R^{+}(e^{-}) = 0$ and $R^{-}(e^{-}) = 0$.

The above scenarios correspond to the way the entire state (example) space was partitioned into the four regions described in Section 5.1. In the proposed ILE approach, the new examples will be determined such that the previous scenario #3 occurs as often as possible. Such an example can be determined by solving a SAT (clause satisfiability) problem or by simply randomly sampling a large enough sample of new examples until one that is classified identically by both Boolean functions is found.

## 6.4 The Proposed Incremental Algorithm

The proposed incremental learning algorithm has some similarity to the GEM system [Reine and Michalski, 1986]. They are similar in the sense that any disagreement between the inferred system and the training examples is not allowed and only the disjunctive clauses triggering the wrong classification are repaired in the proposed algorithm. Nonetheless, they differ in the way new training examples are selected and used to reconstruct the current system. For instance, in the GEM system new information is submitted to a generalization process only when a set of misclassified examples has been collected. In contrast, in this chapter this knowledge (i.e., the group of the two Boolean functions) is repaired by considering examples identified as "undecided." This guarantees either the positive or the negative Boolean function will be altered. Furthermore, the approach presented here differs from the GEM system because we always maintain two Boolean functions as described earlier.

In the GEM system the methodology for repairing only the portion(s) of the function followed the procedures described in [Michalski and Larson, 1978]. In this chapter, however, this repair was divided into the following two mutually exclusive subproblems which capture all possibilities: (i) Repair of a Boolean function that incorrectly rejects a positive example and (ii) repair of a Boolean function that incorrectly accepts a negative example. For both subproblems we assume that the inferred Boolean function is in DNF. The CNF case can be developed in a similar manner. However, it seems that for this kind of problems the DNF case is more intuitive to follow.

### 6.4.1 Repairing a Boolean Function that Incorrectly Rejects a Positive Example

From the definition of the DNF form given as (2.1) in Chapter 2, a Boolean function $F$ accepts an example if and only if at least one of its clauses accepts it. Recall that now each clause is a conjunction. Alternatively, a Boolean function rejects an example if and only if all of its clauses reject it. Next, suppose that the current system, denoted as Boolean function $F$, incorrectly rejects the positive example $e^+$. Then, the following relation should be obviously satisfied:

$$F(e^+) = c_1 \vee c_2 \vee c_3 \vee \cdots \vee c_n = 0, \tag{6.1}$$

where $c_i$ (for $i = 1, 2, 3, \ldots, n$) is the $i$-th clause of $F$.

The previous discussion naturally raises the question of how to decide the clause, among the $n$ clauses $c_i$ (for $i = 1, 2, 3, \ldots, n$) in relationship (6.1), one should alter such that the new positive example will be accepted by the modified Boolean function. The algorithm depicted in Figure 6.2 addresses this problem.

This algorithm indicates (in Step 3) that two extreme strategies can be implemented. The first strategy is to select for change (repair) the clause that is the *most* generalizing one (denoted as the MGC clause), while the second strategy is to select for repair the *least* generalizing clause (denoted as LGC) in Figure 6.2. These two strategies represent two extreme scenarios. They rank the clauses according to their generalization capability and then select the two extreme cases. In this way, it is hoped that one can study all possibilities.

The generalizability of a clause is assessed in terms of two parameters. One is the number of positive examples accepted by that clause. This is represented as $|E^+(c_i)|$ in Figure 6.2. The higher this number is, the higher the generalizability of a clause is assumed to be. The second parameter is the size of the clause denoted as $A(c_i)$. As size, we consider its length or the number of the attributes that define it. The fewer the attributes, the more general the clause is (for the DNF case). For these reasons, the MGC (*Most Generalizing Clause*) in Figure 6.2 is defined as the clause with the maximum $|E^+(c_i)|/A(c_i)$ value. Similarly, the LGC (*Least Generalizing Clause*) is the one that corresponds to the minimum $E^+(c_i)/A(c_i)$ value.

It should be stated at this point that if one ranks the clauses of a Boolean function according to their generalizability power, then one may expect that the clause that ranks the highest (i.e., the MGC clause) has also the most potential to effect the behavior of the Boolean function when that clause is altered. After all, by definition an LGC clause plays a smaller role. Therefore, it is reasonable to expect that focusing attention on the MGC would lead to better results. However, this is not possible to assess quantitatively without some kind of computational experiments. The computational results reported in the next section indicate that the previous hypothesis seems to be indeed the case.

In the same figure the notation OCAT($E^+(c_k)$, $E^-$) denotes a Boolean function inference problem that has as positive examples the members of the set $E^+(c_k)$ and as negative examples the members of the set $E^-$. This notation is used in the remaining of this chapter. The effectiveness of these two selection criteria is further studied

**Input:** The training sets $E^+$ and $E^-$. A Boolean function $F$ in DNF that accepts all the positive examples while it rejects all negative ones. This function is comprised of $n$ clauses (in DNF) denoted as $c_i$ (for $i = 1, 2, 3, \ldots, n$). A new positive example that is incorrectly rejected by $F$.

**Output**: A modified Boolean function $F'$ (in DNF) that accepts all positive examples in $E^+ \cup \{e^+\}$ and rejects all negative examples in $E^-$.

**begin**

*Step* 1: Let $E^+(c_i)$ (for $i = 1, 2, 3, \ldots, n$) be the set of the members of $E^+$ which are accepted by clause $c_i$;

*Step* 2: Let $A(c_i)$ be the number of attributes in $c_i$;

*Step* 3: Select a clause $c_k$ (for some $k$; $1 \leq k \leq n$) according to a clause selection criterion (i.e., MGC or LGC, as described below);

*Step* 4: Let $F \leftarrow F - c_k$;

*Step* 5: Let $E^+(c_k) \leftarrow E^+(c_k) \cup \{e^+\}$;

*Step* 6: Let $f$ be the function (in DNF) that solves the subproblem OCAT($E^+(c_k)$, $E^-$);

*Step* 7: Set $F' \leftarrow F \vee f$;

**end;**

**Clause Selection Criteria:**

1. Most Generalizing Clause (MGC): A clause $c_i$ with the max $|E^+(c_i)|/A(c_i)$ value.
2. Least Generalizing Clause (LGC): A clause $c_i$ with the min $|E^+(c_i)|/A(c_i)$ value.

Note that $x$ indicates the cardinality of set $x$.

**Figure 6.2.** Proposed Strategy for Repairing a Boolean Function which Incorrectly Rejects a Positive Example (for the DNF case).

empirically later in terms of the size and the accuracy of the produced Boolean functions and also the required CPU times.

A key step for achieving an effective and efficient incremental learning solution is the size of the subproblem OCAT($E^+(c_k)$, $E^-$) in Step 6 of the algorithm depicted in Figure 6.2. This is a key concept because if it is assumed that $|E^+(c_k)| \ll |E^+|$ (where $|x|$ is the size of the set $x$), then it is reasonable to assume that the CPU time for solving the subproblem OCAT($E^+(c_k)$, $E^-$) would be significantly shorter than the time required to solve the complete (and much bigger) problem OCAT($E^+ \cup \{e^+\}$, $E^-$). The branch-and-bound (B&B) approach that is used to solve such a problem in Chapter 3 is an NP-complete approach. The faster heuristics described in Chapter 4 are of polynomial time complexity. At this point, it is also important to notice that by using either of the two clause selection criteria, it may happen that one or more clauses might be added to the function $F$, then increasing in this way its size. According to [Michalski and Larson, 1978], this situation

**Input:** The negative example $e^-$ that is incorrectly accepted by the function $F$ (in DNF).
The two training sets $E^+$ and $E^-$.
**Output:** A Boolean function $F''$ that accepts all examples in $E^+$ and rejects all examples in $E^- \cup e^-$.
**begin**
*Step* 1: Let $C$ be the set of clauses $c_i$ (for $i = 1, 2, 3, \ldots, m$) that incorrectly accept $e^-$;
*Step* 2: Let $F \leftarrow F - C$;
*Step* 3: Let $E^+(C)$ be set of the members of $E^+$ which are accepted by $C$;
*Step* 4: Let $f$ be the Boolean function in DNF form that solves the subproblem OCAT($E^+(C)$, $E^- \cup \{e^-\}$);
*Step* 5: Let $F'' \leftarrow F \vee f$;
**end;**

**Figure 6.3.** Repair of a Boolean Function that Erroneously Accepts a Negative Example (for the DNF case).

can be anticipated because the utilization of an ILE approach often results in more complex systems.

### 6.4.2 Repairing of a Boolean Function that Incorrectly Accepts a Negative Example

The algorithm in Figure 6.3 addresses the second scenario for repairing a Boolean function $F$ (in DNF). This scenario occurs when the Boolean function incorrectly accepts as positive a new example that has been classified by the oracle as negative (recall that this example is now denoted as $e^-$). In this case the current function $F$ erroneously satisfies the condition

$$F(e^-) = c_1 \vee c_2 \vee c_3 \vee \cdots \vee c_n = 1. \tag{6.2}$$

The function in (6.2) shows that at least one of the $n$ clauses incorrectly accepts the negative example $e^-$. The main problem in this scenario is how to select the clause(s) to repair so that the updated function, denoted as $F''$, will reject the negative example $e^-$ while maintaining the correctness for the other examples (positive and negative).

The algorithm in Figure 6.3 solves the problem implied in relation (6.2) by first identifying the set of clauses $C$ that incorrectly accept the negative example $e^-$, and then by forming a subset of positive examples (denoted as $E^+(C)$), which is comprised of the examples in $E^+$ that are accepted by the clauses in $C$. The set of negative examples is formed by $E^- \cup \{e^-\}$. As with the subproblem in the first scenario, the potential of the algorithm in Figure 6.3 is based on solving the smaller size Boolean function inference problem denoted as OCAT($E^+(C)$, $E^- \cup \{e^-\}$) in Step 4. This is an important issue because if $|E^+(C)| \ll |E^+|$ (where $|x|$ is the size of the set $x$), then the CPU time requirement for solving this subproblem is most likely significantly shorter than that for solving the entire Boolean function inference problem OCAT($E^+$, $E^- \cup \{e^-\}$).

### 6.4.3 Computational Complexity of the Algorithms for the ILE Approach

An inspection of the algorithms in Figures 6.2 and 6.3 indicates that in the worst case, the entire problems OCAT($E^+ \cup \{e^+\}$, $E^-$) and OCAT($E^+$, $E^- \cup \{e^-\}$) will have to be executed. It should be emphasized here that the Boolean function inference problems described so far can be solved with *any* function construction method and not only the OCAT approach. In the experiments described in the next section these problems were solved by using the fast heuristic (i.e., RA1) of polynomial time described in Chapter 4. Another alternative would be to use the revised B&B approach described in Chapter 3. Other methods could be used as well. The heuristic described in Chapter 4 is essentially of $O(nm^3)$ time complexity (where $n$ is the number of binary attributes and $m$ is the *total number* of training examples). On the other hand, the B&B algorithm is an NP-complete approach. Thus, the proposed incremental learning approach takes the time complexity of the learning algorithm used to solve the smaller size Boolean function inference subproblems.

The next section describes an extensive empirical study of the relative effectiveness of the ILE and NILE approaches when they are combined with the RA1 heuristic for inferring a Boolean function. This empirical study examines the relative performance of the two criteria for clause selection described in Figure 6.2 (i.e., the MGC and LGC criteria). As measurements of performance we used the CPU time, the accuracy of the derived systems (i.e., how accurately they classified the remaining available examples), and the size of the derived systems. The various approaches were analyzed in terms of the *sign test* [Barnes, 1994] in order to determine any difference in the performance of pairs of algorithms. The sign test is a nonparametric test that compares paired observations of two populations. The number of positive and negative signs of the comparisons is used to make inference on the two populations of observations.

## 6.5 Experimental Data

Table 6.1 shows the numbers of documents from the TIPSTER collection that was used in the experimentation. The TIPSTER collection is comprised of numerous documents extracted from various sources. As was mentioned in Section 6.1, this collection of documents is often used to evaluate the performance of information retrieval (IR) and machine learning systems.

These documents were randomly extracted from the four classes of the collection. The numbers in each class were determined from the RAM limitations of the PC we used in the experiments. The computer used was a Pentium II PC with a 400 MHz CPU running the Windows 95 operating system. The computer programs for this study were written in Turbo Pascal 1.5 for Windows [Borland, 1991].

**Table 6.1.** Number of Documents Randomly Selected from Each Class.

| **Class:** | DOE | AP | WSJ | ZIPFF | Total |
|---|---|---|---|---|---|
| **No. of Documents:** | 1,407 | 336 | 624 | 530 | 2,897 |

NOTES: DOE, AP, and WSJ stand for the U.S. Department of Energy, the Associated Press, and the Wall Street Journal, respectively; ZIPFF is a collection of technical documents on various topics.

In order to simulate two mutually exclusive classes, the following three class-pairs (DOE vs. ZIPFF), (AP vs. DOE), and (WSJ vs. ZIPFF) were formed. These three class-pairs were randomly selected from all possible pair combinations. Furthermore, to comply with the notation presented in earlier sections, the first class of each class-pair was denoted as $E^+$, while the second class was denoted as $E^-$ (these class designations were set randomly). The sizes shown in Table 6.1 were dictated by the limitations of the computing resources available for that study.

The conversion of these documents into binary vectors followed the methodology discussed in [Salton, 1989], [Cleveland and Cleveland, 1983], [Meadow, 1992], [Hsiao and Haray, 1970], [Chen, 1996], and [Chen, *et al.*, 1994]. It should be mentioned here that similar examples, also derived from large collections of text documents, were used in some of the studies described in Chapter 13 of this book.

## 6.6 Analysis of the Computational Results

The computational experiments were conducted as follows. First a collection of examples was formed under one of the target class-pairs as defined in the previous section (i.e., in Table 6.1). The test examples were derived by analyzing the text documents in the previous TIPSTER categories. Next, an example was retrieved from the class-pair collection and was presented to the learning algorithm, along with its actual class membership. We used three learning algorithms as follows. The first one was the OCAT approach combined with the RA1 heuristic as described in Figure 4.1. The second algorithm is the ILE approach coupled with the *Most Generalizing Clause* (MGC) selection criterion and also the RA1 heuristic. The third ILE algorithm was similar to the second one, but now the *Least Generalizing Clause* (LGC) selection criterion is used instead of the MGC one.

The results are depicted in Figures 6.4 to 6.12. In these figures the thickest lines correspond to results under the plain OCAT approach, the thick lines to results under the incremental OCAT (i.e., under IOCAT) when it is combined with the MGC criterion, and the thin lines to results under the IOCAT when it is combined with the LGC criterion. In the horizontal axes the term "documents" is used along with the term "examples," as examples correspond to documents from the TIPSTER collection.

The results are grouped into three subsections with three figures in each subsection. We present the plots for each class-pair individually, in order to maintain

**Table 6.2.** Number of Training Documents to Construct a Clause that Classified All 510 Documents.

| | **Class-Pair** | | | |
|---|---|---|---|---|
| | **DOE vs. ZIPFF** | **AP vs. DOE** | **WSJ vs. ZIPFF** | **Mean** |
| OCAT | 335 | 311 | 363 | 336 |
| IOCAT (MGC) | 244 | 311 | 288 | 281 |
| IOCAT (LGC) | 277 | 319 | 303 | 299 |

NOTE: MGC and LGC stand for the Most and the Least Generalizing Clause selection criterion, respectively.

some subtle differences that were observed in these results. The first subsection deals with the accuracy of the derived system (i.e., the combinations of the "positive" and "negative" Boolean functions). The second subsection deals with the number of clauses in the derived Boolean functions, while the third subsection deals with the CPU time required by each approach.

### 6.6.1 Results on the Classification Accuracy

In these experiments the different learning processes started with the same initial random collection of 50 examples and were followed by incrementing the training examples one at a time according to the corresponding methods. The accuracy was determined as the number of correct classifications on the remaining unseen examples from the population of the examples in each class-pair (as shown in Table 6.1). Table 6.2 summarizes the number of training examples these learning algorithms needed to extract Boolean functions (sets of rules) that could classify all the available examples correctly for each of the three class-pairs from the TIPSTER collection.

The data in this table suggest that the Boolean functions constructed by the two IOCAT approaches used significantly fewer examples than the plain OCAT approach before the extracted Boolean functions could correctly classify the entire population (i.e., the training plus the unseen) of examples. Furthermore, an inspection of Figures 6.4, 6.5, and 6.6 indicates that the speed of correct classifications for IOCAT (thick and thin lines) was faster (steeper) than that for the plain OCAT approach (thickest line which is below the previous two lines).

Next, in Table 6.3, the 0 (zero) positive signs from the comparison of the difference OCAT – IOCAT(MGC) for class-pair DOE vs. ZIPF indicates that the ICAT(MGC) approach was always more accurate than the plain OCAT approach in all the paired observations. In contrast, the datum for IOCAT(MGC) – IOCAT(LGC) for class-pair AP vs. DOE indicates that 210 positive signs were obtained. In this case, the high number of positive signs shows that the MGC was a better performer than the LGC. As was stated in Section 6.4.1 this was anticipated since the MGC

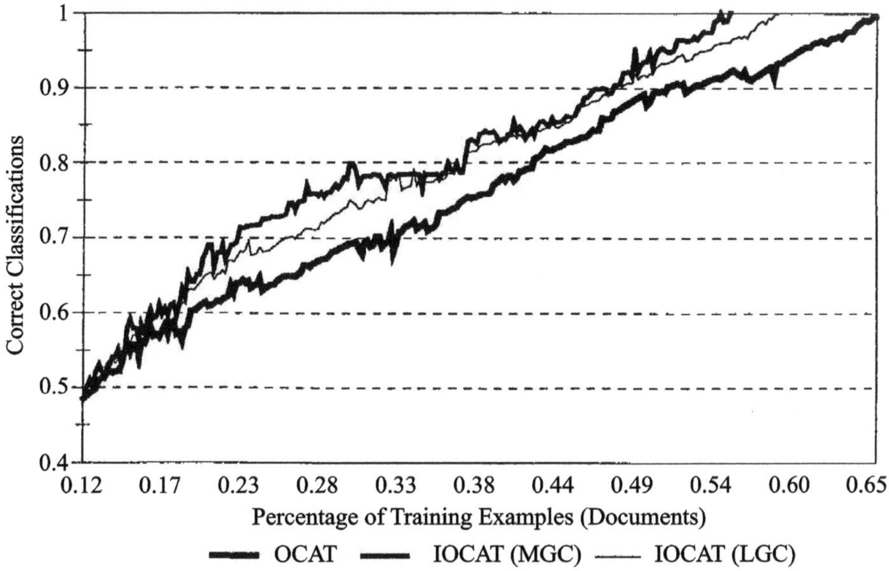


**Figure 6.4.** Accuracy Results for the Class-Pair (DOE vs. ZIPFF).

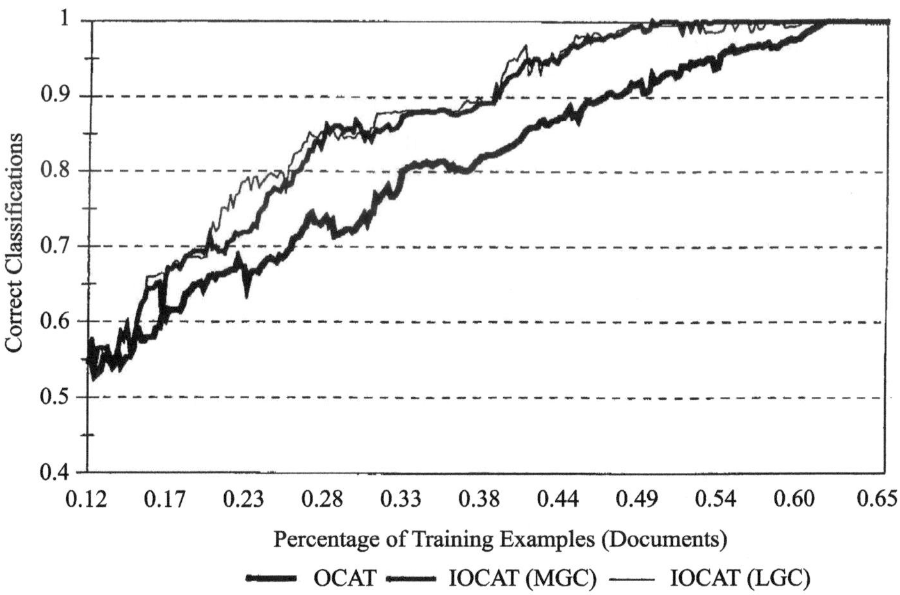


**Figure 6.5.** Accuracy Results for the Class-Pair (AP vs. DOE).

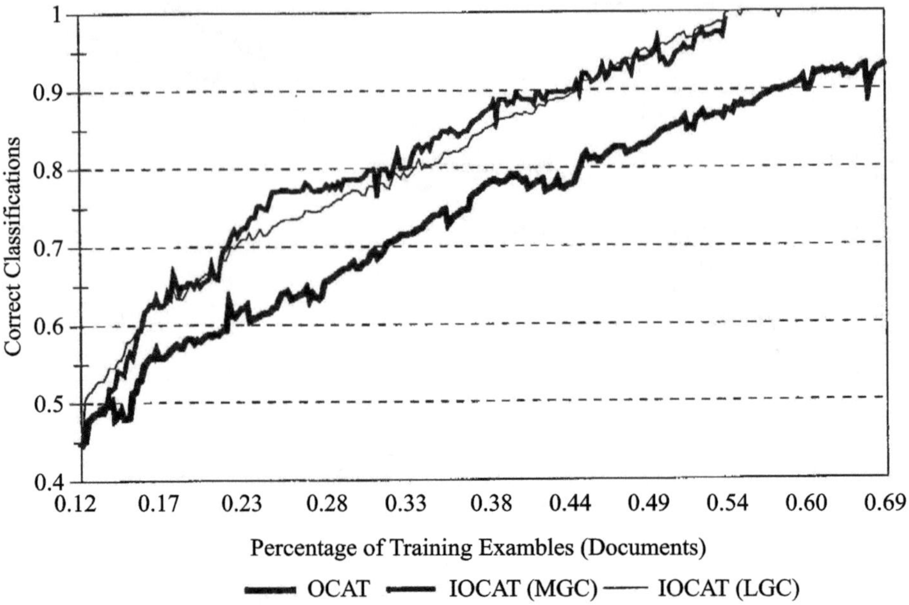


**Figure 6.6.** Accuracy Results for the Class-Pair (WSJ vs. ZIPFF).

**Table 6.3.** Statistical Comparison of the Classification Accuracy Between OCAT and IOCAT.

| | Number of Positive Signs | | |
|---|---|---|---|
| | **DOE vs. ZIPFF** | **AP vs. DOE** | **WSJ vs. ZIPFF** |
| OCAT – IOCAT (MGC) | 0¶ | 1¶ | 5¶ |
| OCAT – IOCAT (LGC) | 1¶ | 2¶ | 0¶ |
| IOCAT (MGC) – IOCAT (LGC) | 92¶ | 210£ | 163£ |

NOTES: MGC and LGC stand for the Most and the Least Generalizing Clause selection criterion, respectively.
All $p$-values were approximated using $\mathrm{N}(np, np(1-p)^{1/2})$. For all ¶ instances, $n = 250$, while for all £ instances $n$ was equal to 237, 240, and 242. These different values of $n$ were needed because all the differences yielding zero were discarded [Barnes, 1994]. The $p$-value for the sign test was equal to 0.50.
¶ denotes a $p$-value close to 0.
£ denotes a $p$-value close to 1.

clause has, by definition, more potential to make an impact in the way the target Boolean function classifies examples.

The small $p$-value of the comparison between the two algorithms indicates that the two versions of IOCAT were much better performers than the plain OCAT approach. Second, the large $p$-values of the comparison of the two versions of IOCAT suggest that the two clause selection criteria performed in a similar manner.

### 6.6.2 Results on the Number of Clauses

The corresponding results are depicted in Figures 6.7, 6.8, and 6.9. These results represent the total number of clauses of the "positive" and the "negative" Boolean functions. As before, some numerical highlights are summarized in Tables 6.4 and 6.5. From these results it is evident that under the plain OCAT approach the two inferred Boolean functions had much fewer clauses (less than 50%) than the functions under the two ILE approaches. The same results also indicate that the IOCAT approach with the MGC selection criterion did better than the IOCAT approach when it was combined with the LGC selection criterion. As with the previous results, this can be attributed to the higher potential of the MGC clause.

An interesting phenomenon in the previous results is the occasional drop in the number of clauses under the two ILE approaches. This occurred when the ILE approach (i.e., either the IOCAT(MGC) or the IOCAT(LGC) approach) had to solve the entire function inference problem. This problem was denoted as subproblem OCAT($E^+$, $E^- \cup \{e^-\}$) in the methodology section. This occasional reconstruction step of the entire function alleviated the problem of occasionally generating too many clauses.

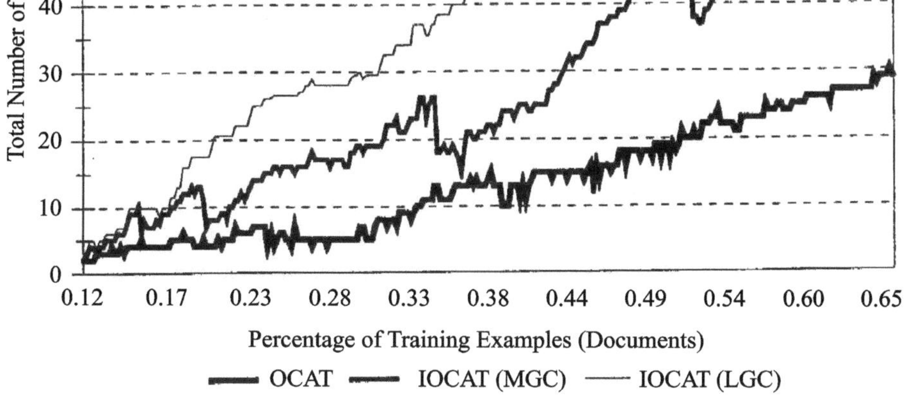


**Figure 6.7.** Number of Clauses for the Class-Pair (DOE vs. ZIPFF).

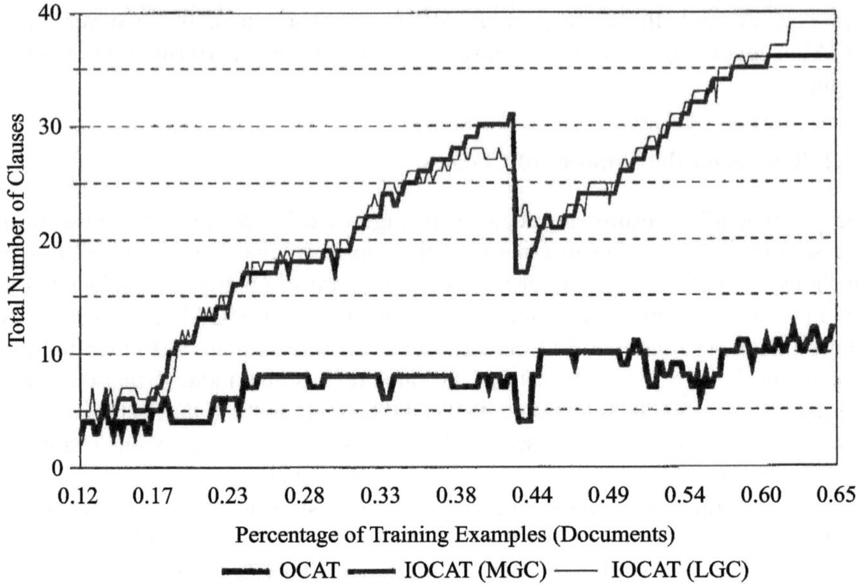


**Figure 6.8.** Number of Clauses for the Class-Pair (AP vs. DOE).

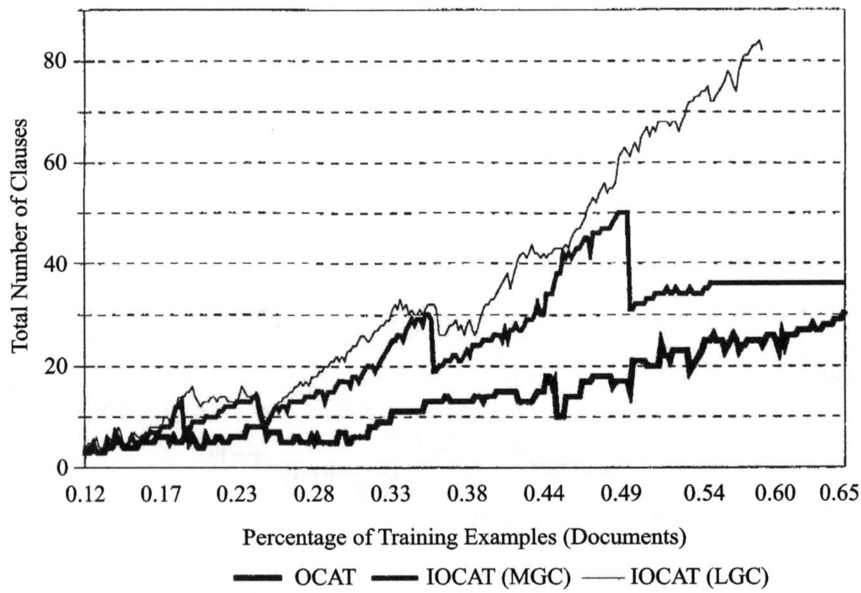


**Figure 6.9.** Number of Clauses for the Class-Pair (WSJ vs. ZIPFF).

**Table 6.4.** Number of Clauses in the Boolean Functions at the End of an Experiment.

| | Class-Pair | | | |
|---|---|---|---|---|
| | **DOE vs. ZIPFF** | **AP vs. DOE** | **WSJ vs. ZIPFF** | **Mean** |
| OCAT | 23 | 10 | 25 | 19 |
| IOCAT (MGC) | 47 | 36 | 36 | 40 |
| IOCAT (LGC) | 37 | 60 | 74 | 57 |

NOTE: MGC and LGC stand for the Most and the Least Generalizing Clause selection criterion, respectively.

**Table 6.5.** Statistical Comparison of the Number of Clauses Constructed by OCAT and IOCAT.

| | Number of Positive Signs | | |
|---|---|---|---|
| | **DOE vs. ZIPFF** | **AP vs. DOE** | **WSJ vs. ZIPFF** |
| OCAT – IOCAT (MGC) | 0 | 1 | 0 |
| OCAT – IOCAT (LGC) | 1 | 2 | 0 |
| IOCAT (MGC) – IOCAT (LGC) | 44¶ | 0 | 0 |

NOTES: MGC and LGC stand for the Most and the Least Generalizing Clause selection criterion, respectively.
All *p-values* were approximated using $\text{N}(np, np(1-p)^{1/2})$. For all ¶ instances, $n = 243$ since all differences yielding zero were discarded [Barnes, 1994].]. The *p-value* for the sign test was equal to 0.50.
¶denotes a *p-value* close to 0.

### 6.6.3 Results on the CPU Times

The results on the CPU times are depicted in Figures 6.10, 6.11, and 6.12, and are summarized in Tables 6.6 and 6.7. As was anticipated, the ILE approaches (i.e., the IOCAT(MGC) and the IOCAT(LGC) approaches) required significantly less CPU time than the plain OCAT approach. This is in agreement with the discussions in the methodology section and also with similar references from the literature. For instance, [Michalski and Larson, 1978] and [Reine and Michalski, 1986] have indicated that incremental approaches always required shorter CPU times than non-incremental approaches.

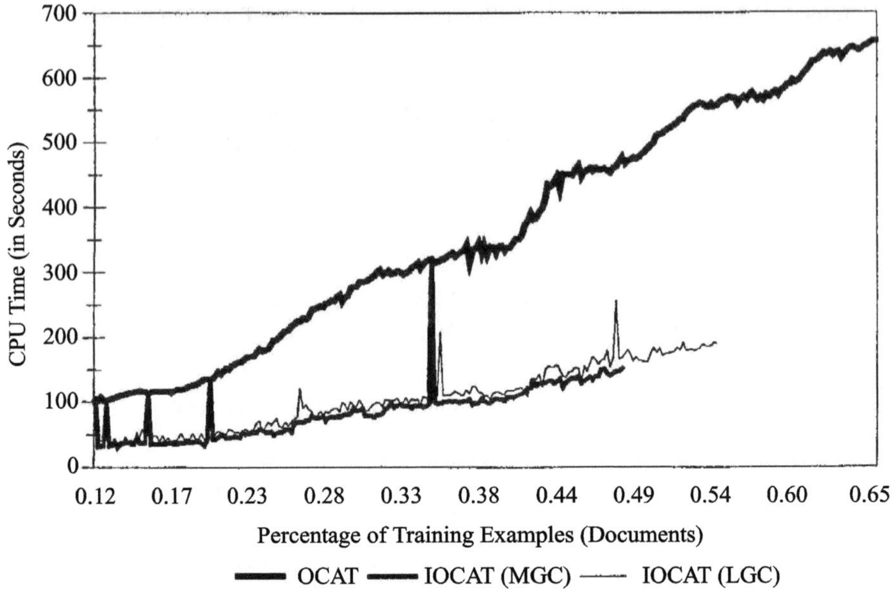


**Figure 6.10.** Required CPU Time for the Class-Pair (DOE vs. ZIPFF).

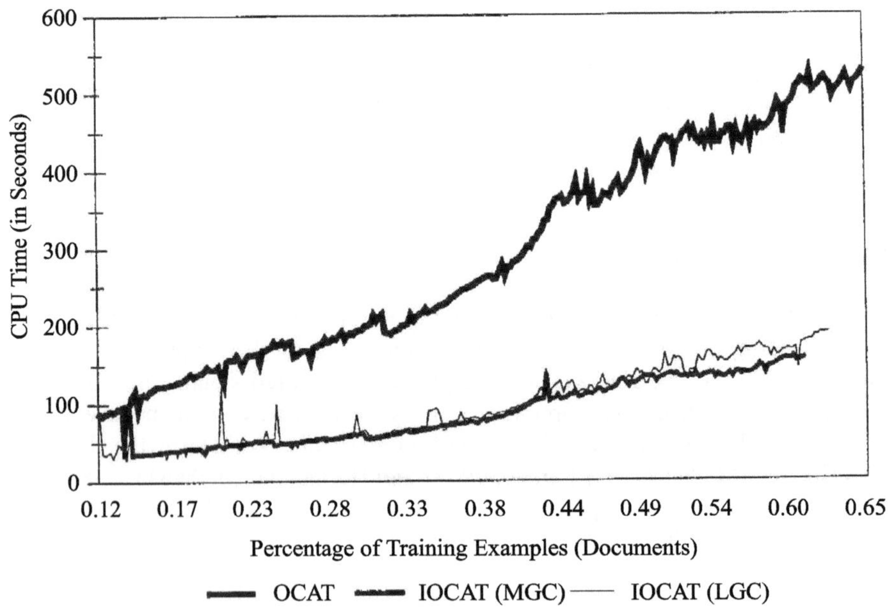


**Figure 6.11.** Required CPU Time for the Class-Pair (AP vs. DOE).

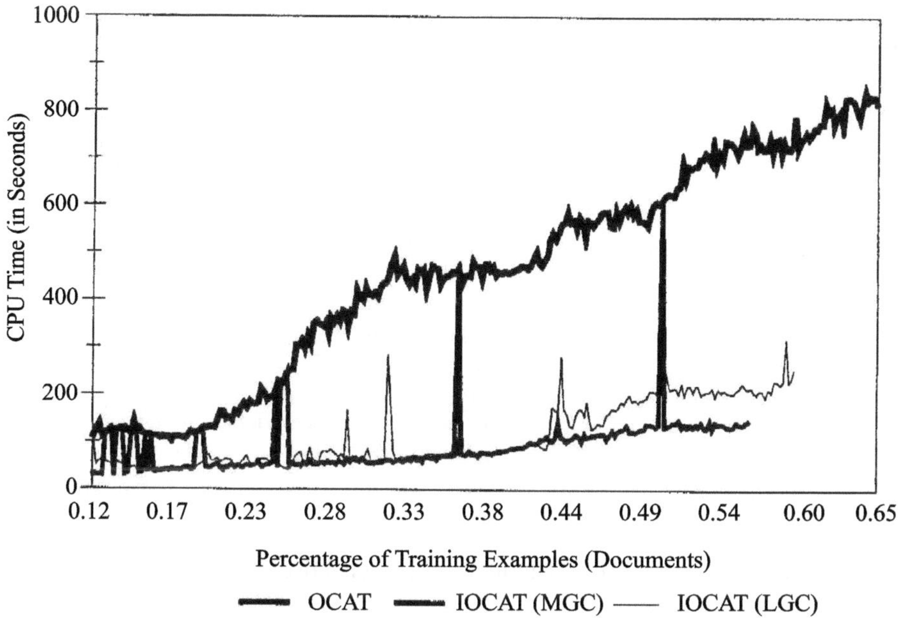


**Figure 6.12.** Required CPU Time for the Class-Pair (WSJ vs. ZIPFF).

**Table 6.6.** The CPU Times (in Seconds) Required to Complete an Experiment.

| | Class-Pair | | | |
|---|---|---|---|---|
| | **DOE vs. ZIPFF** | **AP vs. DOE** | **WSJ vs. ZIPFF** | **Mean** |
| OCAT | 468.532 | 514.584 | 729.430 | 570.848 |
| IOCAT (MGC) | 150.896 | 156.374 | 142.966 | 150.088 |
| IOCAT (LGC) | 178.885 | 168.371 | 215.932 | 187.729 |

NOTE: MGC and LGC stand for the Most and the Least Generalizing Clause selection criterion, respectively.

When the two versions of the IOCAT approach are compared, the small $p$-values shown in Table 6.7 reveal that the MGC selection criterion always required shorter CPU times than the LGC one. As before, this is most likely caused because the MGC can impact the behavior of the target Boolean function more significantly than the LGC.

As with the results regarding the total number of the derived clauses, here too the plots have some spikes. These spikes correspond to occasions when an inference problem had to be solved in its entirety. This occurred when the set $E^+(C)$ was identical to the $E^+$ set in Steps 3 and 4 in Figure 6.3. A similar situation also occurred

**Table 6.7.** Statistical Comparison of the CPU Time to Reconstruct/Modify the Boolean Functions.

| | Number of Positive Signs | | |
|---|---|---|---|
| | **DOE vs. ZIPFF** | **AP vs. DOE** | **WSJ vs. ZIPFF** |
| OCAT – IOCAT (MGC) | 269 | 262 | 257 |
| OCAT – IOCAT (LGC) | 273 | 272 | 274 |
| IOCAT (MGC) – IOCAT (LGC) | 29¶ | 10¶ | 26¶ |

NOTES: MGC and LGC stand for the Most and the Least Generalizing Clause selection criterion, respectively.
All $p$-values were approximated using $N(np, np(1-p)^{1/2})$. For all ¶ instances, $n = 243, 217$, and 247 since all differences yielding zero were discarded [Barnes, 1994]. The $p$-value for the sign test was equal to 0.50.
¶ denotes a $p$-value close to 0.

in the experiments reported in [Utgoff, 1997] when a decision tree had to be rebuilt from the beginning.

## 6.7 Concluding Remarks

This chapter proposed an approach for inferring a Boolean function in an incremental learning environment. In such an environment, it was assumed that some training examples are available and are divided into two mutually exclusive classes. Also a "positive" and a "negative" Boolean function are available and they satisfy the requirements of the initial training data. As new examples become available, either of the two Boolean functions may need to be modified (if it misclassifies new observations) to satisfy the requirements of the existing and also the new training data. The algorithms proposed in this chapter modify a Boolean function in a localized/surgical manner, unless it is determined that the function inference problem needs to be solved in its entirety.

The proposed function modification procedures were combined with an existing algorithm for inferring a Boolean function from two classes of examples. That algorithm is the OCAT (One Clause At a Time) approach as described in Chapter 2. However, *any* Boolean function inference algorithm can be used with the proposed incremental learning approaches.

An extensive empirical study was also undertaken to better assess the numerical properties of the new approaches and how they compare with the nonincremental OCAT approach. As data for the empirical study, we used binary examples that were defined by analyzing text documents from the TIPSTER collection.

The results of this investigation suggest that the proposed approaches are both effective and efficient. In these tests the new approaches returned Boolean functions that were more accurate than the corresponding functions returned by the nonincremental OCAT approach. Furthermore, they did so in a significantly small fraction of the CPU time required by the nonincremental approach. However, as was anticipated, the Boolean functions returned by the incremental approaches had more (slightly more than twice) clauses than under the nonincremental approach.

In summary, the results in this chapter strongly suggest that if a learning task involves frequent incremental processing of large collections of examples (defined on a large set of binary attributes), then the proposed incremental learning algorithms are an effective and efficient alternative to more time-consuming nonincremental approaches. However, the quest for developing even faster incremental learning algorithms is still a very active endeavor in the field of data mining and knowledge discovery from data sets.

# Chapter 7

# A Duality Relationship Between Boolean Functions in CNF and DNF Derivable from the Same Training Examples

## 7.1 Introduction

This chapter discusses a useful relationship between the CNF and DNF forms of the Boolean functions derivable from the same training data. This relationship can benefit approaches which attempt to solve large Boolean function inference problems and use either the CNF or the DNF form in representing a Boolean function.

Suppose that a given learning algorithm depends heavily on the number of the positive (the negative) examples. If the number of the positive (the negative) examples is larger than the number of the negative (the positive) examples, then it is more efficient to solve a slightly different problem. That is, apply the same algorithm as before, but instead of using the original training examples, now use their *complements* (negations). However, now the complements of the original positive (negative) examples should be treated as the new negative (positive) examples.

If the original algorithm derives a CNF (DNF) expression, then in this way it will derive a DNF (CNF) expression which will satisfy the constraints of the original training data. The previous situation is very *similar* to the strategy of solving the *dual* of an LP (linear programming) problem. Recall that if the primal LP problem has many constraints and few variables, then the Simplex approach is faster for the dual problem (which will have fewer constraints and many variables).

Let $e$ be an example (either positive or negative). Then, $\hat{e}$ is defined as the *complement* (negation) of example $e$. For instance, if $e = (1, 0, 0)$, then $\hat{e} = (0, 1, 1)$. The following definition introduces the concept of the complement of a set of examples. Let $E$ be a set of examples (either positive or negative). Then, $\hat{E}$ is defined as the *complement of the set* $E$.

## 7.2 Generating Boolean Functions in CNF and DNF Form

Recall from Chapter 2 (Section 2.3) that the general form of a Boolean function written in CNF or DNF is defined as (7.1) or (7.2), respectively.

E. Triantaphyllou, *Data Mining and Knowledge Discovery via Logic-Based Methods*, Springer Optimization and Its Applications 43, DOI 10.1007/978-1-4419-1630-3_7,

$$\bigwedge_{j=1}^{k}\left(\bigvee_{i\in\rho_j} a_i\right) \tag{7.1}$$

and

$$\bigvee_{j=1}^{k}\left(\bigwedge_{i\in\rho_j} a_i\right), \tag{7.2}$$

where $a_i$ is either $A_i$ or $\bar{A}_i$ and $\rho_j$ is the set of indices.

In other words, a CNF expression is a conjunction of disjunctions, while a DNF expression is a disjunction of conjunctions. For instance, the expression $(A_1 \vee A_2) \wedge (\bar{A}_1 \vee \bar{A}_2)$ is a Boolean function in CNF while $(A_1 \wedge A_2) \vee (\bar{A}_1 \wedge \bar{A}_2)$ is a Boolean function in DNF.

Then the following theorem [Triantaphyllou and Soyster, 1995a] states an important property which exists when Boolean functions in CNF and DNF are inferred from sets of positive and negative examples. The proof of this theorem is based on the simple observation that an example $e$ is accepted (rejected) by a *conjunction* $(\wedge_{i\in\rho} a_i)$ if and only if the example $\hat{e}$ is rejected (accepted) by the *disjunction* $(\vee_{i\in\rho} a_i)$.

**Theorem 7.1.** *Let $E^+$ and $E^-$ be the disjoint sets of positive and negative examples, respectively. A Boolean function in CNF given as (7.1) satisfies the constraints of the $E^+$ and $E^-$ sets if and only if the Boolean function in DNF given as (7.2) satisfies the constraints of $\hat{E}^-$ (considered as the positive examples) and $\hat{E}^+$ (considered as the negative examples).*

## 7.3 An Illustrative Example of Deriving Boolean Functions in CNF and DNF

Suppose that the following are two sets of positive and negative training examples:

$$E^+ = \begin{bmatrix} 1 & 0 & 1 \\ 1 & 0 & 0 \\ 0 & 1 & 1 \end{bmatrix} \quad \text{and} \quad E^- = \begin{bmatrix} 0 & 0 & 1 \\ 1 & 1 & 1 \end{bmatrix}.$$

When the SAT approach, as described in Section 2.5, is used on the previous data, the resulting satisfiability problem has 31 clauses and 66 Boolean variables (it is assumed that $k = 2$). Since there are more positive examples than negative ones, the *complemented* problem is smaller. It has 26 clauses and 58 variables. The complemented sets are as follows:

$$\hat{E}^- = \begin{bmatrix} 1 & 1 & 0 \\ 0 & 0 & 0 \end{bmatrix} \quad \text{and} \quad \hat{E}^+ = \begin{bmatrix} 0 & 1 & 0 \\ 0 & 1 & 1 \\ 1 & 0 & 0 \end{bmatrix}.$$

When the SAT approach is applied on the sets $\hat{E}^-$ and $\hat{E}^+$ (treating the first set as the positive examples and the second set as the negative examples), then the following DNF system is derived: $(A_1 \wedge A_2) \vee (\bar{A}_1 \wedge \bar{A}_2)$. Therefore, according to Theorem 7.1, the following CNF expression satisfies the requirements of the original $E^+$ and $E^-$ data: $(A_1 \vee A_2) \wedge (\bar{A}_1 \vee \bar{A}_2)$.

Similarly, when the OCAT approach is applied on the previous $\hat{E}^-$ and $\hat{E}^+$ data, then the following CNF system is derived: $(A_1 \vee \bar{A}_2) \wedge (\bar{A}_1 \vee A_2)$. According to Theorem 7.1, the following DNF expression satisfies the requirements of the original $E^+$ and $E^-$ data: $(A_1 \wedge \bar{A}_2) \vee (\bar{A}_1 \wedge A_2)$. The above results can easily be verified on the previous data.

## 7.4 Some Computational Results

The previous theorem was next applied to two algorithms which infer a Boolean function from examples. These are the OCAT approach described in Chapters 2 and 3 and the SAT approach described in Chapter 2.

The SAT approach was first proposed in [Kamath, *et al.*, 1992]. Let $m_1$ and $m_2$ be the numbers of examples in the $E^+$ and $E^-$ sets, respectively, and $n$ the number of attributes. When the SAT approach is used to derive a Boolean function in DNF (as described in Section 2.5), then it *preassumes* the value of $k$; the number of conjunctions in the DNF expression. In Section 2.5 it was shown that the SAT formulation is based on $k(n(m_1 + 1) + m_2) + m_1$ clauses (constraints of an integer programming problem), and $k(2n(1 + m_1) + nm_2 + m_1)$ Boolean variables. If this SAT problem is infeasible, then the conclusion is that there is no DNF system which has $k$ or less conjunctions and satisfies the requirements imposed by the examples. Thus, the value of the parameter $k$ needs to be increased until feasibility is reached.

Table 7.1 presents some computational results when the OCAT approach is used on random test problems with $n = 30$ attributes. In this table, $|E^{\circ}|$ represents the *total* number of training examples and $|E^+|$ represents the number of positive examples. To understand the table, consider the third line of the first column. This represents a problem with 5 positive examples and 95 negative examples which required 2 CPU seconds on an IBM ES/3090-600S mainframe computer to derive a set of CNF clauses (which accepts all 5 positive examples and rejects the 95 negative examples). For a given problem size, e.g., $|E^{\circ}| = 100$, note how the CPU time increases as the number of positive examples, denoted as $|E^+|$, increases.

The savings in applying Theorem 7.1 for the SAT approach can be easily determined. Compute the difference in the number of clauses and variables when the parameters $m_1$ and $m_2$ are interchanged. This difference for the number of clauses is equal to

$$(m_1 - m_2)[k(n - 1) + 1],$$

while the difference for the number of variables is equal to

$$(m_1 - m_2)k(n + 1).$$

Hence, when $m_1 > m_2$ (i.e., when the positive examples exceed the negative examples), the number of clauses and the number of variables can be reduced by a

**Table 7.1.** Some Computational Results When $n = 30$ and the OCAT Approach Is Used.

| $\|E^{\circ}\|$ | $\|E^{+}\|$ | Time[a] | $\|E^{\circ}\|$ | $\|E^{+}\|$ | Time | $\|E^{\circ}\|$ | $\|E^{+}\|$ | Time |
|---|---|---|---|---|---|---|---|---|
| 100 | 1 | 1 | 300 | 2 | 1 | 500 | 7 | 15 |
| 100 | 3 | 1 | 300 | 3 | 2 | 500 | 13 | 8 |
| 100 | 5 | 2 | 300 | 14 | 29 | 500 | 16 | 38 |
| 100 | 5 | 3 | 300 | 14 | 25 | 500 | 20 | 19 |
| 100 | 7 | 2 | 300 | 17 | 107 | 500 | 34 | 73 |
| 100 | 7 | 6 | 300 | 22 | 102 | 500 | 35 | 194 |
| 100 | 7 | 3 | 300 | 22 | 70 | 600 | 8 | 16 |
| 100 | 8 | 2 | 300 | 24 | 12 | 600 | 11 | 15 |
| 100 | 9 | 2 | 300 | 36 | 243 | 600 | 23 | 184 |
| 100 | 15 | 7 | 300 | 71 | 524 | 600 | 44 | 41 |
| 200 | 1 | 1 | 400 | 5 | 3 | 600 | 49 | 315 |
| 200 | 2 | 1 | 400 | 7 | 10 | 600 | 83 | 300 |
| 200 | 4 | 2 | 400 | 10 | 10 | 700 | 13 | 26 |
| 200 | 5 | 2 | 400 | 15 | 7 | 700 | 18 | 30 |
| 200 | 6 | 2 | 400 | 16 | 8 | 700 | 19 | 60 |
| 200 | 10 | 6 | 400 | 17 | 23 | 700 | 19 | 15 |
| 200 | 11 | 38 | 400 | 36 | 282 | 700 | 56 | 467 |
| 200 | 18 | 18 | 400 | 47 | 97 | 800 | 64 | 739 |
| 200 | 19 | 4 | 400 | 49 | 400 | 900 | 72 | 836 |
| 200 | 51 | 212 | 500 | 6 | 13 | 1,000 | 47 | 80 |

NOTE: The time is in CPU seconds on an IBM ES/3090-600S machine (such as the mainframe computer at Penn State University in the 1980s and 1990s).

factor proportional to $(m_1 - m_2)$ and a term of the order of $kn$ when the positive and negative examples are interchanged. For instance, if $m_1 - m_2 = 50$, $k = 10$, and $n = 20$, then the reduction in the number of clauses and variables needed for the SAT approach would be equal to 9,550 and 10,500, respectively.

## 7.5 Concluding Remarks

In this chapter we examined an interesting and rather simple relationship that exists between Boolean functions in CNF and DNF derivable from disjoint sets of positive and negative examples. The formulations described in this chapter can benefit *any algorithm* which derives CNF or DNF expressions from positive and negative examples. This relationship was demonstrated on two learning algorithms. Furthermore, these formulations can lead to increased efficiency for solving large-scale learning problems of the type described in this chapter.

# Chapter 8

# The Rejectability Graph of Two Sets of Examples

## 8.1 Introduction

This chapter is based on the findings presented in [Triantaphyllou and Soyster, 1996] and presents the motivation and definition of a special graph which can be easily derived from positive and negative examples. To understand the motivation for introducing this graph, consider a situation with $n = 5$ attributes. Suppose that the vector $v_1 = (1, 0, 1, 0, 1)$ is a *positive example* while the two vectors $v_2 = (1, 0, 1, 1, 1)$ and $v_3 = (1, 1, 1, 0, 1)$ are *negative examples*. For the positive example $v_1$, note that $A_1, \bar{A}_2, A_3, \bar{A}_4$, and $A_5$ are true (or, equivalently, $\bar{A}_1, A_2, \bar{A}_3, A_4$, and $\bar{A}_5$ are false). Similar interpretations exist for the remaining two examples $v_2$ and $v_3$.

Denote by *ATTRIBUTES*$(v)$ the set of the attributes that are true (have value "1") for a particular (either positive or negative) example $v$. With this definition, one obtains from the above data:

$$\mathit{ATTRIBUTES}(v_1) = \mathit{ATTRIBUTES}((1, 0, 1, 0, 1)) = \{A_1, \bar{A}_2, A_3, \bar{A}_4, A_5\}$$

$$\mathit{ATTRIBUTES}(v_2) = \mathit{ATTRIBUTES}((1, 0, 1, 1, 1)) = \{A_1, \bar{A}_2, A_3, A_4, A_5\}$$

$$\mathit{ATTRIBUTES}(v_3) = \mathit{ATTRIBUTES}((1, 1, 1, 0, 1)) = \{A_1, A_2, A_3, \bar{A}_4, A_5\}.$$

Next consider a single CNF clause (i.e., a disjunction), denoted as $C$, of the general form

$$C = \bigvee_{i=1}^{m} a_i \quad (\text{where } a_i \text{ is either } A_i \text{ or } \bar{A}_i).$$

The clause $C$ *accepts* an example $v$ (i.e., $v$ is a positive example of $C$) if and only if at least one of the attributes in the set *ATTRIBUTES*$(v)$ is also one of the attributes in the expression

$$\bigvee_{i=1}^{m} a_i .$$

Otherwise, the example $v$ is *not accepted* (i.e., $v$ is a negative example of $C$). For instance, if the clause $C$ is defined as $C = (\bar{A}_2 \vee A_4)$, then the examples $v_1$ and $v_2$ are accepted by $C$, while the example $v_3$ is *not* accepted.

E. Triantaphyllou, *Data Mining and Knowledge Discovery via Logic-Based Methods*, Springer Optimization and Its Applications 43, DOI 10.1007/978-1-4419-1630-3_8,

Now observe that there is *no single CNF clause* which can *simultaneously* reject the two negative examples $v_2$ and $v_3$, while at the same time accept the positive example $v_1$. This is true because any clause which simultaneously rejects the two examples $v_2$ and $v_3$, should not contain any of the attributes in the *union* of the two sets $ATTRIBUTES(v_2)$ and $ATTRIBUTES(v_3)$. But, if none of the attributes of the set $\{A_1, A_2, \bar{A}_2, A_3, A_4, \bar{A}_4, A_5\} = ATTRIBUTES(v_2) \cup ATTRIBUTES(v_3)$ is present in the clause, then it is *impossible* to accept the positive example $v_1 = (1, 0, 1, 0, 1)$. Therefore, given any clause which accepts the positive example $v_1$, the previous two negative examples $v_2$ and $v_3$ cannot also be *rejected* by such clause.

From the above realizations it follows that given any three examples $v_1$, $v_2$, and $v_3$, the examples $v_2$ and $v_3$ are rejectable by a single clause (disjunction), subject to the example $v_1$ (i.e., such a clause would accept $v_1$), if and only if the following condition is true:

$$ATTRIBUTES(v_1) \not\subseteq ATTRIBUTES(v_2) \cup ATTRIBUTES(v_3).$$

In general, given a set of positive examples $E^+$, two negative examples $v_1$ and $v_2$ are rejectable by a single clause if and only if the condition in the following theorem [Triantaphyllou and Soyster, 1996] is satisfied:

**Theorem 8.1.** *Let $E^+$ be a set of positive examples and $v_1$, $v_2$ be two negative examples. There exists a CNF clause which accepts all the positive examples and rejects both negative examples $v_1$ and $v_2$ if and only if*

$$ATTRIBUTES(v_i) \not\subseteq ATTRIBUTES(v_1) \cup ATTRIBUTES(v_2),$$

*for each positive example $v_i \in E^+$.*

## 8.2 The Definition of the Rejectability Graph

The previous theorem follows directly from the previous considerations. Given two collections of positive and negative examples, denoted as $E^+$ and $E^-$, respectively, Theorem 8.1 motivates the construction of a graph $G = (V, E)$ as follows:

$$V = \{V_1, V_2, V_3, \ldots, Vm_2\},$$

where $m_2$ is the cardinality of $E^-$ (i.e., each vertex of $G$ corresponds to one negative example in $E^-$), and

$$e \in E, \quad \text{where} \quad e = (V_i, V_j),$$

if and only if the $i$-th and the $j$-th examples in $E^-$ are rejectable by a single clause subject to the examples in $E^+$. That is, such a clause would accept all the positive examples. In this notation $E$ is the set of the edges of the graph and it should not be confused with $E^+$ or $E^-$.

We denote this graph as the *rejectability graph* (or *the R-graph*) of $E^+$ and $E^-$. The previous theorem indicates that it is computationally straightforward to construct this graph. If there are $m_2$ negative examples, then the maximum number of edges is $m_2 \times (m_2 - 1)/2$. Therefore, the rejectability graph can be constructed by performing $m_2 \times (m_2 - 1)/2$ simple rejectability examinations.

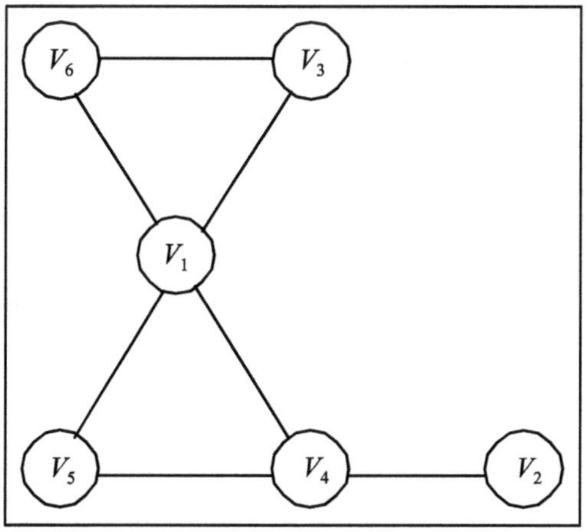


**Figure 8.1.** The Rejectability Graph of $E^+$ and $E^-$.

## An Illustrative Example

Consider the following $E^+$ and $E^-$ sets (also given earlier in other chapters and are repeated here):

$$E^+ = \begin{bmatrix} 0 & 1 & 0 & 0 \\ 1 & 1 & 0 & 0 \\ 0 & 0 & 1 & 1 \\ 1 & 0 & 0 & 1 \end{bmatrix} \quad \text{and} \quad E^- = \begin{bmatrix} 1 & 0 & 1 & 0 \\ 0 & 0 & 0 & 1 \\ 1 & 1 & 1 & 1 \\ 0 & 0 & 0 & 0 \\ 1 & 0 & 0 & 0 \\ 1 & 1 & 1 & 0 \end{bmatrix}.$$

Since there are 6 negative examples, there are $6 \times (6 - 1)/2 = 15$ possible pairwise comparisons (i.e., single rejectability tests). For instance, the first ($v_1$) and third ($v_3$) negative examples correspond to the vertices $V_1$ and $V_3$, respectively. Next one can observe that because

$$ATTRIBUTES(v_1) \cup ATTRIBUTES(v_3) = \{A_1, A_2, A_3, A_4, \bar{A}_2, \bar{A}_4\}$$

and

$$ATTRIBUTES(v_i) \not\subseteq \{A_1, A_2, A_3, A_4, \bar{A}_2, \bar{A}_4\}, \text{ for each } v_i \in E^+,$$

it follows that there is an edge which connects the vertices $V_1$ and $V_3$ in the rejectability graph. The rejectability graph $G$, which corresponds to the previous two sets of examples, is presented in Figure 8.1. ■

### 8.2.1 Properties of the Rejectability Graph

The rejectability graph $G$ of two sets of positive and negative examples possesses a number of interesting properties. Two of these properties refer to its cliques. A *clique*

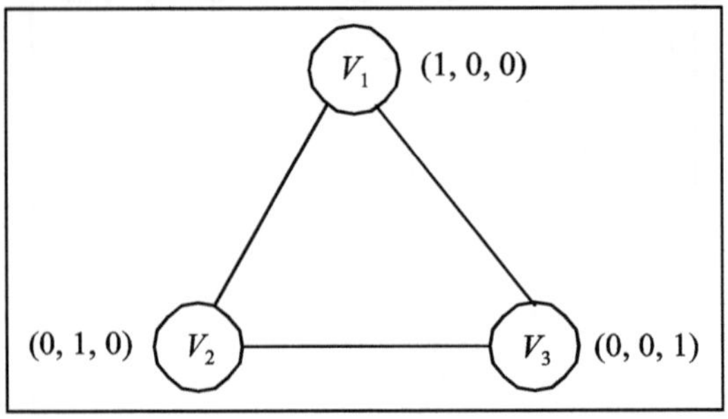


**Figure 8.2.** The Rejectability Graph for the Second Illustrative Example.

of a graph is a subgraph in which all the nodes are connected with each other. The *minimum clique cover number* (denoted as $k(G)$) is the smallest number of cliques needed to cover the vertices of $G$ (see, for instance, Golumbic [1980]). The following theorem [Triantaphyllou and Soyster, 1996] refers to any clique of the rejectability graph.

**Theorem 8.2.** *Suppose that two sets $E^+$ and $E^-$ with positive and negative examples, respectively, are given and $\beta$ is a subset of k negative examples from $E^-$ ($k \leq$ size of set $E^-$) with the property that the subset can be rejected by a single CNF clause which also accepts each of the positive examples in $E^+$. Then, the vertices corresponding to the k negative examples in the rejectability graph G form a clique of size k.*

*Proof.* Consider any two examples $v_1$ and $v_2$ in the subset of negative examples $\beta$. Since all the members in $\beta$ can be rejected by a single clause, obviously the examples $v_1$ and $v_2$ can be rejected by this single clause. From the definition of the rejectability graph $G$, it follows that there is an edge connecting the corresponding two nodes in $G$. Clearly this situation is true for any pair of examples in the subset $\beta$. Therefore, the vertices which correspond to the $k$ negative examples in $\beta$ form a clique in $G$ of size $k$. ■

The previous theorem states that any set of negative examples which can be rejected by a single clause corresponds to a clique in the rejectability graph. However, *the converse is not true*. That is, not every clique in the rejectability graph corresponds to a set of negative examples which can be rejected by a single clause. To see this consider the following illustrative example.

## An Illustrative Example

Consider the following sets $E^+$ and $E^-$ of positive and negative examples, respectively:

$$E^+ = \begin{bmatrix} 1 & 1 & 1 \end{bmatrix} \quad \text{and} \quad \begin{bmatrix} 1 & 0 & 0 \\ 0 & 1 & 0 \\ 0 & 0 & 1 \end{bmatrix}.$$

It can be easily verified that any pair of the three negative examples in $E^-$ can be rejected by a single clause which also accepts the positive example in $E^+$. For instance, the first and second negative examples are rejected by the clause $(A_3)$, which also accepts the positive example in $E^+$. Similarly, the first and third negative examples can be rejected by clause $(A_2)$, while clause $(A_1)$ rejects the second and third negative examples. In all cases, these clauses accept the single positive example in $E^+$. Therefore, the corresponding rejectability graph is a triangle (i.e., a clique with three nodes, see also Figure 8.2). However, a clause which would reject all the three negative examples should *not include* any attributes from the following set:

$$\begin{aligned} &ATTRIBUTES(v_1) \cup ATTRIBUTES(v_2) \cup ATTRIBUTES(v_3) \\ &\quad = ATTRIBUTES((1, 0, 0)) \cup ATTRIBUTES((0, 1, 0)) \cup ATTRIBUTES((0, 0, 1)) \\ &\quad = \{A_1, A_2, A_3, \bar{A}_1, \bar{A}_2, \bar{A}_3\}. \end{aligned}$$

Obviously, no such clause exists as there are no attributes left to be included in such clause when $n = 3$. Therefore, a minimum size set of CNF clauses which satisfy the requirements of the current examples could be $(A_3) \vee (A_2)$, which is of size 2. ■

### 8.2.2 On the Minimum Clique Cover of the Rejectability Graph

Consider two sets of positive and negative examples $E^+$ and $E^-$, respectively. Let $\bar{G}$ be the *complement* of the rejectability graph $G$ of the two sets of examples. Recall that the complement of a graph is constructed as follows: The complement graph has exactly the same vertices as the original graph. There is an edge between any two vertices if and only if there is no edge between the corresponding vertices of the original graph. Next, define $\omega(\bar{G})$ as the *size of the maximum clique* of the graph $\bar{G}$ and $k(G)$ as the minimum clique cover number of the rejectability graph $G$. Let $r$ be the minimum number of CNF clauses required to reject all the examples in $E^-$, while accepting all the examples in $E^+$. Then, the following theorem [Triantaphyllou and Soyster, 1996] states a *lower bound* (i.e., the minimum clique cover $k(G)$) on the *minimum number* of clauses required to reject all the negative examples in $E^-$, while accepting all the positive examples in $E^+$.

**Theorem 8.3.** *Suppose that $E^+$ and $E^-$ are the sets of the positive and negative examples, respectively. Furthermore, let $G$ and $\bar{G}$ be the corresponding rejectability graph and its complement, respectively. Then, the following relation is true: $r \geq k(G) \geq \omega(\bar{G})$.*

This theorem states that a lower bound on the minimum number of CNF clauses which can be inferred from the sets of positive and negative examples is the minimum clique cover of the rejectability graph (i.e., the value of $k(G)$). Another lower bound is the size of the maximum clique of the complement of the rejectability graph (i.e., the value of $\omega(\bar{G})$). Of these two bounds, the former is tighter since $k(G) \geq \omega(\bar{G})$.

At this point it should be stated that according to Theorem 8.3 the gap between $r$ and $k(G)$ could be positive. The same is also true with the gap between $k(G)$ and $\omega(\bar{G})$. Therefore, there is a potential for the gap between $r$ and $\omega(\bar{G})$ to be large (since the value of $\omega(\bar{G})$ can be arbitrarily large, see for instance, [Bollobás, 1979]).

Finding $k(G)$ is NP-complete. The same is true for the determination of $\omega(\bar{G})$. In Carraghan and Pardalos [1990] a survey of algorithms which can find the maximum clique in any graph is presented. The same authors also present a very efficient algorithm which uses a partial enumeration approach which had outperformed any other known algorithm at that time. In that treatment random problems with 3,000 vertices and over one million edges were solved in rather short times (less than one hour on an IBM ES/3090-600S computer). Some other related developments regarding the maximum clique of a graph can be found in [Pardalos and Xue, 1994], [Babel and Tinhofer, 1990], [Babel, 1994], [Balas and Xue, 1993], [Balas and Niehaus, 1994], [Zhang, Sun, and Tsang, 2005], and [Solnon and Fenet, 2006].

## 8.3 Problem Decomposition

The rejectability graph provides some intuitive ways for decomposing the inference of a Boolean function with few (ideally minimum number of) clauses or some bounds on the minimum number of clauses a function can have into a set of smaller problems. Such decompositions can be obtained through partitions of the rejectability graph. These decomposition approaches can be used with *any data mining method* and not only ones that are based on mathematical logic (i.e., they can be used with neural networks, decision trees, support vector machines, and so on). We consider two main approaches:

- *Decomposition via Connected Components and*
- *Decomposition via the Construction of a Clique Cover.*

### 8.3.1 Connected Components

In this case, one inspects the rejectability graph for a *natural* decomposition. A *connected component* of a graph is a maximal subgraph in which there is a path of edges between any pair of vertices. The following corollary is derived from Theorem 8.2 and illustrates an important relation of the connected components of $G$ and the clauses which can be inferred from two collections of positive and negative examples.

**Corollary 8.1.** *Suppose that $E^+$ and $E^-$ are the sets of the positive and negative examples, respectively. Then, any subset of negative examples in $E^-$ which is rejected by a single CNF clause, subject to the examples in $E^+$, corresponds to a subset of vertices of the rejectability graph $G$ which belong to the* ***same connected component*** *of the graph $G$.*

Pardalos and Rentala in [1990] present an excellent survey of algorithms which determine the connected components of a graph. Furthermore, they also propose a *parallel algorithm* which runs on an IBM ES/3090-400E computer (with four processors). That algorithm determines the connected components in super linear time.

The importance of Corollary 8.1 emerges when the sets of positive and negative examples are very large. First, one constructs the rejectability graph $G$. Next, one determines all the connected components of the rejectability graph by applying an algorithm (such as the one described in Pardalos and Rentala [1990]) for finding the connected components. Then, one solves the smaller clause inference problems which are formed by considering *all the positive examples* and the *negative examples which correspond* to the vertices of the individual and distinct connected components in $G$.

In other words, if a graph has two or more connected components, then one can decompose the original problem into separate problems and *the aggregation of the optimal solutions (based on the minimum number of CNF clauses) of the separate problems is an optimal solution to the original problem.* One can observe that each such subproblem (in the CNF case) is comprised of the negative examples for that component and *all* the positive examples, i.e., the positive examples are identical for each subproblem. Obviously, the number of negative examples corresponds to the number of vertices of the connected component of each subproblem.

### 8.3.2 Clique Cover

The second approach is also motivated by partitioning the vertices of the rejectability graph into mutually disjoint sets. However, in this second approach, vertices are subdivided via a sequential construction of cliques. First, the maximum clique of the rejectability graph is determined. The negative examples which correspond to the vertices of the maximum clique, along with *all* the positive examples, form the first subproblem of this decomposition. Next, the maximum clique of the *remaining* graph is derived. The second subproblem is formed by the negative examples which correspond to the vertices of the second clique and all the positive examples. This process continues until all the negative examples (or, equivalently, all the vertices in the rejectability graph) are considered.

We note that this sequence of cliques does not necessarily correspond to a minimum clique cover of the rejectability graph. This procedure is simply a *greedy* approach which *approximates* a minimum clique cover. Furthermore, it is possible that a single subproblem (in which all the vertices in the rejectability graph form a clique) may yield *more than one* clause (as was the case with the illustrative example in Section 8.2.1).

It should be noted at this point that the clique cover derived by using the above greedy approach may not always yield a minimum clique cover. Therefore, the number of cliques derived in that way *cannot* be used as a lower bound on the number of clauses derivable from positive and negative examples. Obviously, if the number of cliques is equal to $\omega(\bar{G})$ (see also Section 8.2.2), then the previous clique cover is minimal. However, even if the previous clique cover is not of minimum size, it can still be very useful as it can lead to a decomposition of the original problem into a sequence of smaller problems. Some computational results described in Section 8.5 provide some insight into the effectiveness of such a decomposition approach.

The approaches for problem decomposition described in this section can be combined into a single approach as follows. One first decomposes the original problem in terms of its connected components. Next, a clique cover, as described above, is derived for the individual problems which correspond to the connected components of the rejectability graph. This approach is further demonstrated in the illustrative example presented in the following section.

## 8.4 An Example of Using the Rejectability Graph

Next consider the following sets of positive and negative examples:

$$E^{+} = \begin{bmatrix} 0&1&0&0&0&1&0&1&1&1\\ 0&1&1&1&1&1&0&0&0&0\\ 0&0&1&0&1&1&1&0&1&0\\ 0&1&0&0&1&1&0&1&1&0\\ 1&0&1&0&0&0&1&0&1&1\\ 1&1&1&0&0&0&0&0&1&1\\ 1&1&0&0&0&0&0&1&1&1\\ 1&0&0&1&0&0&1&1&0&1\\ 0&1&1&0&1&1&0&0&1&0\\ 0&0&1&1&0&1&1&0&0&1\\ 1&1&1&1&0&0&0&0&0&1\\ 1&0&1&0&1&0&1&0&1&0\\ 1&1&1&0&1&0&0&0&1&0 \end{bmatrix} \quad \text{and}$$

$$E^{-} = \begin{bmatrix} 1&0&0&1&1&0&1&1&0&0\\ 0&0&0&1&1&1&1&1&0&0\\ 1&0&1&1&0&0&1&0&0&1\\ 0&0&0&0&1&1&1&1&1&0\\ 1&1&0&1&1&0&0&1&0&0\\ 1&1&0&0&1&0&0&1&1&0\\ 0&0&0&0&0&1&1&1&1&1 \end{bmatrix}.$$

One may use *any method* for inferring clauses from two disjoint classes of examples. In this illustrative example we use the OCAT approach. An application of the OCAT approach, combined with the RA1 heuristic, to this illustrative example yields the following CNF system (Boolean function) of four clauses:

$$(A_2 \vee A_3 \vee \bar{A}_5) \wedge (\bar{A}_1 \vee A_3 \vee \bar{A}_5) \wedge (A_1 \vee A_2 \vee A_4 \vee A_5)$$
$$\wedge\, (\bar{A}_1 \vee A_2 \vee \bar{A}_3 \vee \bar{A}_4).$$

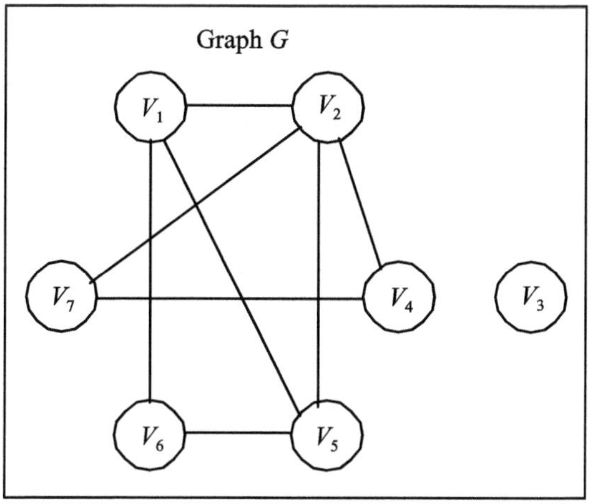


**Figure 8.3.** The Rejectability Graph for $E^+$ and $E^-$.

Of course the question addressed in this section is whether it is possible to derive another system with *fewer* clauses.

To help answer the previous question, we apply Theorem 8.3 to this illustrative example. Since there are 13 positive and 7 negative examples, the construction of the rejectability graph requires 21 ($= 7(7-1)/2$) simple rejectability examinations. When Theorem 8.1 is applied on these data, the rejectability graph $G$ shown in Figure 8.3 is derived. For instance, there is an edge between vertices $V_1$ and $V_6$ because the first and sixth negative examples can be rejected by a single disjunction without violating the constraints imposed by the positive examples in $E^+$ (i.e., this disjunction accepts all the positive examples). A similar interpretation holds for the remaining edges in graph $G$.

The rejectability graph in the current illustrative example has *two connected components* (see also Figure 8.3). One component is comprised by the six vertices $V_1$, $V_2$, $V_4$, $V_5$, $V_6$, $V_7$ while the second component has only the vertex $V_3$. Therefore, the original problem can be partitioned into *two independent* clause inference subproblems.

Both subproblems have the same positive examples. The first subproblem has the same negative examples as in $E^-$ *except* for the *third* negative example. The second problem has *only* the third negative example. The lower bound for the minimum number of CNF clauses required to appropriately classify the 20 examples is derived from the sum of the lower bounds for the two separate components. Since the rejectability graph of the second subproblem contains only a single vertex, the size of the minimum clique cover is one. A minimum clique cover is also obvious for the first subproblem, namely, the two sets $\{V_1, V_5, V_6\}$ and $\{V_2, V_4, V_7\}$. Hence, a minimum clique cover is two for the second subproblem. Thus, an overall lower bound for the minimum number of CNF clauses required is 3 ($= 2 + 1$). Therefore, it may well be possible that only three clauses are needed to appropriately classify all 20 examples.

As was also mentioned in Section 2.5 there is another clause inference approach which can be used to determine a minimum size set of clauses. This method, denoted as SAT (for satisfiability), has been proposed in Kamath, *et al.* [1992]. In that approach one first specifies an upper limit on the number of clauses to be considered, say $k$. That is, the value of $k$ must be *preassumed*. Next a clause satisfiability (SAT) model is formed and solved using an interior point method developed by Karmakar and his associates [1992]. If the clause satisfiability problem is feasible, the conclusion is that it is possible to correctly classify all the examples with at least $k$ clauses. If this SAT problem is infeasible, then one must increase $k$ until feasibility is reached. In this manner, the SAT approach can yield a system (Boolean function) with the minimum number of clauses. It is very important to observe at this point that computationally it is much harder to prove that a given SAT problem is infeasible than it is feasible. Therefore, trying to determine a minimum size Boolean function by using the SAT approach may be computationally too difficult. In this illustrative example, the SAT approach with $k = 3$, is feasible and returns the Boolean function with the following three clauses:

$$(A_2 \vee A_2 \vee A_3 \wedge (\bar{A}_1 \vee A_2 \vee \bar{A}_3 \vee \bar{A}_4) \wedge (\bar{A}_1 \vee A_3 \vee \bar{A}_5).$$

However, when the value $k = 2$ is used, the corresponding SAT formulation is infeasible. Therefore, the above set of clauses is optimal in the sense of this chapter. The last statement also follows from Theorem 8.3 since there exists a minimum size clique cover of 3 and a set of clauses has been derived with exactly this number of members. The previous optimal solution can also be derived much faster by applying the OCAT approach to the subproblems which correspond to the three cliques that fully cover the rejectability graph $G$. Since one of these cliques is of size one, that subproblem is a trivial one.

## 8.5 Some Computational Results

In this section we provide some computational insight into two issues. The first is the role and usefulness of the lower bound of Theorem 8.3. The second issue is on the potential benefit of using the decomposition approach which is based on a clique cover. The first issue is important when one is interested in minimizing the size of the inferred Boolean function. The lower limit described in Theorem 8.3 (i.e., the value of $\omega(\bar{G})$) can also give an idea of how far from optimality a given solution might be. The second issue is important when one wishes to control the CPU time requirement (for instance, when solving large problems). Although the clique decomposition approach results in solving a sequence of smaller Boolean inference problems, there is a new time burden because one now also needs to determine a sequence of maximum cliques. Therefore, it is not immediately obvious that the decomposition approach will be less demanding in terms of the CPU time. A second issue is the size of the systems derived via the clique decomposition. The systems derived via the decomposition approach may be larger in size than otherwise. For these reasons, three series of test problems were performed.

The **first series of test problems** is based on the fifteen systems introduced in Kamath, *et al.*, [1992]. These systems use sets of clauses defined on 8, 16, and 32 attributes and are depicted in Table 3.3. The four systems with 8 attributes are labeled as 8A1, 8A2, 8A3, and 8A4 in Table 8.1 (the secondary classification is based on the number of examples used as input). The same convention was also used for the rest of the systems in Table 8.1. Each of the fifteen sets is used as a "hidden logic" for classifying randomly generated examples, i.e., for determining the sets $E^+$ and $E^-$. Then, the OCAT algorithm was used to generate a set of CNF clauses which correctly classify each of the positive and negative examples.

As described earlier, the rejectability graph and the sequential (greedy) generation of maximum cliques were determined too. Next, Theorem 8.3 was used to establish a lower bound on the required number of inferred clauses. Note that the SAT results were determined by using a VAX 8700 machine running the 10th Edition of UNIX, written in FORTRAN and C [Kamath, *et al.*, 1992]. The OCAT results were derived by using an IBM/3090-600S machine (which is approximately 3 to 4 times faster than a VAX 8700 machine) and the code was written in FORTRAN. This table is complementary to Table 3.1, which also refers to the same computational experiments.

The results of these experiments are provided in Table 8.1. These results include computations for SAT, OCAT, and the lower bound, as determined according to Theorem 8.3. Note that in these tests we did not decompose the problems by using the connected component approach described in Section 8.3 (no code was available to us at that time). The SAT results represent a *feasible* solution, not necessarily a solution with the minimum number of clauses. (To obtain the minimum number of clauses, one must iteratively reduce the value of $k$ until infeasibility occurs, which is a very time consuming process.)

Consider, for instance, the first case (problem 8A1) depicted in Table 8.1. Ten random examples were generated for system 8A (as defined in Table 3.3). With $k$ fixed at 3 the SAT algorithm returned a feasible solution with three clauses. The lower bound from Theorem 8.3 is equal to 2. Hence, it may be possible to correctly classify the ten examples with only two clauses. However, in these tests the value of $k$ was not iteratively reduced to determine its minimum value.

The results of the first set of tests indicate just how well OCAT performs. Of the 24 problems with 16 or fewer attributes, OCAT generated a set of clauses *exactly* at the lower bound in 20 cases. In the other 4 cases, OCAT exceeded the lower bound by only 1 clause. For the 17 problems with 32 attributes, OCAT averaged about 1.23 more clauses than the lower bound.

It is also noteworthy to observe that the performance of SAT, OCAT, and the lower bound according to Theorem 4.3 are not dramatically affected by the number of examples. As expected, as the number of examples increases, the number of required clauses also increases. This is illustrated, for instance, by test problem 32E. The OCAT approach generated 2 clauses with 50 and 100 examples while 3 clauses

**Table 8.1.** Solution Statistics for the First Series of Tests.

| Problem Characteristics | | | SAT Solution | | OCAT Solution | $\omega(\bar{G})$: Lower Bound |
|---|---|---|---|---|---|---|
| Problem ID | No. of Examples | Clauses in "Hidden Logic" | $k$ | No. of Clauses | No. of Clauses | Via Theorem 8.3 |
| 8A1 | 10 | 3 | 3 | 3 | 3 | 2 |
| 8A2 | 25 | 3 | 6 | 3 | 3 | 2 |
| 8A3 | 50 | 3 | 6 | 3 | 3 | 3 |
| 8A4 | 100 | 3 | 6 | 3 | 3 | 3 |
| 8B1 | 50 | 3 | 3 | 2 | 2 | 2 |
| 8B2 | 100 | 3 | 6 | 3 | 3 | 3 |
| 8B3 | 150 | 3 | 10 | 3 | 3 | 3 |
| 8B4 | 200 | 3 | 6 | 3 | 3 | 3 |
| 8C1 | 50 | 3 | 10 | 2 | 2 | 2 |
| 8C2 | 100 | 3 | 10 | 3 | 3 | 3 |
| 8D1 | 50 | 3 | 10 | 3 | 3 | 3 |
| 8D2 | 100 | 3 | 10 | 3 | 3 | 3 |
| 8E1 | 50 | 3 | 10 | 3 | 3 | 3 |
| 8E2 | 100 | 3 | 10 | 3 | 3 | 3 |
| 16A1 | 100 | 4 | 15 | 4 | 4 | 3 |
| 16A2 | 300 | 4 | 6 | 4 | 4 | 4 |
| 16B1 | 200 | 4 | 8 | 5 | 4 | 4 |
| 16B2 | 400 | 4 | 4 | 4 | 4 | 4 |
| 16C1 | 100 | 4 | 20 | 5 | 4 | 4 |
| 16C2 | 400 | 4 | 4 | 4 | 4 | 4 |
| 16D1 | 200 | 4 | 10 | 4 | 4 | 4 |
| 16D2 | 400 | 4 | 4 | 4 | 4 | 4 |
| 16E1 | 200 | 4 | 15 | 5 | 5 | 4 |
| 16E2 | 400 | 4 | 4 | 4 | 4 | 4 |
| 32A1 | 250 | 3 | 3 | 3 | 3 | 3 |
| 32B1 | 50 | 3 | 3 | 3 | 2 | 1 |
| 32B2 | 100 | 3 | 3 | 3 | 3 | 2 |
| 32B3 | 250 | 3 | 3 | 3 | 3 | 2 |
| 32B4 | 300 | 3 | 3 | 3 | 3 | 2 |
| 32C1 | 50 | 3 | 3 | 3 | 2 | 1 |
| 32C2 | 100 | 3 | 3 | 3 | 2 | 1 |
| 32C3 | 150 | 3 | 3 | 3 | 3 | 1 |
| 32C4 | 1000 | 3 | 3 | 3 | 3 | 3 |

**Table 8.1.** Continued.

| Problem Characteristics | | | SAT Solution | | OCAT Solution | $\omega(\bar{G})$: Lower Bound |
|---|---|---|---|---|---|---|
| **Problem ID** | **No. of Examples** | **Clauses in "Hidden Logic"** | ***k*** | **No. of Clauses** | **No. of Clauses** | **Via Theorem 8.3** |
| 32D1 | 50 | 4 | 4 | 4 | 3 | 1 |
| 32D2 | 100 | 4 | 4 | 4 | 3 | 2 |
| 32D3 | 400 | 4 | 4 | 4 | 4 | 2 |
| 32E1 | 50 | 3 | 3 | 2 | 2 | 1 |
| 32E2 | 100 | 3 | 3 | 3 | 2 | 1 |
| 32E3 | 200 | 3 | 3 | 3 | 3 | 2 |
| 32E4 | 300 | 3 | 3 | 3 | 3 | 2 |
| 32E5 | 400 | 3 | 3 | 3 | 3 | 2 |

were needed for 200, 300, and 400 examples. As more examples are generated, the set of inferred clauses becomes a better approximation of the underlying "hidden logic," which for test problem 32E is comprised of 3 clauses.

The **second and third series** of computational experiments were executed as follows. First, a Boolean function with $K_0$ clauses (in the CNF form) was determined randomly. In order to have a good balance of positive and negative examples, each attribute in any clause was present with probability 15% to 25%. Also, if attribute $A_i$ was selected to be in a clause, then attribute $\bar{A}_i$ was not allowed to be present and vice versa. That was done in order to avoid constructing clauses which would accept all examples (i.e., to avoid tautologies). A system derived in this way was considered as the "hidden logic" system in these experiments. The number of attributes was set to be equal to 10 and 30 (i.e., $n = 10$ or 30). These systems are indexed by ID numbers, such as 1A10, 2A10, ..., 1B10, 2B10, etc. In the above coding scheme the first digit indicates the test number, "A" or "B" indicates whether $n$, the number of attributes, was equal to 10 or to 30, respectively, and the last two digits indicate the number of clauses in the "hidden logic" (see also Tables 8.2 and 8.3).

Next, collections of 400 and 600 examples (400 examples were considered when $n = 10$ and 600 examples when $n = 30$) were randomly generated and classified according to the previous clauses. The computational results are presented in Tables 8.2 and 8.3 (for $n = 10$ and $n = 30$, respectively). At first, the branch-and-bound (B&B) algorithm described in Chapter 3 of this book was applied to the original problem consisting of $m_1$ positive and $m_2$ negative examples. The CPU time (in seconds on an IBM 3090-600E computer) and the number of clauses derived this way are presented in the columns labeled "$Time_1$" and "$K_1$", respectively.

Since in these experiments the "hidden logic" is also known to us, the original system can be compared with the derived system by asking both systems to classify 10,000 randomly generated examples. The corresponding accuracy rates are presented in the column labeled "$A_1$." The next phase in these experiments was to apply

**Table 8.2.** Solution Statistics When $n = 10$ and the Total Number of Examples Is Equal to 400.

| Problem | | | | w.o. Decomposition | | | With Clique Decomposition | | | | | | |
|---|---|---|---|---|---|---|---|---|---|---|---|---|---|
| ID | $m_1$ | $m_2$ | $K_0$ | $K_1$ | $Time_1$ | $A_1$ | $K_2$ | Limit | Cliques | $Time_2$ | $Time_3$ | $Time_4$ | $A_2$ |
| 1A10 | 247 | 153 | 10 | 10 | 4.96 | 0.97 | 10 | 10 | 10 | 0.03 | 0.04 | 0.07 | 0.96 |
| 2A10 | 258 | 142 | 10 | 9 | 2.09 | 0.98 | 9 | 9 | 9 | 0.03 | 0.03 | 0.06 | 0.98 |
| 3A10 | 198 | 202 | 10 | 10 | 3.34 | 0.98 | 10 | 9 | 9 | 0.03 | 0.07 | 0.10 | 0.98 |
| 4A10 | 253 | 147 | 10 | 10 | 2.54 | 0.96 | 10 | 10 | 10 | 0.03 | 0.05 | 0.08 | 0.97 |
| 5A10 | 288 | 112 | 10 | 10 | 5.06 | 0.97 | 11 | 10 | 11 | 0.04 | 0.03 | 0.07 | 0.97 |
| **Mean:** | **248.8** | **151.2** | **10** | **9.8** | **3.60** | **0.97** | **10** | **9.6** | **9.8** | **0.03** | **0.04** | **0.08** | **0.97** |
| 1A20 | 193 | 207 | 20 | 14 | 14 | 0.96 | 18 | 14 | 15 | 0.05 | 0.25 | 0.3 | 0.95 |
| 2A20 | 266 | 134 | 20 | 18 | 55.22 | 0.94 | 22 | 18 | 19 | 0.08 | 0.04 | 0.12 | 0.95 |
| 3A20 | 329 | 71 | 20 | 12 | 44.76 | 0.95 | 16 | 12 | 14 | 0.07 | 0.01 | 0.08 | 0.94 |
| 4A20 | 291 | 109 | 20 | 16 | 57.51 | 0.95 | 20 | 15 | 17 | 0.08 | 0.02 | 0.10 | 0.93 |
| 5A20 | 133 | 267 | 20 | 13 | 4.29 | 0.97 | 16 | 13 | 14 | 0.03 | 200.6 | 200.6 | 0.97 |
| **Mean:** | **242.4** | **157.6** | **20** | **14.6** | **35.16** | **0.95** | **18.4** | **14.4** | **15.8** | **0.06** | **40.18** | **40.24** | **0.95** |
| 1A30 | 315 | 85 | 30 | 17 | 103.76 | 0.93 | 18 | 16 | 17 | 0.08 | 0.01 | 0.09 | 0.94 |
| 2A30 | 326 | 74 | 30 | 21 | 376.06 | 0.93 | 24 | 20 | 22 | 0.11 | 0.01 | 0.12 | 0.93 |
| 3A30 | 306 | 94 | 30 | 18 | 64.55 | 0.93 | 17 | 17 | 17 | 0.07 | 0.02 | 0.09 | 0.93 |
| 4A30 | 287 | 113 | 30 | 17 | 38.29 | 0.93 | 21 | 16 | 18 | 0.09 | 0.03 | 0.12 | 0.92 |
| 5A30 | 293 | 107 | 30 | 22 | 267.25 | 0.92 | 25 | 19 | 21 | 0.10 | 0.03 | 0.13 | 0.92 |
| **Mean:** | **305.4** | **94.6** | **30** | **19** | **169.98** | **0.93** | **21** | **17.6** | **19** | **0.09** | **0.02** | **0.11** | **0.93** |

**Table 8.2.** Continued.

| Problem | | | | w.o. Decomposition | | | With Clique Decomposition | | | | | | |
|---|---|---|---|---|---|---|---|---|---|---|---|---|---|
| ID | $m_1$ | $m_2$ | $K_0$ | $K_1$ | $\text{Time}_1$ | $A_1$ | $K_2$ | Limit | Cliques | $\text{Time}_2$ | $\text{Time}_3$ | $\text{Time}_4$ | $A_2$ |
| 1A40 | 317 | 83 | 40 | 26 | 316.21 | 0.89 | 24 | 20 | 22 | 0.11 | 0.01 | 0.12 | 0.9 |
| 2A40 | 252 | 148 | 40 | 22 | 159.09 | 0.9 | 24 | 21 | 22 | 0.08 | 0.05 | 0.13 | 0.89 |
| 3A40 | 286 | 114 | 40 | 20 | 129.13 | 0.91 | 26 | 19 | 21 | 0.10 | 0.04 | 0.14 | 0.92 |
| 4A40 | 328 | 72 | 40 | 19 | 220.66 | 0.92 | 20 | 17 | 18 | 0.09 | 0.01 | 0.10 | 0.93 |
| 5A40 | 271 | 129 | 40 | 22 | 139.31 | 0.9 | 27 | 20 | 24 | 0.11 | 0.06 | 0.17 | 0.9 |
| **Mean:** | **290.8** | **109.2** | **40** | **21.8** | **192.87** | **0.9** | **24.2** | **19.4** | **21.4** | **0.10** | **0.03** | **0.13** | **0.91** |
| 1A50 | 256 | 144 | 50 | 27 | 260.39 | 0.86 | 30 | 23 | 27 | 0.11 | 0.07 | 0.18 | 0.87 |
| 2A50 | 265 | 135 | 50 | 25 | 159.03 | 0.88 | 28 | 21 | 24 | 0.11 | 0.07 | 0.18 | 0.87 |
| 3A50 | 270 | 130 | 50 | 23 | 128.13 | 0.86 | 27 | 20 | 22 | 0.10 | 0.05 | 0.15 | 0.86 |
| 4A50 | 285 | 115 | 50 | 26 | 356.04 | 0.88 | 26 | 21 | 25 | 0.10 | 0.04 | 0.14 | 0.88 |
| 5A50 | 296 | 104 | 50 | 21 | 112.31 | 0.86 | 26 | 20 | 22 | 0.10 | 0.02 | 0.12 | 0.88 |
| **Mean:** | **274.4** | **125.6** | **50** | **24.4** | **203.18** | **0.87** | **27.4** | **21** | **24** | **0.10** | **0.05** | **0.15** | **0.87** |

**Table 8.3.** Solution Statistics When $n = 30$ and the Total Number of Examples Is Equal to 600.

| Problem | | | | w.o. Decomposition | | | With Clique Decomposition | | | | | | |
|---|---|---|---|---|---|---|---|---|---|---|---|---|---|
| ID | $m_1$ | $m_2$ | $K_0$ | $K_1$ | $Time_1$ | $A_1$ | $K_2$ | Limit | Cliques | $Time_2$ | $Time_3$ | $Time_4$ | $A_2$ |
| 1B10 | 520 | 80 | 10 | 8 | 16.02 | 0.97 | 14 | 3 | 4 | 117.21 | 5.22 | 122.43 | 0.93 |
| 2B10 | 485 | 115 | 10 | 9 | 20.68 | 0.96 | 19 | 3 | 5 | 41.38 | 17,960.00 | 18,001.00 | 0.89 |
| 3B10 | 563 | 37 | 10 | 7 | 28.07 | 0.96 | 10 | 2 | 3 | 35.11 | 0.01 | 35.12 | 0.94 |
| 4B10 | 554 | 46 | 10 | 6 | 5.15 | 0.99 | 10 | 3 | 3 | 3.70 | 0.12 | 3.82 | 0.97 |
| 5B10 | 589 | 11 | 10 | 4 | 10.97 | 0.98 | 5 | 2 | 2 | 6.24 | 0.00 | 6.24 | 0.98 |
| 6B10 | 570 | 30 | 10 | 5 | 9.64 | 0.98 | 7 | 2 | 3 | 1.59 | 0.00 | 1.59 | 0.95 |
| 7B10 | 546 | 54 | 10 | 6 | 78.10 | 0.97 | 9 | 2 | 3 | 2.46 | 0.02 | 2.48 | 0.95 |
| 8B10 | 563 | 37 | 10 | 6 | 39.84 | 0.97 | 9 | 2 | 3 | 17.84 | 0.02 | 17.86 | 0.95 |
| 9B10 | 569 | 31 | 10 | 6 | 38.61 | 0.97 | 7 | 2 | 3 | 45.97 | 0.00 | 45.97 | 0.96 |
| 10B10 | 515 | 85 | 10 | 8 | 57.23 | 0.98 | 20 | 3 | 4 | 202.88 | 259.06 | 461.93 | 0.94 |
| **Mean:** | **547.4** | **53** | **10** | **6.5** | **30.43** | **0.97** | **11** | **2.4** | **3.3** | **47.44** | **1,822.50** | **1,869.90** | **0.95** |
| 1B30 | 586 | 14 | 30 | 5 | 22.95 | 0.97 | 6 | 2 | 2 | 10.44 | 0.00 | 10.44 | 0.97 |
| 2B30 | 583 | 17 | 30 | 4 | 1.76 | 0.99 | 5 | 3 | 3 | 0.21 | 0.00 | 0.21 | 0.98 |
| 3B30 | 533 | 67 | 30 | 10 | 383.66 | 0.96 | 17 | 3 | 4 | 546.33 | 5.89 | 552.22 | 0.92 |
| 4B30 | 574 | 26 | 30 | 5 | 66.23 | 0.97 | 7 | 3 | 3 | 1.67 | 0.00 | 1.67 | 0.96 |
| 5B30 | 559 | 41 | 30 | 10 | 1,380.84 | 0.93 | 13 | 3 | 4 | 97.77 | 0.07 | 97.85 | 0.93 |
| 6B30 | 499 | 101 | 30 | 9 | 538.02 | 0.93 | 14 | 3 | 5 | 55.77 | 2,375.00 | 2,431.00 | 0.93 |
| 7B30 | 513 | 87 | 30 | 9 | 41.06 | 0.95 | 17 | 3 | 4 | 62.71 | 3.9 | 66.61 | 0.91 |
| 8B30 | 537 | 63 | 30 | 12 | 1,501.81 | 0.91 | 17 | 3 | 3 | 903.19 | 0.47 | 903.66 | 0.91 |
| 9B30 | 553 | 47 | 30 | 11 | 1,791.30 | 0.89 | 14 | 3 | 3 | 417.47 | 0.09 | 417.56 | 0.92 |
| 10B30 | 551 | 49 | 30 | 11 | 10,714.65 | 0.91 | 16 | 3 | 3 | 834.61 | 0.1 | 834.71 | 0.91 |
| **Mean:** | **549** | **51** | **30** | **8.6** | **1,644.23** | **0.94** | **12.6** | **2.8** | **3.4** | **293.01** | **238.55** | **531.57** | **0.93** |

**Table 8.3.** Continued.

| Problem | | | | w.o. Decomposition | | | With Clique Decomposition | | | | | | |
|---|---|---|---|---|---|---|---|---|---|---|---|---|---|
| ID | $m_1$ | $m_2$ | $K_0$ | $K_1$ | $\text{Time}_1$ | $A_1$ | $K_2$ | Limit | Cliques | $\text{Time}_2$ | $\text{Time}_3$ | $\text{Time}_4$ | $A_2$ |
| 1B50 | 586 | 14 | 50 | 4 | 1.57 | 0.97 | 4 | 2 | 2 | 0.46 | 0.01 | 0.47 | 0.98 |
| 2B50 | 584 | 16 | 50 | 5 | 64.47 | 0.96 | 7 | 2 | 2 | 20.99 | 0.01 | 21.00 | 0.96 |
| 3B50 | 575 | 25 | 50 | 5 | 13.49 | 0.97 | 8 | 2 | 3 | 10.45 | 0.01 | 10.46 | 0.95 |
| 4B50 | 550 | 50 | 50 | 6 | 28.34 | 0.97 | 7 | 2 | 3 | 7.65 | 0.02 | 7.67 | 0.97 |
| 5B50 | 503 | 97 | 50 | 13 | 1,465.44 | 0.89 | 17 | 3 | 4 | 277.57 | 281.24 | 558.81 | 0.86 |
| 6B50 | 512 | 88 | 50 | 13 | 2,181.36 | 0.91 | 21 | 3 | 4 | 804.24 | 550.02 | 1,354.26 | 0.86 |
| 7B50 | 489 | 111 | 50 | 19 | 7,415.04 | 0.82 | 27 | 3 | 4 | 2,438.29 | 4,410.85 | 6,849.14 | 0.83 |
| 8B50 | 501 | 99 | 50 | 15 | 5,135.15 | 0.89 | 24 | 3 | 5 | 411.08 | 286.33 | 697.41 | 0.87 |
| 9B50 | 496 | 104 | 50 | 18 | 11,735.38 | 0.88 | 27 | 3 | 5 | 2,422.45 | 1,970.34 | 4,392.79 | 0.80 |
| 10B50 | 517 | 83 | 50 | 15 | 7,987.36 | 0.87 | 23 | 3 | 5 | 2,594.78 | 125.63 | 2,720.41 | 0.82 |
| **Mean:** | **531** | **68.7** | **50** | **11.3** | **3,602.76** | **0.91** | **16.5** | **2.6** | **3.7** | **898.8** | **762.44** | **1,661.24** | **0.89** |

the clique decomposition approach. The original problem with the $m_1$ and $m_2$ examples was decomposed into a number of smaller problems according to the clique cover approach described at the end of Section 8.4. Note that the natural decomposition via the connected component approach was not used in these tests. We only used the decomposition approach imposed by the sequence of the maximum cliques (as described in Sections 8.3 and 8.4).

The number of cliques which was generated for this purpose is depicted under the column labeled "*Cliques*." The value of $\omega(\bar{G})$ is depicted under the column labeled "*Limit*." The values under "*Time*$_2$" and "*Time*$_3$" present the CPU time required by the revised B&B algorithm (only) described in Chapter 3 and the calculation of the cliques, respectively. The values under "*Time*$_4$" (total CPU time when the clique decomposition is used) are the sum of the values in the columns labeled "*Time*$_2$" and "*Time*$_3$." The number of clauses of the proposed systems is under the column labeled "$K_2$." Finally, the values under "$A_2$" are the accuracy rates when the system proposed by the decomposition approach and the original system are compared.

The computational experiments shown in Table 8.2 were performed for groups of random "hidden logics" with 10, 20, 30, 40, and 50 clauses. These results indicate that the limit provided by Theorem 8.3 (i.e., the values of $\omega(\bar{G})$ in column "*Limit*") is rather tight. For instance, when $K_0 = 40$, the average number of clauses derived by using B&B without decomposition was equal to 21.8 versus 19.4 being the average lower limit. That is, the B&B approach returned systems of very small or even probably of minimal size.

A "hidden logic" was generated randomly with a predetermined number of CNF clauses, say $K_0$. Suppose that two collections of positive and negative examples are generated such that all positive examples are accepted by the $K_0$ clauses, while each negative example is rejected by at least one of the previous clauses. If $r$ denotes the minimum number of CNF clauses which satisfy the requirements of the previous collections of positive and negative examples, then from the previous consideration the following relation follows to be true: $K_0 \geq r$.

When one examines the sizes of the systems returned when the clique decomposition approach was used, it can be observed that most of the time the decomposition approach returned systems with at most 10% (on the average) more CNF clauses than without decomposition. However, this was done with a *fraction* of the CPU time when compared with no decomposition.

For instance, when $K_0 = 40$, then on the average, the decomposition approach returned systems with 24.2 clauses (versus systems with 21.8 clauses without decomposition) by consuming, on the average, 0.13 CPU second versus 192.87 CPU seconds without the clique decomposition. That is, in the above test problems one obtains a system with at most 10% more clauses, but in return, realizes a gain of an almost 1,500 times speedup on CPU time!

The results in Table 8.3 are similar to those in Table 8.2. However, the lower bound (i.e., the value of $\omega(\bar{G})$) described in Theorem 8.3 is considerably less tight. This is seen by the size of the gap between the values in column "$K_1$" (or "$K_2$") and column "*Limit*." Also, now the CPU times are significantly much higher, since we are dealing with larger and more difficult problems. At the same time, the CPU times are more unpredictable.

For instance, observe that when $K_0 = 10$, test problem 2B10 took 17,960.00 CPU seconds for computing the required cliques. This time is obviously excessive when compared with the times in the rest of the problems in these experiments. The CPU times became more variable when the "hidden logics" had more clauses (which naturally resulted in harder problems).

However, even now the results suggest that the proposed clique decomposition approach may *significantly reduce* the CPU requirements in solving large problems with only a moderate increase on the number of the derived clauses. Finally, it is remarkable to observe that in terms of the accuracy rates the systems derived by using the decomposition approach are almost equally accurate as the systems derived without the clique decomposition (which are computed at much higher CPU time requirements).

From the above analyses and computational results it becomes evident that the rejectability graph provides at least two benefits:

(i) When the value of $\omega(\bar{G})$ is rather high, the value of $\omega(\bar{G})$ can serve as a tight lower bound on the minimum number of clauses derivable from positive and negative examples. Of course, the clique cover can still be used for decomposing a large inductive inference/data mining problem;
and

(ii) When the value of $\omega(\bar{G})$ is low (and hence might be a large gap between $\omega(\bar{G})$ and $r$), the rejectability graph can still be useful in decomposing a large inductive inference/data mining problem because it may lead to a significant reduction of the CPU time.

These decompositions are based on constructing a sequence of cliques. This operation depends on the algorithm used to determine the maximum clique in a graph. Present algorithms are rather efficient when the graph is sparse (as is the case with the rejectability graphs in the problems described in Table 8.2). In our computational experiments we used the clique algorithm described in [Carraghan and Pardalos, 1990]. This algorithm is considered to be very good for sparse graphs and it is often used as a benchmark in the literature.

However, when the graphs become dense, this clique algorithm becomes too slow. Very often (around 20%–30% of the time) we had to abort tests running for the results in Table 8.3, because the clique construction phase of our program would take too long (more than 5 hours on an IBM 3090-600E mainframe computer). We believe that other current algorithms may be more efficient for dense graphs. Such algorithms are the maximum clique algorithm described in [Pardalos and Rogers, 1992] or the ones developed by Balas and his associates [Balas and Xue, 1993], [Balas and Niehaus, 1994] or even some newer ones [Zhang, Sun, and Tsang, 2005], and [Solnon and Fenet, 2006].

One issue became profoundly apparent in this investigation. Future developments in determining a maximum clique in a graph, will directly benefit the efficient solution of large inference problems by employing the rejectability graph and the decomposition approaches described in this chapter.

## 8.6 Concluding Remarks

The paramount importance of data mining and knowledge discovery from data sets creates an immense demand for being able to process large collections of data. It also increases the pressure on creating Boolean expressions which have a small number of clauses. The rejectability graph, which was introduced and discussed in this chapter, provides the means for establishing a lower bound on the number of CNF or DNF clauses which can be inferred from positive and negative examples.

This graph also provides an effective and intuitive way for partitioning the original data and, thus, solve large-scale learning problems. Furthermore, the rejectability graph suggests a time efficient approach for decomposing the original problem into a sequence of smaller problems and still infer a compact Boolean expression from the partial solutions of the smaller problems.

It should be emphasized here that any other data mining approach (such as neural networks, decision trees, support vector machines, and so on) could be used in conjunction with these graph-based decomposition. This approach could benefit such methods when they solve large-size problems (and not necessarily only when one needs to infer a small-size system of some kind). Of course, of critical importance here is to have the time savings due to decompositions be significantly larger than the CPU time needed to determine these decompositions.

The previous findings were discussed in terms of two learning algorithms. The first algorithm is a greedy approach (i.e., the OCAT approach) and is based on the branch-and-bound algorithm described in Chapter 2 of this book. The second approach is based on formulating a satisfiability problem [Kamath, *et al.*, 1992] and then solving it by using an interior point method [Karmakar, *et al.*, 1991] (see also Chapter 2).

Finally, it is possible for the rejectability graph to have even more interesting properties than the ones described in this chapter. For instance, one possible extension might be to define a rejectability graph even when the data are not binary or have missing values. More research in this direction may reveal more connections between graph theory and the learning from examples problem.

# Part II

# Application Issues

# Chapter 9

# The Reliability Issue in Data Mining: The Case of Computer-Aided Breast Cancer Diagnosis

## 9.1 Introduction

Almost any use of a data mining and knowledge discovery method on a data set requires some discussion on the accuracy of the extracted model on some test data. This accuracy can be a general description of how well the extracted model classifies test data. Some studies split this accuracy rate into two rates: the false-positive and false-negative rates. This distinction might be more appropriate for most real-life applications. For instance, it is one thing to wrongly diagnose a benign tumor as malignant than the other way around. Related are some of the discussions in Sections 1.3.4, 4.5, and 11.6.

Usually, such studies suggest that higher accuracy rates may be achieved by using more and more data points as training data. This chapter studies in depth the reliability issue of models derived by the use of data mining approaches. It shows that even *billions* of observations, may still not be sufficient to accurately capture the behavior of a system under consideration.

It uses a real-life application which is based on the diagnosis of breast cancer. Although the results presented in this chapter cannot be directly generalized to all data mining and knowledge discovery application areas, one may argue that these results are reflective of what happens in many other areas as well. The discussions presented in this chapter are based on the research results reported in [Kovalerchuk, *et al.*, 2000].

## 9.2 Some Background Information on Computer-Aided Breast Cancer Diagnosis

Breast cancer is the most common cancer in women in the U.S. For instance, there were an estimated 182,000 cases in 1995 [Wingo, *et al.*, 1995]. The most effective tool in the battle against breast cancer is screening mammography. However, several retrospective analyses have found diagnostic error rates ranging from 20%

E. Triantaphyllou, *Data Mining and Knowledge Discovery via Logic-Based Methods*, Springer Optimization and Its Applications 43, DOI 10.1007/978-1-4419-1630-3_9,

to 43% [Bird, *et al.*, 1992], [Burhenne, *et al.*, 1994]. Also, of the breast biopsies performed due to suspicious mammograms, 70%–89% will be found benign [Hall, 1988]. Elmore, *et al.*, in [1994] studied the variability of radiologists' interpretation of a set of mammograms. They observed an average intraobserver variability of approximately 8% in addition to a 19% interobserver variability for the diagnosis of cancer, for which the variability in management recommendations was 25%. They also found that 9 out of 10 radiologists recognized fewer than 3% of the mammograms which they screened 5 months earlier, while 1 out of 10 claimed to have recognized about 25% of the cases [Elmore, *et al.*, 1994]. These startling statistics and other discussions on computer-aided diagnosis (CAD) [Kopans, 1994], [Gurney, 1994], [Boone, 1994] clearly demonstrate the need for (and the possible magnitude of) improvements in the reliability of breast cancer diagnosis.

Today, with the proliferation of powerful computers, a great effort is directed toward developing computerized methods that can assist radiologists in breast cancer diagnosis. Currently, such methods include neural networks, nearest neighbor methods, discriminant analysis, cluster analysis, decision trees, and linear programming-based methods (see, for instance, [Gale, *et al.*, 1987], [Getty, *et al.*, 1988], [Swets, *et al.*, 1991], [D'Orsi, *et al.*, 1992], [Wu, *et al.*, 1993], [Vyborny, 1994], [Vyborny and Giger, 1994], [Mangasarian, 1993] and the references mentioned in Section 2.2). These methods extract general diagnostic models which are based on a sample of specific cases. Thus, the better the available data represent the underlying models, the more accurate the predictions based on the inferred models can become. Therefore, these methods rely on obtaining representative samples.

Often the available training data are insufficient to achieve desirable prediction accuracy. In other words, the available knowledge is often insufficient to make confident recommendations. According to Johnson [1991], the use of Bayesian models in medical diagnosis can be controversial, if not unethical, because the fundamental requirement of strict randomness rarely occurs and it can rarely be tested with the available training data. This critical issue is further elaborated in Sections 9.3 and 9.4.

Monotonicity of the data is a frequent property (although at different degree) that has not been adequately utilized by traditional approaches. This property has the potential to significantly improve the reliability of breast cancer diagnosis (and in many other areas too). The monotonicity approach described in this chapter does not assume a particular model and in this sense maintains a general representation power. However, it should be stated at this point that if the existence of an appropriate parametric model (as described by Duda and Hart in [1973]) can be established, then its application might lead to a higher degree of confidence than the use of the monotonicity approach described in this chapter. Note that in Chapters 10 and 11 of this book the monotonicity property and some related data mining algorithms are discussed in more detail.

This chapter is organized as follows. Section 9.3 introduces some reliability criteria of computer-aided breast cancer diagnosis. The same section also uses these criteria to analyze the reliability of some published diagnostic results which are based on neural networks. Section 9.4 is devoted to the representation/narrow vicinity hypothesis. Section 9.5 presents the results of the validation of this hypothesis on 11 mammographic and related clinical attributes. The last section summarizes the main results of this study and formulates some directions for future research. Two appendices to this chapter provide more details, or links to more discussions, on the techniques used in the previous sections.

## 9.3 Reliability Criteria

Most experts would agree that the validity and accuracy (reliability) of a computer-aided diagnostic system should be reasonably high for clinical applications. To explore the issue of reliability, assume that we have the 11 binary (0 or 1 value) diagnostic attributes described in Appendix I of this chapter. This example is a rather simple one since it assumes only 11 attributes which are binary valued and the entire setting is assumed to be deterministic (i.e., a given case always belongs to the same class). However, this illustrative setting is still sufficient to provide the main motivation of the key concepts described in this chapter. By using the previous 11 attributes, each medical case can be expressed as a combination of binary values defined on these attributes. For instance, the ordered sequence (01111100011) describes the case with "0" value for the 1-st, 7-th, 8-th, and 9-th attributes and with "1" value for the rest of them. Furthermore, by considering the definitions in Appendix I of this chapter, the above binary vector means that:

(1) The number of calcifications/cm$^2$ is small (value 0);
(2) the volume (in cm$^3$) is small (value 1);
(3) the total number of calcifications is large (value 1);
(4) the irregularity in the shape of individual calcifications is marked (value 1);
(5) the variation in the shape of calcifications is marked (value 1);
(6) the variation in size of the calcifications is marked (value 1);
(7) the variation in density of the calcifications is mild (value 0);
(8) the density of the calcifications is mild (value 0);
(9) no ductal orientation is present (value 0);
(10) the comparison with previous exam is "pro cancer/biopsy" (value 1);
(11) the associated findings are "pro cancer/biopsy" (value 1).

Note that the grade "small" was deliberately coded differently for the number of calcifications/cm$^2$ and the volume (in cm$^3$). This step enabled us to take advantage of the monotonicity property (as described later), which is of critical importance to the effectiveness of the proposed method.

Next, a computer-aided diagnostic (CAD) system which is based on the previous 11 binary attributes should be able to categorize new cases represented by binary vectors. Each such case is assumed to be either in the "highly suspicious for malignancy" class or in the "not highly suspicious for malignancy" class and in only one of them. That is, in mathematical terms a CAD system operates as a discriminant function, say $f(x_1, x_2, \dots, x_n)$, which is defined on the space of $n$ attributes denoted as $x_1, x_2, \dots, x_n$. In order to help fix ideas, assume that a discriminant function for the current illustrative example was constructed from a sample of 80 training cases (each of which is either highly suspicious for malignancy or not highly suspicious for malignancy). It should be noted here that many, if not the majority of, published studies consider sample sizes of about 80 cases each [Gurney, 1994]. Next, suppose that the function $f$ discriminates the entire state space (which in this illustrative scenario is of size $2^{11} = 2{,}048$) by categorizing 78% (i.e., 1,597) of the cases as suggestive of cancer and the remaining 22% (i.e., 451) as negative for cancer.

Some key definitions follow next. The **state space** expresses all possible combinations of attributes. Note that the concept of the state space is different from that of the **population**. In fact, the population is a subset of the state space, because the actual population may not exhibit all the attribute combinations. That is, some cases (examples) may not occur in reality and can be eliminated from further consideration. As a result, the **sample set** (i.e., the **training data**) consists of elements (binary vectors in our illustrative example) drawn, with replacement, from the population rather than from the state space. Another result is that a sample may not represent the population and the state space equally. For instance, if a population includes 1,843 unique cases (i.e., 90% of the state space of 2,048 cases), then a sample of 80 vectors covers at most 3.9% of the state space and 4.3% of the population.

Therefore, in this illustrative scenario, it is assumed that 80 different cases were used to represent 2,048 cases. At this point one may wish to ask the question: "Is the function, which was inferred using a training sample of no more than 4.3% of the population, sufficiently reliable to recommend surgery for new patients?" While this function can be an interesting one, its statistical significance is questionable for a reliable diagnosis of cancer. If one considers multivalued, instead of binary attributes, then a sample of 80 (which is a common sample size in such published studies) becomes a miniscule portion of the entire state or population space. This statistical weakness becomes even more dramatic if one considers more attributes.

Next we define some key parameters for dealing with the reliability issue. Let us denote the **number of unique cases (examples) in a sample** as $S$, the **size of the state space** as $N$, and the **population size** as $P$. Obviously, the following relationship is always true: $N \geq P \geq S$. It should be observed that the sample size might not be the same as the number of training cases, since the sample size is the number of unique cases. For instance, patient #1 and patient #2 may correspond to the same combination, say (01111100011). Therefore, the size of the sample set may be smaller than the number of cases in the sample. For example, the 15,000 mammograms of breasts without malignancy (unpublished data provided to us by the Woman's Hospital in Baton Rouge, Louisiana, in 1995) can be represented by fewer than 300 combinations of 11 attributes. Thus, the number of cases here is about 50 times greater than $S$.

Next we define the **index of potential reliability** by the ratio $S/N$ and the **index of actual reliability** by the ratio $S/P$. In practice, it is very difficult to accurately estimate the size of the population, and consequently the index of actual reliability $S/P$. Obviously, if one has a sample which covers the entire population, then the index of actual reliability is equal to 1. On the other hand, if one has a proper subset of the population, then the size of the population cannot be determined directly.

Note that it is possible to have different levels of reliability for different diagnostic classes within the same training set. In order to demonstrate this, we next compute the indices of potential reliability for the previous 11 attributes. Suppose that there are $N_1 = 1{,}600$ "highly suspicious for malignancy" vectors in a state space (which is of size $2^{11} = 2{,}048$). Then the remaining $N_0 = 448 (= 2{,}048 - 1{,}600)$ cases correspond to "not highly suspicious for malignancy" vectors. Next, suppose that there are $S_1 = 50$ unique "highly suspicious for malignancy" cases in a training set. Similarly, let the class "not highly suspicious for malignancy" have $S_0 = 400$ unique cases in this training set. Then the indices of potential reliability for the respective groups are $S_1/N_1 = 0.03125 (= 50/1{,}600)$ and $S_0/N_0 = 0.89286 (= 400/448)$, respectively.

In the light of the previous reliability indices, we next consider the neural network (NN) approach described in [Wu, *et al.*, 1993]. These authors constructed two different feedforward neural networks (NNs) which contained two layers of processing elements (PEs). Both NNs contained 43 input units, each corresponding to an extracted radiographic attribute, and a single PE on the output layer representing the diagnosis (which was 0 for benign and 1 for malignancy). The two NNs differ merely by the number of PEs on the hidden layer; the first one used 10 while the second one used only 5 PEs. For each NN independently, they trained the PEs by back propagating their errors. For the training process, a set of 133 cases was selected from a mammography atlas. In addition, 60 other cases (of which 26 were malignant and 34 were benign) were randomly selected to evaluate the accuracy of the trained neural network. An experienced radiologist extracted the 43 attributes from each case and rated each attribute on a scale from 0 to 10.

One can compute the potential reliability, as expressed by the sample/state space ratio, for the training data. The state space was defined on 43 attributes, each using 11 grades. This state space corresponds to a total number of $11^{43}$ different cases. Thus, $S/N = 133/11^{43} = 2.21 \times 10^{-43}$. This means that the available sample is $2.21 \times 10^{-41}\%$, an extraordinarily miniscule fraction, of the total possible number of different vectors on the state space. Note that for a particular number of training cases, the reliability index depends on the size of the attribute set. As a result, 133 cases may be an insufficient number of cases for the previous state space, while say 32 cases could be sufficient for a reliable diagnosis in a smaller state space. Suppose that one has only 5 binary diagnostic attributes. Then, the state space consists of $32 (= 2^5)$ combinations. If all the 32 training cases represent unique vectors, then the size of the sample is 32 and the sample/state space ratio is equal to $1.00 (= 32/32)$, which is much better than the previous value of $2.21 \times 10^{-41}\%$. This example illustrates that the relative number of cases (i.e., the indices of reliability) is crucial, while a large number of cases may not be as valuable. Therefore, the question which is naturally raised here is:

> "Can a relatively small number of training cases be considered reliable in order to assist in accurately diagnosing new (and thus unknown) cases?"

Some neural network studies (e.g., [Baum and Haussler, 1989]) suggest that the number of cases should be no less than 10 times the number of connections (i.e., the parameters needed to be estimated), to reliably train a network. This measure is similar to our index of potential reliability expressed as $S/N$. Gurney in [1994] showed that this relatively weak requirement is not fulfilled in breast cancer CAD methods. For example, the largest neural network considered in [Wu, *et al.*, 1993] has 43 input units, 10 hidden layer PEs, and a single output PE. Thus, this network has $440(= 43 \times 10 + 10 \times 1)$ connections which is less than $532(= 4 \times 133)$, where 133 is the number of cases used to estimate the weights for these connections.

Boone in [1994] disputed the 10:1 requirement on the number of cases versus the number of connections. He compared the neural network with biological networks (e.g., radiologists) and argued that for biological networks the ratio is much worse, over $10^{10}$ times less, than the neural network studies criticized by Gurney [1994]. Thus, Boone asked: "Is there any reason that we should hold a computer to higher standards than a human?" Maybe not, but we should ask for both systems: "Is learning based on a small training subset sufficiently reliable to distinguish suspicious from non-suspicious (for malignancy) cases given the vast diversity of mammographic images?"

Machine learning theory (see, for instance, [Schapire, 1992], [Machine Learning, 1995a; 1995b], and [Computational Learning Theory, 1995a; 1995b]) addresses, among other issues, the problem of concept learning. It has been shown that there are relatively simple concepts that no algorithm is capable of learning in a reasonable amount of time (i.e., in polynomial time). The Probably Approximately Correct (PAC) learning theory, as introduced by Valiant in [1984] (see also [Angluin, 1988], and [Haussler and Warmath, 1993]), provides a popular model of learnability (see also Section 2.2).

The machine learning literature provides a plethora of families of relatively simple concepts which cannot be learned reliably in this sense (see also the previous references). Therefore, the question of reliability is among the most fundamental questions of scientific rigor and practical data mining and knowledge discovery applicability to mammographic diagnosis. In this chapter we explore the following two key questions:

(i) "Are accessible and relatively small samples sufficiently representative for learning?" and
(ii) "how can a broad range of mammographic attributes be evaluated?"

## 9.4 The Representation/Narrow Vicinity Hypothesis

The problems of the sample/space ratio and the sample/population ratio reliability criteria are part of a general problem of many data mining methods. Data mining methods such as neural networks, methods based on mathematical logic, decision trees, and so on, generalize from prototypes (i.e., training sets). That is, these approaches propose models that can discriminate new cases which were not among the prototypes (training cases). A common fundamental hypothesis, supporting small samples in data mining, is the hypothesis that a small sample is representative of the entire population. This **representation hypothesis** is best stated by Miller, *et al.* in [1992] (on page 462) as follows: "The training data must still form a representative sample of the set of all possible inputs if the network is to perform correctly." The same authors also suggested that: "The principal problems which must be addressed when producing a complete network application are: collecting and classifying sufficient training and testing data, choosing a valid data presentation strategy and an appropriate network architecture."

Therefore, without confirming the representation hypothesis as it applies to mammography, data mining with small samples may be of questionable reliability. In addressing this issue we study a restrictive version of the representation hypothesis, namely, the hypothesis of narrow vicinity (or the NV hypothesis):

> *All real possible cases are in a* **narrow vicinity** *of an accessible small training sample.*

If the NV hypothesis can be accepted, then we may generalize the training sample (of $S$ vectors) to the actual population (of $P$ vectors), but not to the remaining $(N - P)$ vectors, which do not represent feasible cases anyway. More formally, the NV hypothesis indicates that the $P/N$ ratio is very small, i.e., the size of the actual population $P$ is significantly less than the size of the state space $N$. If, for instance, $S = 80$, $P = 160$, and $N = 2{,}048$, then the $P/N$ ratio is equal to 0.078. Also, the index of the actual reliability $S/P(= 0.50)$ is significantly greater than the index of potential reliability $S/N(= 0.039)$. Therefore, the NV hypothesis provides the grounds to generalize from a small training subset. In Appendix II of this chapter we describe some methods which can allow one to estimate the $P/N$, $S/N$, and $S/P$ ratios without having the actual population for some typical mammographic and clinical attributes.

The problem of narrow vicinity is graphically illustrated in Figure 9.1. This figure shows the areas (i.e., the narrow vicinities) surrounding the points that were used to train a hypothetical CAD system. The small ovals and rectangles represent test cases from the "noncancer" and "cancer" diagnostic classes, respectively. Next, suppose that the actual borderline is the thick line near the "noncancer" training data and that linear discriminant analysis provided the dashed line. Then, the dotted rectangles will be misclassified by the estimated discriminant line. This illustrative example indicates that the extrapolation of training cases which are far away from their narrow vicinities may lead to dramatically inaccurate conclusions.

Our concerns about insufficient training data and the violation of the NV hypothesis were confirmed for our actual data set. We used 156 actual cases, of which 77 were malignant, and 79 were benign, provided to us by the Woman's Hospital in Baton Rouge, Louisiana, in 1995. The cases were defined on the 11 attributes of clustered calcifications with the diagnostic classes "malignant" and "benign" as described in Appendix I. A raw version of this data set can be found at the following URL: **http://www.csc.lsu.edu/trianta** (i.e., the webpage of the author) and is also briefly reviewed in Chapter 15 of this book. We analyzed these data by using Fisher's linear discriminant analysis [Fisher, 1938], [Fisher, 1936], and [Johnson and Wichern, 1992]).

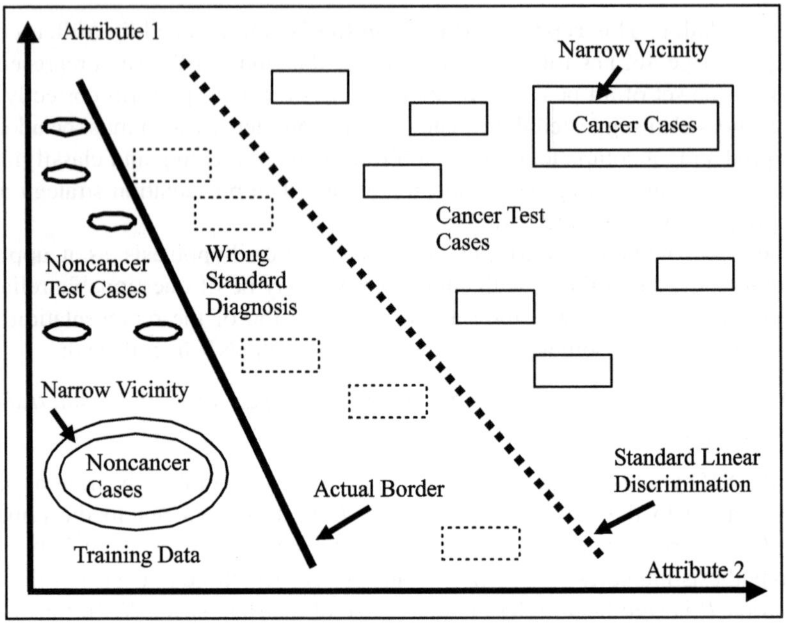


**Figure 9.1.** Comparison of the Actual and Computed Borders Between Diagnostic Classes (*a Conceptual Representation*).

By using linear discriminant analysis (LDA) one can estimate the line that minimizes the misclassification probability (given that this linear combination of the attributes follows a normal distribution and the classes have the same variance–covariance matrix). For the Woman's Hospital data, the line provided by LDA was able to correctly classify only 76% of the 156 cases. That is, a significant portion of the malignant cases were classified as benign and vice versa. Note that the discriminant analysis framework is not capable of handling much more complex classification systems. For example, if the variance–covariance matrices are unequal, then the classification model becomes quadratic and it may lead to some strange results in dimensions higher than 2. This situation indicates the need for an entirely new framework of assumptions.

The classification models (patterns) should be derived from the narrow vicinities of the available points. However, if one focuses only on the narrow vicinities of the available points, then there is a possibility to have too few data points and thus the derived results may not be statistically significant on an 11-dimensional space. These observations are in direct agreement with the sample/population space ($S/P$) ratio problem discussed earlier. Furthermore, this brief analysis indicates the need for developing new inference approaches capable of dealing with the previous methodological weaknesses, which may wrongly extrapolate away from the observed points. Related to this topic is also the discussion in Section 11.6.

## 9.5 Some Computational Results

A detailed description of the proposed method, and the specific steps, can be found in [Kovalerchuk, Triantaphyllou, and Vityaev, 1995], [Kovalerchuk, Triantaphyllou, and Ruiz, 1996], [Kovalerchuk, Triantaphyllou, Deshpande, and Vityaev, 1996] and also in Chapter 10 which describes some recent developments in this area. This method is also briefly summarized for this particular medical application in Appendix II of this chapter.

At first, let us note that the previous percentages of 78% and 22% of "cancer" and "noncancer" cases, respectively, are close to the actual percentages given in Section 9.3. About 80% of all possibilities in the state space indicate suspicion for cancer and recommendation for biopsy/short-term follow-up. A more detailed analysis has also shown that the borders of the biopsy/nonbiopsy regions are near the bottom of the state space (i.e., close to the vector containing all zeros).

We have found that our state space, as defined on the 11 binary diagnostic attributes, consists of 7.42% of possible cases (binary vectors) for which "biopsy/short-term follow-up is not necessary," and 92.58% of the vectors for which "biopsy/short-term follow-up is necessary." Similarly, this state space consists of 86.7% of "highly suspicious for malignancy" cases and 13.3% of "not highly suspicious for malignancy" cases (see also Table 9.1).

In order to understand the actual implications of the above issues one needs to consider the information derived by using actual historic cases. Suppose that one wishes to determine the above borders and percentages by using some sampled data $S$, which include cases of all examined patients at a hospital during a single year. At the Woman's Hospital in Baton Rouge, Louisiana (unpublished data, 1995) there are 15,000 new cases with complete data each year. Approximately 0.2% of these patients have cancer and 99.8% have no cancer. Approximately 1.1% of these 15,000 women will undergo biopsy/short-term follow-up while the remaining 98.9% will receive routine follow-up.

The situation in the state space is almost the reverse of the real-life situation found in the Woman's Hospital experience, namely, 0.2% and 99.8%. These numbers indicate that in a population of 15,000 mammograms we will have just 34 cases with cancer. Let us take this sample to discriminate 1,775 vectors representing suspicious findings (i.e., 86.7% of the total vectors) and the remaining 13.3% (i.e., 273 vectors) suggestive of benign lesions. Here the sample/space ($S/N$) ratio for cancer is equal to $34/1{,}775 = 0.019$ (i.e., 1.9%) and for noncancer we have a ratio of $15{,}000/273 = 54.94$ (i.e., 5,495%). Thus, we have a large surplus sample of patterns (i.e., training examples or data points) which are not representative of cancer and a very small sample representing highly suspicious findings indicating the presence of cancer (see also Table 9.1 and Figures 9.2 and 9.3). Moreover, 1.9% is an upper estimate for the $S/N$ ratio because different cases can be represented by the same combination of attributes.

The analysis presented in Figures 9.2 and 9.3 shows that, in general, the narrow vicinity (NV) hypothesis is not valid for mammographic evaluation. Recall that this is exactly the hypothesis implicitly used by all traditional pattern recognition/data

**Table 9.1.** Comparison of Sample and Class Sizes for Biopsy and Cancer (from Woman's Hospital in Baton Rouge, Louisiana, Unpublished Data, 1995).

| | Sample Size | Class Size in State Space | Sample/Space Ratio |
|---|---|---|---|
| **Total size** | 15,000 | 2,048 | 7.32 |
| Cases with Cancer | 34 | 1,775 | 0.02 |
| Percent | 0.20 | 86.70 | |
| Cases without Cancer | 14,966 | 273 | 54.82 |
| Percent | 99.80 | 13.30 | |
| Number of Biopsies | 165 | 1,896 | 0.09 |
| Percent | 1.10 | 92.58 | |
| Number of Nonbiopsies | 14,835 | 152 | 97.60 |
| Percent | 98.90 | 7.42 | |

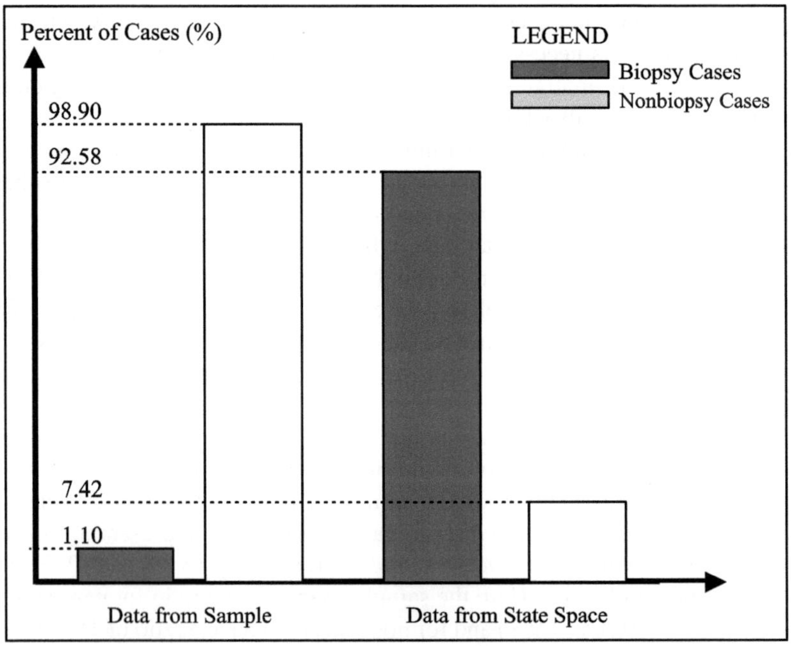


**Figure 9.2.** Relations Between Biopsy Class Size and Sample.

mining methods in breast cancer diagnosis! The diagnostic parameters which we used are typical for mammographic diagnosis [Wu, *et al.*, 1993].

The introduction of digital mammography has spawned a great deal of research in artificial intelligence/data mining techniques applied to breast cancer diagnosis. These methods range from $K$-nearest neighbor models (e.g., [Hojjatoleslami, and Kittler, 1996]) to the application of genetic algorithms (GAs) (e.g., [Sahiner, *et al.*,

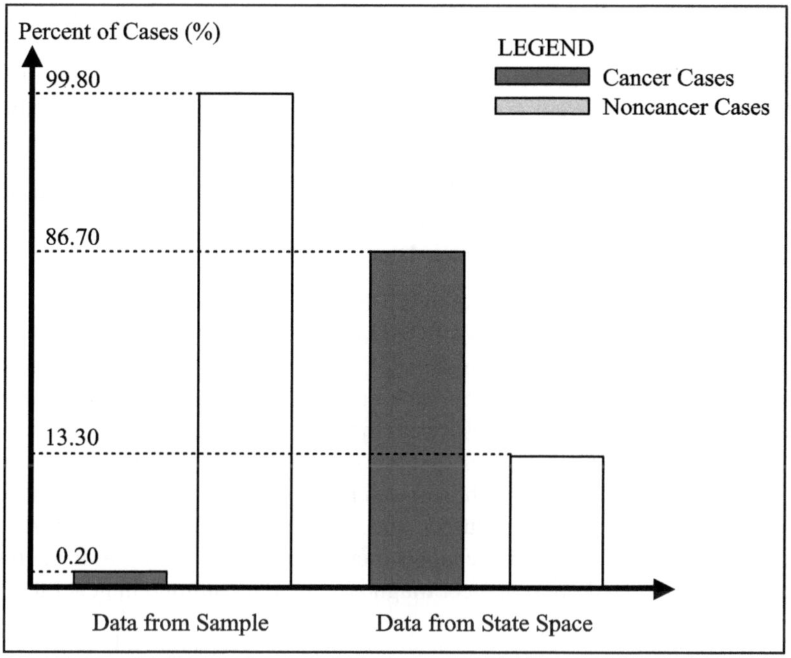


**Figure 9.3.** Relations Between Cancer Class Size and Sample.

1996]). However, the majority of these studies are concerned with the application of neural networks to extract attributes and classify tumors. For some closely related developments in this particular area the interested reader may want to consult the work reported in [Chan, *et al.*, 1995], [Floyed Carey, *et al.*, 1994], and [Zhang, *et al.*, 1994].

## 9.6 Concluding Remarks

This chapter argued that the development of reliable CAD/data mining methods for breast cancer diagnosis requires more attention on the problem of the selection of training and testing data and processing methods. Strictly speaking, all CAD/data mining methods are still very unreliable in spite of the apparent, and possibly fortuitous, high accuracy of cancer diagnosis reported in the literature. Our computations clearly show that a standard random selection of test cases [Wu, *et al.*, 1993] does not give a true picture of the accuracy/reliability issue of breast cancer diagnosis. The *Receiver Operator Characteristic* (ROC)-based analysis (see, for instance, [Kegelmeyer, *et al.*, 1994], [Jiang, *et al.*, 1996], [Huo, *et al.*, 1996], and [D'Orsi, *et al.*, 1992]) which is used to evaluate the accuracy of diagnosis suffers from this weakness.

There are several approaches and methods which can be used to improve this situation (some of them were used in our studies). Nevertheless, the low reliability of CAD/data mining should be recognized by the scientific community and their application should be reconsidered. It should be emphasized here that the problem is not only in the methods themselves. Not only are they implemented in situations where they are inappropriate, but they also provide a false sense of security since the literature tends to inflate their reliability. Reliable diagnosis can be obtained if:

(a) Research on the attribute space has shown that the chosen training data actually represent the border between diagnostic classes, and
(b) the mathematical method to be used can extract this border.

The main advantage of the methods which we used is that they allow one to identify and evaluate the reliability of CAD methods. Standard random selection of test cases (training examples) often does not adequately represent the critical border points (see also Figure 9.1). An approach based on the concept of monotonicity of the data allows one to select test cases near the border of diagnostic classes, i.e., critical points for verifying those cases that are regarded as the borderline between the benign and the malignant cases. The last and most important point is that the applied method can improve gains in accuracy and construct reliable diagnostic (discriminant) functions. In summary, this chapter focused on the following four main issues related to computerized breast cancer diagnosis and data mining, in general:

(i) The state space of all possible attributes may be astronomically large, and as a result, the size of the population (which is a subset of the state space) may be of the same magnitude.
(ii) Most samples represent a tiny fraction of the possible population space. Therefore, results obtained by traditional approaches, although they might be correct on some test cases, are not statistically significant, unless the *representation/narrow vicinity (NV) hypothesis* can be accepted.
(iii) The representation/narrow vicinity (NV) hypothesis may not always be valid.
(iv) Fortunately, real-life data may exhibit the monotonicity property. Approaches which explicitly use this critical property may alleviate some of the previous problems. The approach proposed in Chapters 10 and 11 and which uses the monotonicity property, may offer an effective and efficient way in overcoming these reliability problems.

Finally, it should be stated that we used the paradigm of breast cancer diagnosis because it is a socially and medically critical subject and because it possesses important characteristics that require a critical appraisal of the reliability issue in a real-life situation. It is expected that most of the findings described in this chapter can be extended to other domains too.

# Appendix I

## Definitions of the Key Attributes

The main study described in this chapter was performed for the binary attributes presented below. We deliberately used nonspecific terms, such as *small*, *large*, *pro cancer*, and *contra cancer* in order to allow us to further refine the language. An approach which uses nonbinary values and which is based on fuzzy logic is described in [Kovalerchuk, *et al.*, 1997] and also in Chapter 16 of this book. Also, the terminology used in the following paragraphs is deliberately simplified because the emphasis is on the technical procedures.

The definitions of indirect diagnostic attributes, along with their meaning, are as follows:

$x_1$ – Amount and volume of calcifications
(0-contra cancer/biopsy;
1-pro cancer/biopsy).

In addition, $x_1$ was considered to be a function $\psi(w_1, w_2, w_3)$ of the attributes $w_1, w_2, w_3$ defined as follows:

$w_1$ – Number of calcifications/cm$^2$
(1-large; 0-small)
$w_2$ – Volume, in cm$^3$ (approximate)
(1-small; 0-large)
$w_3$ – Total number of calcifications
(1-large; 0-small)

$x_2$ – Shape and density of calcifications
(0-contra cancer/biopsy;
1-pro cancer/biopsy).

Note that we considered $x_2$ as a function $\psi(y_1, y_2, y_3, y_4, y_5)$ of $y_1, y_2, y_3, y_4, y_5$, which are determined as follows:

$y_1$ – Irregularity in the shape of individual calcifications
(1-marked; 0-mild)
$y_2$ – Variation in the shape of calcifications
(1-marked; 0-mild)
$y_3$ – Variation in the size of calcifications
(1-marked; 0-mild)
$y_4$ – Variation in the density of calcifications
(1-marked; 0-mild)
$y_5$ – Density of the calcifications
(1-marked; 0-mild)
$x_3$ – Ductal orientation
(1-marked; 0-mild)

$x_4$ – Comparison with previous exam
(0-contra cancer/biopsy;
1-pro cancer/biopsy)

$x_5$ – Associated findings
(0-contra cancer/biopsy;
1-pro cancer/biopsy)

Thus, we used a state space consisting of the 11 binary attributes $w_1$, $w_2$, $w_3$, $y_1$, $y_2$, $y_3$, $y_4$, $y_5$, $x_3$, $x_4$, $x_5$. Attributes $x_1$ and $x_2$ were used to construct a hierarchy of attributes, as described next in Appendix II.

# Appendix II

## Technical Procedures

### 9.A.1 The Interactive Approach

Next, let us consider how one can validate the narrow vicinity (NV) hypothesis when a small sample set is available. If one has a large sample set available, then he/she does not need the NV hypothesis. On the other hand, with a small sample, one does not have to directly validate this key hypothesis. We developed a new methodology to overcome these difficulties. The main idea is to extend insufficient clinical cases with information from an experienced radiologist. Another approach is mentioned in [Miller, *et al.*, 1992] (on page 462): "One obvious solution to the problem of restricted training and testing data is to create simulated data using either a computer based or physical model." We used experienced experts as a "human" model to generate new examples.

One can ask a radiologist to evaluate a particular case when a number of attributes take on a set of specific values. A typical query in our experiments had the following format:

> "If attribute 1 has value $V_1$, attribute 2 has value $V_2, \ldots,$ attribute $n$ has value $V_n$, then should biopsy/short term follow-up be recommended or not? Or, does the above setting of values correspond to a highly suspicious case or not?"

The above queries can be defined with artificially constructed vectors (as will be explained below) or with artificially generated new mammograms by modifying existing ones. In this way one may increase a sample size but not as much as may be necessary. Roughly speaking, the technical weakness now is the same as before. That is, it is practically impossible to ask a radiologist to generate many thousands of artificial (i.e., synthetic) mammographic cases.

One can overcome these difficulties in two ways. First, if the attributes can be organized in a hierarchical manner, then a proper exploitation of this structure can lead to a significant reduction of the needed queries. Second, if the property of monotonicity, as explained below, is applicable, then the available data can be generalized to cover a larger training sample. The specific mathematical steps of how to achieve the above two goals are best described in Chapters 10 and 11 of this book.

At this point it should be stated that the issue of monotonicity in Boolean functions has been studied extensively by Hansel [1966]. Hansel proposed what has become a famous theorem (also known as Hansel's lemma) on the worst-case complexity of learning monotone Boolean functions. However, his theorem had not been translated into English until recently. There are numerous references to it in the non-English literature (Hansel wrote his paper in French). This theorem is one of the finest results of the long-term efforts in monotone Boolean functions that began with Dedekind in [1897]. The monotonicity property and related data mining issues are discussed further in Chapters 10 and 11.

### 9.A.2 The Hierarchical Approach

One can construct a hierarchy of *medically interpretable* attributes from a very generalized level to a less generalized level. For example, we considered the generalized binary attribute "Shape and density of calcification" with grades (0-"contra cancer/biopsy" and 1-"pro cancer/biopsy") denoted by $x_2$. On the second level we considered the attribute $x_2$ to be some function $\psi$ of five other attributes: $y_1, y_2, \ldots, y_5$. That is, $x_2 = \phi(y_1, y_2, \ldots, y_5)$, where

$y_1$ – irregularity in the shape of the individual calcifications,
$y_2$ – variation in the shape of the calcifications,
$y_3$ – variation in the size of the calcifications,
$y_4$ – variation in the density of the calcifications,
$y_5$ – density of the calcifications.

For illustrative purposes we will consider the above attributes as being binary valued with grades: (1) for "marked" and (0) for "minimal" or, equivalently, (1)-"pro cancer/biopsy" and (0)-"contra cancer/biopsy."

### 9.A.3 The Monotonicity Property

If we can identify regularities in advance, then it is possible to decrease the number of calls to a radiologist required to classify (diagnose) particular vectors (examples of clinical cases). Monotonicity is one such regularity and it may greatly reduce the number of diagnoses while maintaining a general hypothesis because many non-monotone regularities can also be represented as a combination of several monotone regularities (see also Chapters 10 and 11 of this book).

In order to clarify how the monotonicity property can be applied to the breast cancer diagnosis problem, consider the evaluation of calcifications in a mammogram. For simplicity and illustrative purposes assume that $x_1$ is the number and the volume occupied by calcifications, in a binary setting, as follows: (0-"contra cancer/biopsy" (or "not marked (i.e., mild);" 1-"pro cancer/biopsy" (or "marked")). As was stated in Appendix I, the following definitions were used:

$x_2$ – {shape and the density of the calcifications}, with values:
0-"contra cancer/biopsy";
1-"pro cancer/biopsy,"
$x_3$ – {ductal orientation}, with values:
0-"contra cancer";
1-"pro cancer,"
$x_4$ – {comparison with previous examination}, with values:
0-"contra cancer/biopsy";
1-"pro cancer/biopsy,"
and $x_5$ – {associated findings}, with values:
0-"contra cancer/biopsy";
1-"pro cancer/biopsy."

Given the above definitions we can represent clinical cases in terms of binary vectors with these five attributes as $(x_1, x_2, x_3, x_4, x_5)$. Next consider the two clinical cases which are represented by the two binary vectors (10100) and (10110). The vector (10100) means that the number and the volume occupied by calcifications is "pro cancer/biopsy" (e.g., $x_1 = 1$) and ductal orientation is "pro cancer/biopsy" (e.g., $x_3 = 1$) for the first case. The vector (10110) shows an extra "pro cancer/biopsy" attribute for the second case, i.e., the comparison with the previous examination is "pro cancer/biopsy" (i.e., $x_4 = 1$).

If a radiologist correctly diagnosed the first clinical case (10100) as malignant, then we can also conclude that the second clinical case (10110) should also be malignant. The latter case has all the "pro cancer/biopsy" characteristics (attributes) of the first case plus an extra one (i.e., $x_4 = 1$). In a similar manner, if we know that (01010) is not considered suspicious for cancer, then a second case, say (01000), should also not be considered suspicious for cancer. This is true because the second case has all the "contra cancer/biopsy" characteristics as the former one and, in addition, a new "contra cancer/biopsy" characteristic. This is indicated by the replacement of one "1" with a "0" value in the fourth attribute. These illustrative examples roughly demonstrate the property of monotonicity in Boolean functions and indicate how our algorithms can explicitly exploit this important property. One can combine a hierarchical approach with monotonicity and generalize accordingly. In this way, some major weaknesses of the traditional pattern recognition/data mining methods can be alleviated.

## 9.A.4 Logical Discriminant Functions

In Chapter 10 of this book we show the mathematical procedures which can be used to derive monotone Boolean discriminant functions. When these procedures are applied to this problem, it can be shown that monotone Boolean discriminant functions for the attributes on the uppermost level of the hierarchy are as follows.

For the "biopsy/short term follow-up" subproblem:

$$f_1(x) = x_2x_4 \vee x_1x_2 \vee x_1x_4 \vee x_3 \vee x_5. \tag{9.1}$$

In the above expression note that, for instance, $x_2x_4$ means $x_2 \wedge x_4$. Similar abbreviations are used throughout this discussion.

Similarly, for the second subproblem (i.e., "highly suspicious for cancer") the extracted function was

$$f_2(x) = x_1x_2 \vee x_3 \vee (x_2 \vee x_1 \vee x_4)x_5. \tag{9.2}$$

Regarding the second level of the hierarchy (which recall has 11 binary attributes) we interactively constructed the following functions (an interpretation of the attributes is presented in Appendix I of this chapter):

$$x_1 = \phi(w_1, w_2, w_3) = w_2 \vee w_1w_3, \tag{9.3}$$

and

$$x_2 = \psi(y_1, y_2, y_3, y_4, y_5) = y_1 \vee y_2 \vee y_3y_4y_5. \tag{9.4}$$

By combining the functions in (9.1) to (9.4) one can obtain the formulas of all the 11 attributes for "biopsy/short term follow-up" as follows:

$$\begin{aligned} f_1(x) = {} & (y_2 \vee y_1 \vee y_3y_4y_5)x_4 \\ & \vee (w_2 \vee w_1w_3)(y_2 \vee y_1 \vee y_3y_4y_5) \\ & \vee (w_2 \vee w_1w_3)x_4 \vee x_3 \vee x_5, \end{aligned}$$

and for "highly suspicious for cancer" cases:

$$\begin{aligned} f_2(x) &= x_1x_2 \vee x_3 \vee (x_2 \vee x_1 \vee x_4)x_5 \\ &= (w_2 \vee w_1w_3)(y_1 \vee y_2 \vee y_3y_4y_5) \\ &\quad \vee x_3 \vee (y_1 \vee y_2 \vee y_3y_4y_5) \\ &\quad \vee (w_2 \vee w_1w_3 \vee x_4)x_5. \end{aligned}$$

The benefit of having these functions is twofold. First, they express patterns as logical expressions (i.e., clauses of Boolean functions) which can allow us to identify the real border between diagnostic classes. Second, they allowed us to compute the size of the classes presented in Table 9.1 and depicted in Figures 9.2 and 9.3.

# Chapter 10

# Data Mining and Knowledge Discovery by Means of Monotone Boolean Functions

## 10.1 Introduction

In all previous discussions the problem was how to infer a *general* Boolean function based on some training examples. Such a Boolean function can be completely inferred if all possible binary examples (states) in the space of the attributes are used for training. Thus, one may never be 100% certain about the validity of the inferred knowledge when the number of training examples is less than $2^n$. The situation is different, however, if one deals with the inference of systems that exhibit *monotonic* behavior. The developments presented in this chapter are based on the award-winning doctoral work of Vetle I. Torvik and in particular on the research results first published in [Torvik and Triantaphyllou, 2002; 2003; 2004; 2006].

Thus, this chapter addresses the problem of learning monotone Boolean functions with the underlying objective to *efficiently acquire simple and intuitive knowledge that can be validated and has a general representation power.* The following key properties strengthen the argument in favor of this objective:

*Key Property 1.* Monotone Boolean functions are inherently frequent in applications.

The following three examples illustrate the versatility of the monotonicity property and how it applies to practical situations. (1) Suppose a computer tends to crash when it runs a particular word processor and web browser simultaneously. Then, the computer will probably crash if it, in addition, runs other software applications. Further, suppose this computer does not tend to crash when it runs a particular CD player and web browser simultaneously. Then, it will probably not crash when it only runs the web browser (or only the CD player). (2) If a keyword search in a database gives interesting hits, then hits for a proper superset of these keywords are also probably going to be interesting. On the other hand, if a keyword search in a database does not give interesting hits, then hits for a proper subset of these keywords are probably not going to be interesting either. (3) With all other factors constant, a student with a high Grade Point Average (GPA) is more likely to be accepted into a particular college than a student with a low GPA.

E. Triantaphyllou, *Data Mining and Knowledge Discovery via Logic-Based Methods*, Springer Optimization and Its Applications 43, DOI 10.1007/978-1-4419-1630-3_10,

Recent literature contains a plethora of phenomena that can be modeled by using monotone Boolean functions. Such diverse phenomena include, but are not limited to, social worker's decisions, lecturer evaluation, and employee selection [Ben-David, 1992], chemical carcinogenicity, tax auditing, and real estate valuation [Boros, *et al.*, 1994], breast cancer diagnosis and engineering reliability [Kovalerchuk, *et al.*, 1996c; 2000], signal processing [Shmulevich, 1997], rheumatology [Bloch and Silverman, 1997], voting rules in the social sciences [Judson, 1999], financial systems [Kovalerchuk and Vityaev, 2006], record linkage in administrative databases [Judson, 2001; 2006], and in bibliographic databases [Torvik, *et al.*, 2004].

*Key Property 2.* Monotone Boolean functions are simple and intuitive.

This property is perhaps the most important one when human interaction is involved since people tend to make very good use of knowledge they can easily interpret, understand, validate, and remember. Due to the increasing computational efficiency and storage capacity, the recent trend has been to increase the knowledge representation power in order to capture more complex knowledge. For example, the popular neural networks are not capable of representing very complex knowledge. Unfortunately, even small neural networks can be hard to interpret and validate.

*Key Property 3.* Monotone Boolean functions can represent relatively complex knowledge and still be validated.

Validating knowledge that is generalized from a set of specific observations (training examples), which may be noisy and incomplete, is based on philosophical arguments and mathematical assumptions. Traditional statistical approaches tend to require specific modeling in small dimensions, to gain a theoretical justification for the final model. This justification is obtained at the cost of eliminating the computational feasibility of learning higher dimensional models. On the other hand, the more general the knowledge representation is, the more one tends to lose the handle on its validation.

In practice, a great deal of effort is put into the knowledge discovery process. Software applications are tested, diseases are researched, search engines are trained to be intelligent, and so on. The inference process generally involves gathering and analyzing data. Gathering the data often involves some sort of labor that far outweighs the computations used to analyze the data in terms of cost. Therefore, the main objective in this chapter is to minimize the cost associated with gathering the data, as long as it is computationally feasible.

Monotone Boolean functions lay the ground for a simple and efficient question-asking strategy, where it may be easy to pinpoint questions whose answers make incomplete knowledge more general or stochastic knowledge more accurate. Due to the underlying monotonicity property, this learning strategy may significantly increase the learning rate, as an unguided learner might not receive the relevant pieces of information early enough in the inference process. Therefore, it is highly desirable not only to be able to pose questions, but also to pose "smart" questions. The main problem addressed in this chapter is how to identify these "smart" questions in order to efficiently infer monotone Boolean functions. This chapter focuses on the case where the monotone Boolean functions are defined on the set of $n$-dimensional Boolean vectors $\{0, 1\}^n$. This does not necessarily limit the application domain as the methodology developed in this chapter can be applied to any finite set of vectors $V \subset R^n$, and as it is shown in Chapter 11, any general Boolean function can be represented by a set of monotone Boolean functions.

This chapter is organized as follows: The background information and the relevant literature is reviewed in Section 10.2. Formal definitions of the problems and their solution methodology are given in Section 10.3. In Section 10.4 some experimental results are provided, for which a summary and discussion are given in Section 10.5. Section 10.6 concludes this chapter.

## 10.2 Background Information

### 10.2.1 Problem Descriptions

Let $V$ denote a finite set of vectors (i.e., examples) defined on $n$ binary attributes. A vector $v \in V$ is said to *precede* another vector $w \in V$, denoted by $v \preceq w$, if and only if (iff) $v_i \leq w_i$ for $i = 1, 2, \ldots, n$. Here, $v_i$ and $w_i$ denote the $i$-th element of vectors $v$ and $w$, respectively. Similarly, a vector $v \in V$ is said to *succeed* another vector $w \in V$, denoted by $v \succeq w$, iff $v_i \geq w_i$ for $i = 1, 2, \ldots, n$. When $v$ precedes (or succeeds) $w$, and the two vectors are distinct (i.e., $v \neq w$), then the vector $v$ is said to *strictly precede* (or *strictly succeed*, respectively) $w$, denoted by $v \prec w$ (or $v \succ w$, respectively). If a vector $v$ either precedes or succeeds $w$, they are said to be *related* or *comparable*. A Boolean function defined on the set of vectors $\{0, 1\}^n$ is simply a mapping to $\{0, 1\}$. A monotone Boolean function $f$ is called *nondecreasing* iff $f(v) \leq f(w) \forall v, w \in \{0, 1\}^n : v \preceq w$, and *nonincreasing* iff $f(v) \geq f(w) \forall v, w \in \{0, 1\}^n : v \preceq w$. This chapter focuses on nondecreasing functions, which are referred to as just *monotone*, as analogous results hold for nonincreasing functions.

Monotone Boolean functions lay the ground for a simple question-asking (i.e., guided learning) strategy, which forms the basis of this chapter. More specifically, the problem of inferring monotone Boolean functions by successive and systematic *function evaluations* (*membership queries* submitted to an oracle) is addressed. A monotone Boolean function can be thought of as a phenomenon, such as breast cancer or a computer crash, together with a set of predictor attributes. An *oracle* can be thought of as an entity that knows the underlying monotone Boolean function and provides a Boolean function value in response to each membership query. As in previous treatments, an oracle may take the shape of a human expert, or it may be the outcome of performing tasks such as running experiments or searching large databases.

This inference problem is broken down by the nature of the oracle: whether it is deterministic or stochastic, and whether it is two-valued or three-valued. The simplest variant considers the guided inference of a deterministic monotone Boolean function defined on at most $n$ Boolean attributes. This case is referred to as Problem 1 which is generalized into two different problems. The first generalization includes a pair of nested monotone Boolean functions and is referred to as Problem 2. Since this problem includes two oracles, it is further broken down into three subproblems 2.1, 2.2, and 2.3, differing only in the manner in which these two oracles are accessed. The second generalization includes stochastic membership queries and is referred to as Problem 3. A simple monotone Boolean function is shown later in Figure 10.3 while a nested pair of such functions is shown in Figure 10.4, where more details are provided.

**Problem 1 (Inferring a Monotone Boolean Function from a Deterministic Oracle).** Initially, the entire set of $2^n$ Boolean vectors in $\{0, 1\}^n$ is considered to be unclassified. That is, the values of the underlying monotone Boolean function $f$ are all unknown and may be 0 or 1. A vector $v$ is then selected from the set of unclassified vectors $U$ and is submitted to an oracle as a membership query. After the vector's function value $f(v)$ is provided by the oracle, the set of unclassified vectors is reduced according to the following monotonicity constraints: $f(w) = 0$, $\forall w \in U : w \preceq v$, when $f(v) = 0$, or the following monotonicity constraints: $f(w) = 1, \forall w \in U : v \preceq w$, when $f(v) = 1$. Here, the relationship $v \preceq w$ holds if and only if $v_i \leq w_i$, for $i = 1, 2, \ldots, n$, where $v_i$ and $w_i$ denote the $i$-th Boolean elements (fields or attributes) of the vectors $v$ and $w$, respectively. Vectors are then repeatedly selected from the unclassified set until they are all classified (i.e., until $U = \{\ \}$). Given the classification of any unclassified vector, other vectors may be concurrently classified if the underlying Boolean function is assumed to be monotone. Therefore, only a subset of the $2^n$ vectors need to be evaluated in order to completely reconstruct the underlying function. Thus, a key problem is to select a sequence of "promising" vectors so as to reduce (or ideally minimize) the total number of queries (or *query complexity*). These queries are needed to completely infer a hidden logic under the assumption that it is a monotone Boolean function.

**Problem 2 (Inferring a Pair of Nested Monotone Boolean Functions from Deterministic Oracle(s)).** A pair of monotone Boolean functions $f_1$ and $f_2$ are called *nested* when the following relationship holds: $f_1(v) \geq f_2(v)$ (or $f_1(v) \leq f_2(v))\forall v \in \{0, 1\}^n$. The case when $f_1 \geq f_2$ is addressed in this chapter as analogous results hold for the case when $f_1 \leq f_2$. A single monotone Boolean function does not capture the idea of a classification intermediate to 0 and 1. However, a pair of nested monotone Boolean functions can do so. For instance, some vectors might belong to a class with a high probability (i.e., when $f_1 = 1$ and $f_2 = 1$), and some might belong to the other class with a high probability (i.e., when $f_1 = 0$ and $f_2 = 0$). Other instances might not be classifiable with a satisfactorily high probability. A pair of nested monotone Boolean functions allows for this intermediate classification (i.e., when $f_1 = 1$ and $f_2 = 0$) to be incorporated. The case $f_1 = 0$ and $f_2 = 1$ is infeasible as the two monotone functions are nested (and $f_1 \geq f_2$). This makes the monotone Boolean function model more powerful.

Since the inference of a pair of nested monotone Boolean functions may include two oracles, it is further broken down into the three subproblems 2.1, 2.2, and 2.3

(described next), differing only in the manner in which the oracle(s) is(are) accessed. These three problems are defined to capture the main inference scenarios that may arise in real-life applications.

**Problem 2.1 (Sequentially Inferring Nested Functions from Two Oracles).** For this problem the two functions are considered to be available via their two respective oracles where the inference situation dictates that, for example, function $f_1$ should be completely reconstructed before the inference of function $f_2$ begins. In other words, the two functions are to be *sequentially inferred*. This approach may simply be the only feasible or reasonable one or it may be dictated by the cost of querying the oracle associated with $f_2$ far surpassing the cost of querying the other oracle.

**Problem 2.2 (Inferring Nested Functions from a Single Three-Valued Oracle).** For this problem the two nested monotone Boolean functions are viewed as a single function $f$ taking on the three values 0, 1, and 2, corresponding to $(f_1, f_2) = (0, 0), (1, 0)$, and $(1, 1)$, respectively. That is, it is a *ternary function*. Recall that $(f_1, f_2)$ cannot take on the values $(0, 1)$ due to the nestedness constraint $f_1 \geq f_2$. The single three-valued function is used to emphasize that the Boolean function values arrive in pairs, for each vector, from a single oracle.

**Problem 2.3 (Inferring Nested Functions from Two Unrestricted Oracles).** This problem is similar to Problem 2.1, in that two oracles are queried separately. Unlike Problem 2.1, no restrictions are put on the manner in which the two oracles are queried. At each inference step, a vector can be submitted to either of the two oracles. In this sense, this is the least restrictive of the three problems, and it is therefore expected that its solution approach will be more efficient.

**Problem 3 (Inferring a Monotone Boolean Function from a Stochastic Oracle).** This problem is identical to Problem 1, except that the membership values are now stochastic in nature. As in Problem 1, vectors are selected from $\{0, 1\}^n$ and are submitted to an oracle as membership queries. Unlike Problem 1, it is assumed that the oracle misclassifies each vector $v$ with an unknown probability $q(v) \in w(0, 1/2)$. That is, for a given monotone Boolean function $f$, the oracle returns 1 for vector $v$ with probability $p(v) = q(v) \times (1 - f(v)) + (1 - q(v)) \times f(v)$, and it returns 0 with probability $1 - p(v)$. It is assumed that the oracle is not misleading the inference process and is better at classifying the vectors than completely random guessing, hence the oracle's misclassification probability is assumed to be less than $1/2$.

The stochastic inference problem involves estimating the misclassification parameter $q(v)$ for each vector $v$, as well as reconstructing the underlying function $f$. These two tasks are based on a maximum likelihood framework. A monotone Boolean function that is the most likely to match the underlying function, given the observed queries, is referred to as the *inferred function* and is denoted by $f^*$. Associated with a function $f^*$ are the estimated misclassification probabilities which are denoted by $q^*(v)$ for each vector $v$.

The inference process consists of two steps that are repeated successively. In the first step, a vector is submitted to the oracle as a query. After a vector's function (class) value is provided by the oracle, both $q^*(v)$ and $f^*$ may have to be updated, according to the following monotonicity property: $p(v) \leq p(w)$ if and only if $v \preceq w, \forall v, w \in w\{0, 1\}^n$. These two steps are repeated until the likelihood of the inferred function $f^*$ matching the underlying function $f$ is high relative to the likelihood of any of the other monotone Boolean functions matching $f$. In other words, the underlying function is considered completely inferred when the maximum likelihood ratio for the inferred function, denoted by $\lambda(f^*)$, reaches a value that is close to 1. Again, the key problem is to select "promising" vectors so as to reduce the total number of queries required in this process.

### 10.2.2 Hierarchical Decomposition of Attributes

In some applications, the attributes may be monotone Boolean functions themselves defined on a set of Boolean attributes at a lower level. Kovalerchuk, *et al.* [1996c; 2000] decomposed five breast cancer diagnostic attributes in a hierarchical manner as follows. This is the same illustrative application as the one used in Chapter 9. Function $f_1(v)$ describes their "biopsy subproblem" and is defined as 1 if a biopsy is recommended for a tumor with the features described by vector $v$, and 0 otherwise. Function $f_2(v)$ describes their "cancer subproblem" and is defined as 1 if a tumor with the features described by $v$ is highly suspicious for malignancy, and 0 otherwise. The first attribute $v_1$ is defined as 1 if the *amount and volume of calcifications* is "pro cancer," and 0 if it is "contra cancer." In reality, this attribute was inferred (through queries to a radiologist) as the following monotone Boolean function: $v_1(x_1, x_2, x_3) = x_2 \vee x_1 x_3$ (see also relation (9.3) in Chapter 9). In the previous expression recall that $x_1 x_3$ stands for $x_1 \wedge x_3$. Here, the extra attributes are defined as follows (see also Appendix I, Chapter 9):

$x_1 = 1$ if the *number of calcifications/cm²* is "large," 0 if "small,"
$x_2 = 1$ if the *volume of calcifications (cm³)* is "small," 0 if "large," and
$x_3 = 1$ if the *total number of calcifications* is "large," 0 if "small."

The second attribute $v_2$ is defined as 1 if the *shape and density of calcifications* is "pro cancer," and 0 if it is "contra cancer." In reality, this attribute was inferred (through queries to a radiologist) as the following monotone Boolean function: $v_2(x_4, x_5, x_6, x_7, x_8) = x_4 \vee x_5 \vee x_6 x_7 x_8$ (see also relation (9.4) in Chapter 9. Also notice that the notation is slightly different). Here, the extra attributes are defined as follows (see also Appendix II, Chapter 9):

$x_4 = 1$ if the *irregularity in the shape of individual calcifications* is "marked," 0 if "mild,"
$x_5 = 1$ if the *variation in the shape of calcifications* is "marked," 0 if "mild,"
$x_6 = 1$ if the *variation in the size of calcifications* is "marked," 0 if "mild,"
$x_7 = 1$ if the *variation in the density of calcifications* is "marked," 0 if "mild," and
$x_8 = 1$ if the *density of calcifications* is "marked," 0 if "mild."

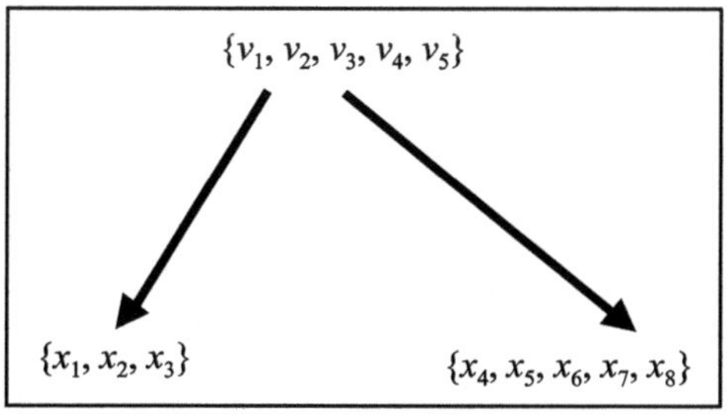


**Figure 10.1.** Hierarchical Decomposition of the Breast Cancer Diagnosis Attributes.

In general, one can construct a hierarchy of the sets of attributes, where each set of attributes corresponds to an independent inference problem. Figure 10.1 shows this hierarchy for the breast cancer diagnostic attributes. The upper level consists of the set $\{v_1, v_2, v_3, v_4, v_5\}$ which is linked to the sets of attributes $\{x_1, x_2, x_3\}$, and $\{x_4, x_5, x_6, x_7, x_8\}$ at the lower level. Here, the attributes $v_1$ and $v_2$ have to be defined before the inference process defined on the set of attributes $\{v_1, v_2, v_3, v_4, v_5\}$ can begin. In general, the inference processes at the lower level have to be completed before the inference processes at the upper levels can begin.

The breast cancer inference problem is defined on the set of Boolean attributes $\{x_1, x_2, x_3, x_4, x_5, x_6, x_7, x_8, v_3, v_4, v_5, f_i\}$. This problem includes a total of $2^{12} =$ 4,096 vectors (states, examples) to choose from. However, it can be approached hierarchically, as three independent problems defined on the sets $\{x_1, x_2, x_3\}$, $\{x_4, x_5, x_6, x_7, x_8\}$, and $\{v_1, v_2, v_3, v_4, v_5, f_i\}$, respectively. These problems include a total of $2^3 + 2^5 + 2^6 = 104$ possible vectors to choose from. Thus, the hierarchical approach to this problem reduces the number of possible vectors to choose from by a factor of $4{,}096/104 \approx 39.4$.

Notice that a single monotone Boolean function is to be inferred for each of the two sets $\{x_1, x_2, x_3\}$, and $\{x_4, x_5, x_6, x_7, x_8\}$. This corresponds to Problem 1 defined on the sets (binary spaces) $\{0, 1\}^3$ and $\{0, 1\}^5$, respectively. In contrast, a pair of nested monotone Boolean functions defined on the set $\{v_1, v_2, v_3, v_4, v_5\}$ are to be sequentially inferred. This corresponds to Problem 2.1 and includes the query domain $\{0, 1\}^6$.

### 10.2.3 Some Key Properties of Monotone Boolean Functions

An ordered set of related vectors $v^1 \preceq v^2 \preceq \ldots \preceq v^p$ is sometimes called a *chain*, while an *antichain* (or *layer*) consists of a set of mutually unrelated vectors. When a set of vectors is partitioned into as few layers as possible, a *layer partition* is formed. Similarly, when a set of vectors is partitioned into as few chains as possible, a *chain partition* is formed. For a particular layer partition, the layers can be ordered as $L^1, L^2, \ldots, L^r$ so that a vector $v^i \in L^i$ cannot succeed another vector $v^j \in L^j$, if $i < j$. Let $\{0, 1\}^n$ denote the set of vectors defined on $n$ Boolean attributes. The layer partition for the set $\{0, 1\}^n$ is unique, while its chain partition is not unique. In fact,

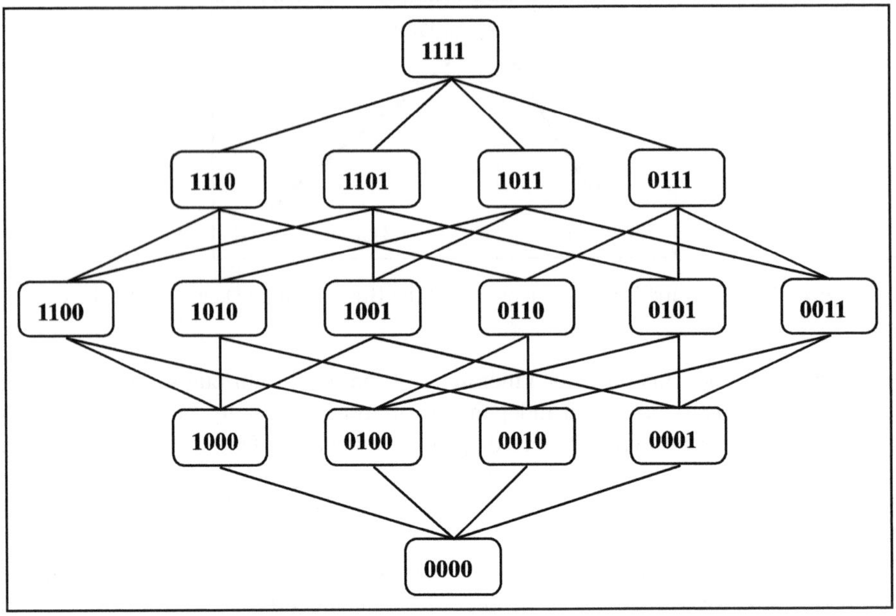


**Figure 10.2.** The Poset Formed by $\{0, 1\}^4$ and the Relation $\preceq$.

the way one partitions $\{0, 1\}^n$ into chains can be used effectively in the inference of monotone Boolean functions. An example is the symmetric chain partition used by Hansel [1966] and Sokolov [1982] as described in Section 10.2.4.

A *directed graph G* is often written in the form $(V, E)$, where $V$ denotes its set of vertices, and $E$ denotes its set of directed edges. Here, a directed edge from vertex $v$ to vertex $w$ is written as $(v, w)$. A directed graph $(V, E)$ is called *cyclic* if it has a sequence of edges that starts and ends with a vector $v : (v, v^1), (v^1, v^2), \ldots, (v^r, v) \in E$. Figure 10.2 shows a *partially ordered set* (or *poset* for short). In general, posets can be formed by a set of vectors $V$ together with the precedence relation $\preceq$, and are written as $(V, \preceq)$. A poset can be viewed as a directed graph where each vertex corresponds to a vector and each directed edge $(v, w)$ represents the precedence relation $v \preceq w$. When drawing a poset as a directed graph, its edges' directions are often omitted without loss of information.

The graph of a poset is acyclic and so all the directions can be forced upwards on a page by ordering the vertices by layers, as in Figure 10.2. Precedence relations that are transitively implied by other relations are considered *redundant*. For example, in Figure 10.2 the precedence relation (0000) $\preceq$ (1100) is redundant because it is implied by the two precedence relations (0000) $\preceq$ (1000) and (1000) $\preceq$ (1100). For the purpose of reducing storage and simplifying the visualization of posets, redundant precedence relations are generally omitted, as in Figure 10.2.

Two posets $P^1$ and $P^2$ are said to be *isomorphic* if there exists a one-to-one mapping of the vectors in $P^1$ to the vectors in $P^2$, where the precedence relations are preserved. That is, if $v^1 \rightarrow v^2$ and $w^1 \rightarrow w^2$, then $v^1 \preceq w^1$ if and only if $v^2 \preceq w^2$, $\forall v^1, w^1 \in P^1$ and $v^2, w^2 \in P^2$. For example, the poset formed by the vectors {0000, 1001, 0100} is isomorphic to the poset formed by the vectors {1110, 1100, 1101}. Here, one possible isomorphic mapping is as follows: $(0000) \rightarrow (1100)$, $(1001) \rightarrow (1110)$, and $(0100) \rightarrow (1101)$.

A vector $v^*$ is called an *upper zero* of a Boolean function $f$ if $f(v^*) = 0$ and $f(v) = 1 \forall v \in \{0, 1\}^n : v \succ v^*$. Similarly, a vector $v^*$ is called a *lower unit* if $f(v^*) = 1$ and $f(v) = 0 \forall v \in \{0, 1\}^n : v \prec v^*$. Lower units, denoted as LU($f$), and upper zeros, denoted as UZ($f$), are also referred to as *border vectors*. For any monotone Boolean function $f$, the set of lower units LU($f$) and the set of upper zeros UZ($f$) *are unique and either of these two sets uniquely identifies* $f$. Boolean functions are often written in disjunctive normal form (DNF) or in conjunctive normal form (CNF) (see also Section 2.4). A DNF or a CNF representation is *minimal* if removing any of its clauses results in a different mapping $\{0, 1\}^n \rightarrow \{0, 1\}$. For any monotone Boolean function $f$ there is a one-to-one relationship between its lower units and its minimal DNF representation, as follows:

$$f(v_1, v_2, \ldots, v_n) = \bigvee_{w \in LU(f)} \left( \bigwedge_{i:w_i=1} v_i \right).$$

Similarly, there is a one-to-one relationship between the upper zeros of a monotone Boolean function $f$ and its minimal CNF representation as follows:

$$f(v_1, v_2, \ldots, v_n) = \bigwedge_{w \in UZ(f)} \left( \bigvee_{i:w_i=0} v_i \right).$$

For instance, the monotone Boolean function defined by its lower units {110, 101} can be written in minimal DNF as $v_1 v_2 \vee v_1 v_3$ (i.e., $(v_1 \wedge v_2) \vee (v_1 \wedge v_3)$). The corresponding upper zeros of the same Boolean function are {011, 100} and its minimal CNF representation is $(v_2 \vee v_3)v_1$ (i.e., $(v_2 \vee v_3) \wedge (v_1)$). Often the operator $\wedge$ is omitted when writing out Boolean functions, as in the previous two examples. Since the lower units and upper zeros are unique to a monotone Boolean function, so are its minimal representations in DNF and CNF. Another nice property of monotone Boolean functions is that they can be written in minimal CNF or DNF without using the NOT (i.e., the negation) operation.

The *set of all monotone Boolean functions* defined on $\{0, 1\}^n$ is denoted by $M_n$. For example, the set of all monotone Boolean functions defined on $\{0, 1\}^2$ is given by $M_2 = \{F, v_1 v_2, v_1, v_2, v_1 \vee v_2, T\}$. Here the functions $T$ and $F$ are defined by $f(v) = 1, \forall v \in \{0, 1\}^n$, and $f(v) = 0, \forall v \in \{0, 1\}^n$, respectively.

Let $m(f)$ denote the *number of border vectors* associated with a Boolean function $f$. It is well known (e.g., [Engel, 1997]) that $m(f)$ achieves its maximum value for a function that has all its border vectors on two of the most populous layers of $\{0, 1\}^n$. That is, the following equation holds:

$$\max_{f \in M_n} m(f) = \binom{n}{\lfloor n/2 \rfloor} + \binom{n}{\lfloor n/2 \rfloor + 1}.$$

In this equation the first term stand for "$n$ choose $\lfloor n/2 \rfloor$" and so on.

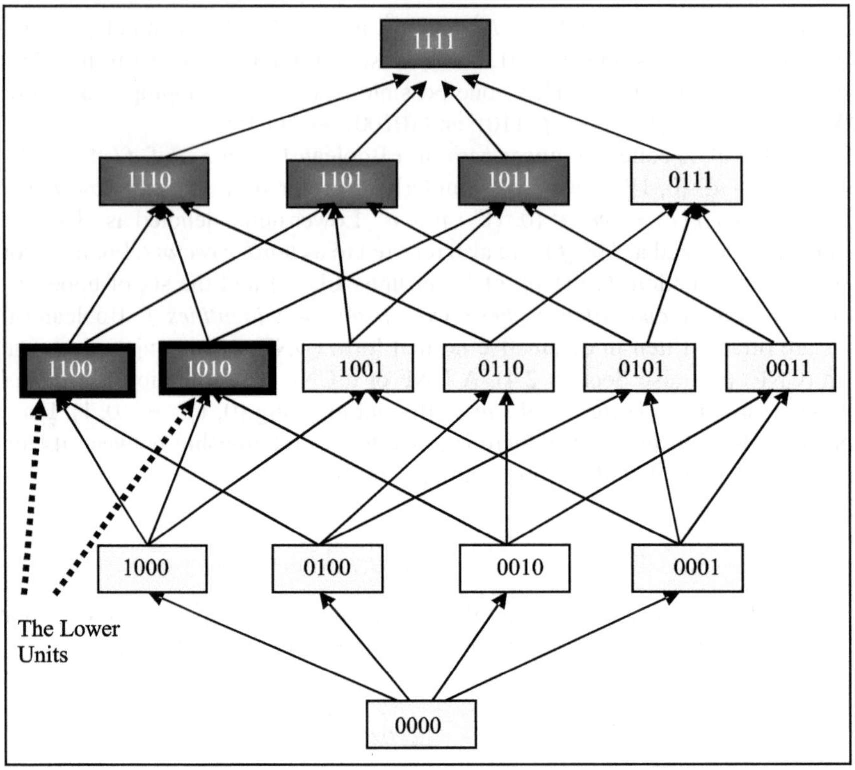


**Figure 10.3.** Visualization of a Sample Monotone Boolean Function and Its Values in $\{0, 1\}^4$ ($f(x) = (x_1 \wedge x_2) \vee (x_1 \wedge x_3)$).

**Table 10.1.** History of Monotone Boolean Function Enumeration.

| | |
|---|---|
| $\Psi(1) = 3$, $\Psi(2) = 6$, $\Psi(3) = 20$ | |
| $\Psi(4) = 168$ | by Dedekind [1897] |
| $\Psi(5) = 7{,}581$ | by Church [1940] |
| $\Psi(6) = 7{,}828{,}354$ | by Ward [1946] |
| $\Psi(7) = 2{,}414{,}682{,}040{,}998$ | by Church [1965] |
| $\Psi(8) = 56{,}130{,}437{,}228{,}687{,}557{,}907{,}788$ | by Wiedemann [1991] |

The borders of any monotone Boolean function $f$ are the only vectors that require evaluations in order to completely reconstruct the function. Hence, the value of $m(f)$ works as a lower bound on the number of queries for Problem 1. The inference problem of a monotone Boolean function reduces to the problem of inferring these border points (vectors).

Figure 10.3 depicts a sample monotone Boolean function when $n = 4$. This function is $f(x) = (x_1 \wedge x_2) \vee (x_1 \wedge x_3)$, where $x_i$ (for $i = 1, 2, 3, 4$) are binary attributes. The dark vectors represent the cases where this function evaluates to true value, while the rest (semi-gray) of the vectors are the cases for which $f(x)$ evaluates to false. This function is completely determined if one knows the composition, for instance, of its two lower units (as shown in Figure 10.3).

The *number of monotone Boolean functions* defined on $\{0, 1\}^n$ is denoted by $\Psi(n)$. That is, $\Psi(n)$ is equal to the size (dimension) of the set $M_n$. All of the known values for $\Psi(n)$ are given in Table 10.1. For larger values of $n$ the best known asymptotic is due to Korshunov [1981]:

$$\Psi(n) \approx \begin{cases} 2^{\binom{n}{n/2}} e^{\binom{n}{n/2-1}\left(\frac{1}{2^{n/2}}+\frac{n^2}{2^{n+5}}-\frac{n}{2^{n+4}}\right)}, & \text{for even } n. \\ 2^{\binom{n}{n/2-1/2}+1} e^{\binom{n}{n/2-3/2}\left(\frac{1}{2^{(n+3)/2}}-\frac{n^2}{2^{n+6}}-\frac{n}{2^{n+3}}\right)+\binom{n}{n/2-1/2}\left(\frac{n}{2^{(n+1)/2}}+\frac{n^2}{2^{n+4}}\right)}, & \\ \text{for odd } n. & \end{cases}$$

The number of *pairs* of nested monotone Boolean functions defined on $\{0, 1\}^n$ is simply $\Psi(n+1)$. This fact can be observed by first constructing the poset connecting the two posets $P_1 = (\{0, 1\}^n, \preceq)$ and $P_2 = (\{0, 1\}^n, \preceq)$ associated with functions $f_1$ and $f_2$, respectively, and then by adding the edges corresponding to the precedence relations $f_1(v) \geq f_2(v), \forall v \in \{0, 1\}^n$. Figure 10.4 depicts the main idea of having a pair of nested monotone Boolean functions.

### 10.2.4 Existing Approaches to Problem 1

Let $\varphi(A, f)$ denote the number of queries performed by an algorithm $A$, when reconstructing (inferring) the monotone Boolean function $f$. A *Teacher* can be thought of as an inference algorithm that knows the function ahead of time. It simply verifies that the function is correct by querying only the border vectors. Thus, $\varphi(\mathit{Teacher}, f) = m(f), \forall f \in M_n$. Recall that $m(f)$ denotes the number of border vectors associated with a function $f$.

For any monotone Boolean function inference algorithm $A$, the value $m(f)$ can be considered as a lower bound on the number of queries. Thus, $\varphi(A, f) \geq m(f), \forall f \in M_n$. It turns out that it is possible to achieve fewer or the same number of queries as the upper bound on $m(f)$, for all monotone Boolean functions defined on $\{0, 1\}^n$. This can be achieved by partitioning the set of vectors into chains as described in Hansel [1966]. In general, there are a total of $\binom{n}{\lfloor n/2 \rfloor}$ chains in $n$ dimensions.

An inference algorithm that searches these chains in increasing length is referred to as Hansel's algorithm (also known as Hansel's theorem or lemma). A key property of the Hansel chains is that once the function values are known for all the vectors in all the chains of length $k$, the function values are unknown for at most two vectors in each chain of the next length $k + 2$. Proof of this property can be found in both [Hansel, 1966] and [Sokolov, 1982]. As a result, Hansel's algorithm results in fewer or the same number of queries as the upper bound on $m(f)$ as follows. When $n$ is odd, the shortest chains contain two vectors each, and there are a total of $\binom{n}{\lfloor n/2 \rfloor}$ chains. In this case, the maximum number of queries used by Hansel's algorithm

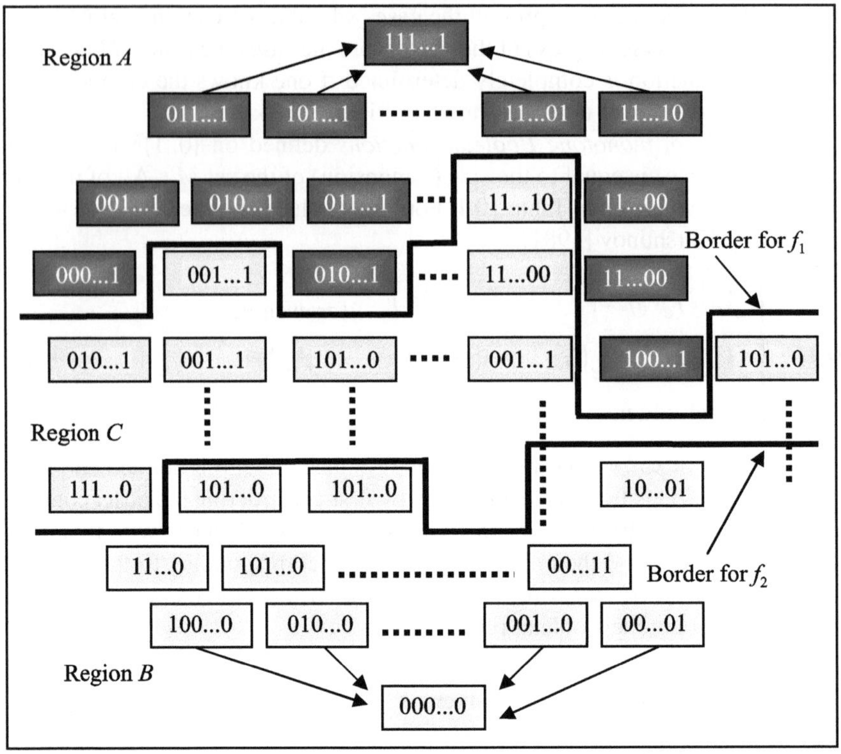


**Figure 10.4.** A Visualization of the Main Idea Behind a Pair of Nested Monotone Boolean Functions.

is $2\binom{n}{\lfloor n/2 \rfloor} = \left(\binom{n}{\lfloor n/2 \rfloor} + \binom{n}{\lfloor n/2 \rfloor+1}\right)$. Similarly, when $n$ is even, there are $\binom{n}{n/2} - \binom{n}{n/2+1}$ chains of length one, and $\binom{n}{n/2+1}$ chains of length greater than one. In this case, the maximum number of queries is $\binom{n}{n/2} - \binom{n}{n/2+1} + 2\binom{n}{n/2+1} = \binom{n}{n/2} + \binom{n}{n/2+1}$. That is, the following inequality holds:

$$\varphi(\text{Hansel}, f) \leq \max_{f \in M_n} m(f) = \binom{n}{\lfloor n/2 \rfloor} + \binom{n}{\lfloor n/2 \rfloor + 1}, \quad \forall f \in M_n.$$

The algorithm described in [Sokolov, 1982] is also based on Hansel chains. In contrast to Hansel's algorithm, it considers the chains in the reverse order (i.e., in decreasing length) and performs binary search within each chain. It turns out that Sokolov's algorithm is much more efficient for functions that have all their border vectors in the longer Hansel chains. As an example, consider the monotone Boolean function $T$. This function has only one border vector $(00\ldots0)$, which is located in the longest chain. For this function, Sokolov's algorithm performs at most $\lfloor \log_2(n) \rfloor + 1$ evaluations, while Hansel's algorithm needs at least $\binom{n}{\lfloor n/2 \rfloor}$ evaluations. For instance, when $n = 20$ this translates into at least 184,756 evaluations performed by Hansel's algorithm and at most 5 evaluations performed by Sokolov's algorithm.

Sokolov's algorithm does not satisfy the upper bound, as the following example shows. Suppose that $n > 4$ and even, and the monotone Boolean function to be inferred is defined by $f(v) = 1 \forall v \in \{0,1\}^n : |v| \geq n/2$, and 0 otherwise. Then the set of border vectors is $\{v, |v| = n/2 \text{ or } n/2 - 1\}$ and $m(f) = \binom{n}{\lfloor n/2 \rfloor} + \binom{n}{\lfloor n/2 \rfloor + 1}$. In Sokolov's algorithm, the first vector $w^1$ submitted for evaluation is a border vector since $|w^1| = n/2$. The second vector $w^2$ is not a border vector because $|w^2| = \lceil 3n/4 \rceil \neq n/2$ and $n/2 - 1$. Therefore, the following inequality holds:

$$\varphi(Sokolov, f) > \left( \binom{n}{\lfloor n/2 \rfloor} + \binom{n}{\lfloor n/2 \rfloor + 1} \right), \text{ for at least one } f \in M_n.$$

In an attempt to provide a unified efficiency testing platform, Gainanov [1984] proposed to compare inference algorithms based on the number of evaluations needed for each border vector. To that end, he presented an algorithm that searches for border vectors one at a time, and we refer to this algorithm as FIND-BORDER. At the core of the algorithm is a subroutine that takes as input any unclassified vector $v$, and finds a border vector by successively evaluating adjacent vectors. This subroutine is also used in the algorithms of Boros, *et al.* [1997], Makino and Ibaraki [1995], and Valiant [1984]. As a result, any inference algorithm $A$ that feeds unclassified vectors to this subroutine satisfies the following upper bound:

$$\varphi(A, f) \leq m(f)(n+1), \forall f \in M_n.$$

For the majority of monotone Boolean functions, the expression $m(f)(n+1)$ is greater than or equal to $2^n$, in which cases the bound is useless.

Earlier work on monotone Boolean function inference (such as [Hansel, 1966], [Sokolov, 1982], [Gainanov, 1984]) focuses on reducing the query complexity. More recent work (like [Boros, *et al.*, 1997], [Makino and Ibaraki, 1997], and [Fredman and Khachiyan, 1996]) considers both the query complexity and the computational complexity. The problem of inferring a monotone Boolean function via membership queries is equivalent to many other computational problems in a variety of fields (see, for instance, [Bioch and Ibaraki, 1995], and [Eiter and Gottlob, 1995]). These applications use algorithms that are efficient in terms of query and computational complexity.

In practice, queries often involve some sort of effort, such as consulting with experts, performing experiments, or running simulations. For such applications, queries far surpass computations in terms of cost. Therefore, this chapter focuses on *minimizing the query complexity* as long as it is computationally feasible.

### 10.2.5 An Existing Approach to Problem 2

Kovalerchuk, *et al.* [1996c; 2000] considered the problem of inferring a pair of nested monotone Boolean functions. Their algorithm, which exhibited a promising efficiency in their cancer diagnosis application, is an extension of Hansel's inference algorithm for a single monotone Boolean function. However, the algorithm performance analysis is far from conclusive as a single application represents a single pair of nested monotone Boolean functions.

### 10.2.6 Existing Approaches to Problem 3

The problem of guided inference in the presence of stochastic errors is referred to as sequential design of experiments in the statistics community. The field of optimal experiment design [Federov, 1972] contains various optimality criteria that are applicable in a sequential setting. The most common vector selection criterion is based on instantaneous variance reduction. Other selection criteria, such as the maximum information gain used in MacKay [1992], and Tatsuoka and Ferguson [1999], have been studied. However, no guided inference studies using a maximum likelihood framework were found in the literature.

The theory of optimal experiment design is most extensive for simple regression models [Federov, 1972]. Fortunately, efficient guided inference for more complex models has been studied, such as the feedforward neural networks in [Cohn, 1996], even though a sound theory has not been established. In fact, the same article reported a convergence problem for which a partial remedy was introduced in [Cohn, 1965].

### 10.2.7 Stochastic Models for Problem 3

Suppose a set of observed vectors $V = \{v^1, v^2, \ldots, v^k\}$ is given. For a given number of queries $m$, let $m_z(v)$ be the number of times the oracle classified vector $v$ as $z$ (for $z = 0$ or 1, and $v \in V$). Associated with a monotone Boolean function $f$, the *number of errors* it performs on the set of observations is denoted as $e(f)$ and it is given by

$$e(f) = \sum_{i=1}^{k} (f(v^i)m_0(v^i) + (1 - f(v^i))m_1(v^i)).$$

It is assumed that the oracle misclassifies each vector $v$ with a probability $q(v) \in (0, 1/2)$. That is, for a given monotone Boolean function $f$, the oracle returns for vector $v$

$$\begin{aligned} &1 \text{ with probability } p(v) = q(v) \times (1 - f(v)) + (1 - q(v)) \times f(v), \text{ and} \\ &0 \text{ with probability } 1 - p(v). \end{aligned}$$

A key assumption is that the misclassification probabilities are all less than $1/2$, otherwise it would not be possible to infer the correct monotone Boolean function. If the sampled values are considered fixed, their joint probability distribution function can be thought of as the likelihood of function $f$ matching the underlying function as follows:

$$L(f) = q^{e(f)}(1 - q)^{m - e(f)}.$$

The likelihood value of a particular monotone Boolean function decreases exponentially as more observations are added and therefore this value is generally very small. However, the likelihood ratio given by

$$\lambda(f^*) = \frac{L(f^*)}{\sum_{f \in F(V)} L(f)}$$

measures the likelihood of a particular function $f*$ relative to the likelihood of all possible monotone Boolean functions $F(V)$, defined on the set of vectors $V$. Note that when the set of vectors $V$ is equal to $\{0, 1\}^n$, the set of all possible monotone Boolean functions $F(V)$ is equal to $M_n$.

The goal of the maximum likelihood problem is to find a monotone Boolean function $f^* \in F(V)$, so that $L(f^*) \geq L(f) \forall f \in F(V)$. Assuming that the misclassification probabilities $q(v)$ are all less than 1/2, this problem is equivalent to identifying a monotone Boolean function $f^*$ that minimizes the number of errors $e(f^*)$ [Boros, *et al.*, 1995]. Note that if $q$ can take on values greater than 2, then the maximum likelihood solution may maximize the number of errors, as demonstrated by Boros, *et al.* [1995]. In this chapter, error maximization is avoided by restricting $q$ to be less than $1/2$; existence of such a solution is shown in [Torvik and Triantaphyllou, 2004].

The error minimization problem can be converted into an integer maximization problem as follows:

$$\begin{aligned} \textbf{min}\, e(f) &= \textbf{min} \sum_{i=1}^{k} (f(v^i)m_0(v^i) + (1 - f(v^i))m_1(v^i)) \\ &= \textbf{min} \left( - \sum_{i=1}^{k} f(v^i)(m_1(v^i) - m_0(v^i)) + \sum_{i=1}^{k} m_1(v^i) \right). \end{aligned}$$

Since the $\sum_{i=1}^{k} m_1(v^i)$ part is a constant, it thus can be removed from the optimization objective. Furthermore, maximizing a particular objective function is equivalent to minimizing the negative of that objective function, resulting in the following simplified integer optimization problem:

$$\begin{aligned} \textbf{max} \quad & \sum_{i=1}^{k} f(v^i)(m_1(v^i) - m_0(v^i)) \\ \textbf{\textit{subject to}}: \quad & f(v^i) \leq f(v^j) \forall v^i, v^j \in V : v^i \preceq v^j, \\ & \text{and } f(v^i) = 0 \text{ or } 1. \end{aligned}$$

This problem is known as a *maximum closure problem*, which can be converted into a *maximum flow problem* [Picard, 1976]. The most efficient algorithms developed for the maximum flow problem use the idea of preflows developed by Karzanov [1974]. For example, the lift-to-front algorithm (e.g., [Cormen, *et al.*, 1997]) takes $O(V^3)$ time. The fact that this problem can be solved in polynomial time is a nice property of the single $q$ parameter model. For two-dimensional problems (i.e., $V \subset R^2$), the minimum number of errors can also be guaranteed via a dynamic programming approach [Bloch and Silverman, 1997].

A more complex error model can potentially maintain as many parameters as the size of the domain $V$. That is, each vector $v$ may have an associated unique parameter $p(v)$. In this case, minimizing the weighted least squares

$$\textbf{min} \quad \sum_{i=1}^{k}(p'(v^i) - p(v^i))(m_1(v^i) + m_0(v^i))$$

$$\textbf{\textit{subject to}}: \quad p(v^i) \leq p(v^j)\forall v^i, v^j \in V : v^i \preceq v^j,$$

where

$$p'(v^i) = \frac{m_1(v^i)}{m_1(v^i) + m_0(v^i)}, \text{ for } i = 1, 2, 3, \ldots, k,$$

yields a maximum likelihood solution [Robertson, *et al.*, 1988]. This is a hard optimization problem, and several algorithms have been developed to solve it optimally and near optimally. The *Pooled Adjacent Violators Algorithm* (PAVA) by Ayer, *et al.* [1955] only guarantees optimality when $(V, \preceq)$ forms a chain poset (also referred to as a simple order). The *Min-Max algorithm* developed by Lee [1983] and the *Isotonic Block Class with Stratification* (IBCS) algorithm by Block, *et al.* [1994] guarantee optimality for the general poset but both algorithms can potentially consume exponential time. Unfortunately, no polynomial algorithm for the general poset was found in the literature.

In addition to the full parametric model, there are models of intermediate parametric complexity. One example is the logistic regression model with nonnegativity constraints on its parameters, as used for record linkage in databases by Judson [2001; 2006]. A monotone decision tree approach can be found in Makino, *et al.* [1999], and a sequential monotone rule induction approach can be found in [Ben-David, 1992; 1995].

It should be noted that the single parameter error model considered in this chapter is somewhat restrictive, in the sense that it does not estimate misclassification probabilities that vary across the vectors. However, one of the goals of this chapter is to efficiently uncover the underlying monotone Boolean function and not necessarily come up with accurate estimates for the individual errors. The fixed misclassification probability assumption does not affect the capability of the inference methodology as will be demonstrated in the subsequent sections. The assumption is simply used to estimate the error rate and the confidence in having inferred the correct function, and a more accurate estimate of the maximum likelihood ratio may require a substantial increase in computational complexity, as for the full parametric model described above.

## 10.3 Inference Objectives and Methodology

### 10.3.1 The Inference Objective for Problem 1

An inference algorithm that performs fewer queries than another algorithm when reconstructing a particular deterministic monotone Boolean function is considered more efficient on that particular function. However, it has not been clear how to compare algorithms on the entire class of monotone Boolean functions defined on $\{0, 1\}^n$.

The main existing algorithms by Hansel [1966], Sokolov [1982], and Gainanov [1984] focus on the upper bounds of their query complexities. Unfortunately, the worst case scenario reflects the algorithm performance on a few specific functions. It does not reflect what to expect when executing the algorithm on an arbitrary monotone Boolean function. For example, algorithms that implement Gainanov's subroutine (which we refer to as FIND-BORDER) indirectly suggest minimizing the upper bound on the number of evaluations per border vector. These algorithms greatly favor the simplest functions (which may only have a single border vector) over the complex functions (with up to $\binom{n}{\lfloor n/2 \rfloor}+\binom{n}{\lfloor n/2 \rfloor+1}$ border vectors). Kovalerchuk, *et al.* [1996c; 2000] demonstrated promising results for a Hansel-based inference algorithm on a real-life application. However, their performance analysis is far from conclusive as a single application represents a single pair of monotone Boolean functions.

With no prior knowledge (other than monotonicity) about the inference application, each function is equally likely to be encountered and should therefore carry the same weight in the objective. The objective for this problem is to develop an algorithm that minimizes the average number of queries over the *entire class* of monotone Boolean functions defined on the set $\{0, 1\}^n$. This objective can be expressed mathematically as follows:

$$Q(n) = \min_{A} \frac{\sum_{f \in M_n} \varphi(A, f)}{\Psi(n)}.$$

The objective $Q(n)$ represents the entire class of monotone Boolean functions $M_n$. As such, it provides a better indication of what to expect when executing an algorithm on an arbitrary monotone Boolean function.

### 10.3.2 The Inference Objective for Problem 2

The approach taken to this problem is analogous to that of Problem 1. The minimum average number of queries for Problem 2.$k$ (for $k = 1, 2$, or 3) can be expressed mathematically as follows:

$$Q_k(n) = \min_{A_k} \frac{\sum_{f_1, f_2 \in M_n : f_2 \le f_1} \varphi(A_k, f_1, f_2)}{\Psi(n+1)},$$

where $\varphi(A_k, f_1, f_2)$ denotes the number of queries performed by algorithm $A_k$, in reconstructing the pair of nested monotone Boolean functions $f_1$ and $f_2$ defined on the set $\{0, 1\}^n$. Furthermore, $A_1$, $A_2$, and $A_3$ denote algorithms designed for Problems 2.1, 2.2, and 2.3, respectively. Recall from Section 10.2.3 that the number of pairs of nested monotone Boolean functions defined on the set $\{0, 1\}^n$ is equal to $\Psi(n + 1)$, that is, the number of monotone Boolean functions defined on the set (binary space) $\{0, 1\}^{n+1}$.

Since these three problems differ in the way the oracles are queried, it should be clarified that a query unit pertains to the membership value from one of the two functions $f_1$ and $f_2$. This definition is intuitive for Problems 2.1 and 2.3, where two oracles are accessed individually. For Problem 2.2, the membership values are provided in pairs from a single *three-valued oracle* (i.e., a *ternary oracle*). To make the definition of $Q_2(n)$ comparable to $Q_1(n)$ and $Q_3(n)$, each query to the three-valued oracle will be counted as two queries.

### 10.3.3 The Inference Objective for Problem 3

The approach taken to Problem 3 is similar to that of Problems 1 and 2. The goal is to minimize the average number of queries needed to completely reconstruct the underlying monotone Boolean function, expressed mathematically as follows:

$$\min_A \frac{\sum_{f \in M_n} \varphi(A, f, q)}{\Psi(n)}.$$

In this expression $\varphi(A, f, q)$ denotes the expected number of queries performed by algorithm $A$ in completely reconstructing the underlying monotone Boolean function $f$ from an oracle with a fixed misclassification probability $q$. Completely reconstructing the underlying function translates into making the likelihood ratio $\lambda(f^*)$ for the inferred function $f^*$ reach a sufficiently high value (e.g., 0.99).

It should be stressed that the misclassification probability $q$ is unknown and ranges from 0 up to $1/2$. However, it is expected that the average number of queries will increase significantly with $q$, since, by definition, it approaches infinity as $q$ approaches $1/2$, and it is finite when $q$ is equal to 0. Therefore, the average over a large range $q$ may not be an accurate prediction of how many queries to expect for a particular application. The average query complexity will therefore be evaluated as a function of $n$ and $q$, even though $q$ is unknown.

### 10.3.4 Incremental Updates for the Fixed Misclassification Probability Model

Suppose the error minimizing function $f^*_{\text{old}}$ and its misclassification parameter $q^*_{\text{old}}$, associated with a set of vectors $V = \{v^1, v^2, \ldots, v^k\}$ and their $m_0(v)$ and $m_1(v)$ values, are given. When a new vector is classified by the oracle (i.e., $m_z(v) \leftarrow m_z(v) + 1$), the function $f^*_{\text{old}}$ and its misclassification parameter $q^*_{\text{old}}$ may have to be updated. Since the new error minimizing function is likely to be close to the old function, it may be inefficient to solve the entire problem over again.

Simply stated, the incremental problem consists of finding $f^*_{\text{new}}$ and consequently $q^*_{\text{new}}$ when $m_z(v) \leftarrow m_z(v) + 1$. If the new classification is consistent with the old function (i.e., $f^*_{\text{old}}(v) = z$), then the old function remains error minimizing (i.e., $f^*_{\text{old}} = f^*_{\text{new}}$). Therefore, the number of errors remains the same and the misclassification estimate is reduced to $q^*_{\text{new}} = e(f^*_{\text{old}})/(m_{\text{old}} + 1)$. Note that this case is the most likely one since it occurs with an estimated probability of $1 - q^*_{\text{old}} \geq 1/2$. If, on the other hand, the new classification is inconsistent with the old function (i.e., $f^*_{\text{old}}(v) = 1 - z$), then the old function may or may not remain error minimizing. The only case in which the old function does not remain error minimizing is when there is an alternative error minimizing function $f^*_a$ on the old data for which $f^*_a(v) = z$. In this case $f^*_a$ is error minimizing for the new data.

The number of possible error minimizing functions may be exponential in the size of the set $V$, and therefore storing all of them may not be an efficient solution to this problem. To avoid this computational burden an incremental algorithm, such as the one described in [Torvik and Triantaphyllou, 2004], can be used.

### 10.3.5 Selection Criteria for Problem 1

When computing the optimal solutions, many different and complex posets are encountered. The optimal vectors of these posets seemed to display two general properties [Torvik and Triantaphyllou, 2002]. First, the optimal vectors tend to be in the *vertical middle*. More specifically, all posets observed in the inference process when $n$ is equal to 4 or less have at least one optimal vector in the most populous (i.e., in the middle) layer. This observation alone is not sufficient to pinpoint an optimal vector. The second property observed is that the optimal vectors also tend to be *horizontal end points*.

Now consider creating a selection criterion based on the ideas of the vertical middle and the horizontal end points. Suppose a subset of unclassified vectors, $V = \{v^1, v^2, \ldots, v^p\}$, is given. Let $K_1(v^i)$ and $K_0(v^i)$ be the numbers of vectors that are concurrently classified when $f(v^i)$ equals 1 and 0, respectively. Invariably selecting a vector $v$ with the minimum $|K_1(v) - K_0(v)|$ value guarantees the minimum average number of queries for inference problems with $n$ strictly less than 5 [Torvik and Triantaphyllou, 2002].

Unfortunately, this selection criterion is not optimal for all the posets generated for $n$ equal to 4. It is only optimal for the subset of posets encountered when using the criterion $\min |K_1 - K_0|$. Another drawback is that it is not optimal for the inference problem when $n$ is equal to 5. However, the criterion is probably close to optimal since the larger posets eventually decompose into smaller posets.

It is important to note that what may look like intuitive criteria (without the consultation of optimal solutions) may lead to poor performance and ambiguous choices. For example, it may seem reasonable to attempt to classify as many vectors as possible for each query (i.e., employ a greedy approach). The two criteria $\max(K_1(v) + K_0(v))$ and $\max(K_1(v) \times K_0(v))$ are consistent with this philosophy (see, for instance, [Judson, 1999; 2006]). However, they are extremely counterproductive to minimizing the average query complexity and should be avoided. As an example, consider the set of vectors in $\{0, 1\}^4$. The criterion $\max(K_1(v) + K_0(v))$ selects either the (0000) or the (1111) vector, which happens to **maximize** the average number of queries. The criterion $\max(K_1(v) \times K_0(v))$ ties the entire set of vectors, and therefore the choice of a vector is ambiguous.

There is a logical explanation for why these two selection criteria are counterproductive. Vectors that are able to concurrently classify more vectors are also more likely to be classified by others. Following this line of thought, the selection criterion $\min(K_1(v) + K_0(v))$ seems reasonable. This criterion is similar to $\min |K_1(v) - K_0(v)|$, but it does not satisfy the same optimality conditions for the inference problem when $n$ is equal to 4.

### 10.3.6 Selection Criteria for Problems 2.1, 2.2, and 2.3

The minimum average number of queries for the unrestricted problem, denoted as $Q_3(n)$, is equal to that of the single function case in one dimension higher, that is, it is equal $Q(n+1)$. Thus, $Q_3(n) = Q(n+1)$. To see this connection consider a pair of nested monotone Boolean functions $f_1$ and $f_2$ defined on $\{0, 1\}^n$. The query domain for the nested case can be viewed as the product $\{0, 1\}^n \times \{f_2, f_1\}$. Each of the vertices in the resulting poset $(\{0, 1\}^{n+1}, \preceq)$ may take on function values of 0 or 1, where the monotonicity property is preserved. In other words, a pair of nested monotone Boolean functions defined on $\{0, 1\}^n$ are equivalent to a single monotone Boolean function defined on $\{0, 1\}^{n+1}$.

The selection criterion $\min|K_1(v) - K_0(v)|$ was shown to be very efficient in minimizing the average number of queries in Problem 1. Therefore, it will be used for the three nested problems with a slight modification. The query domain for the nested case is made up of the set of vectors $\{0, 1\}^n \times \{f_2, f_1\}$. For a vertex labeled $(vf_i)$, let $K_z(v, f_i)$ be the number of vertices that are concurrently classified when the value of $f_i(v)$ is queried and the answer is $f_i(v) = z$, for $z = 0$ or 1. When the access to the oracles is unrestricted (i.e., we have Problem 2.3), vertices are selected based on the criterion $\min|K_1(v, f_i) - K_0(v, f_i)|$. This criterion is equivalent to the criterion $\min|K_1(v) - K_0(v)|$ for the single function case. The only change is in the notation since the oracle that is to provide the answer has to be identified for Problem 2.3.

For sequential oracles (i.e., Problem 2.1), queries of the form $f_2(v)$ are infeasible until all of the queries of the form $f_1(v)$ are classified. In this case, the criterion used during the first phase is $\min|K_1(v, f_1) - K_0(v, f_1)|$, after which the criterion $\min|K_1(v, f_2) - K_0(v, f_2)|$ is used.

For the three-valued oracle (i.e., Problem 2.2), the queries are of the form $(f_1(v), f_2(v))$ and are selected by using the criterion $\min|K_{11}(v) - K_{00}(v)|$. Here the value of the function $K_{zz}(v)$ equals the number of vertices concurrently classified when vertex $v$ is queried and the result of the query is $f_1(v) = f_2(v) = z$, for $z = 0$ or 1. Once there are no pairs of vertices of the form $(f_1(v), f_2(v))$ left unclassified, the criterion $\min|K_1(v, f_i) - K_0(v, f_i)|$ is used for the remaining of the query selections.

### 10.3.7 Selection Criterion for Problem 3

The status of the inference process will be considered to be in one of three stages. Stage 1 starts with the first question and lasts until a deterministic monotone Boolean function is obtained. During Stage 1 only vectors that may take on either 0 or 1 value are queried. As a result, no (identifiable) errors are observed in Stage 1, and thus the monotone Boolean function inferred during Stage 1 is deterministic. This function, however, may or may not be the correct one. In fact, the probability that it is the correct function is equal to the probability that no misclassifications were made: $(1-q)^m$, where $m$ is the number of questions used during Stage 1 and $q$ is the true misclassification probability. This probability decreases rapidly with $m$,

**Table 10.2.** A Sample Data Set for Problem 3.

| $V$ | $m_1(v)$ | $m_0(v)$ | $m_1(v) - m_0(v)$ |
|---|---|---|---|
| 111 | 0 | 1 | −1 |
| 110 | 3 | 5 | −2 |
| 101 | 4 | 1 | 3 |
| 11 | 3 | 1 | 2 |
| 100 | 4 | 5 | −1 |
| 10 | 2 | 0 | 2 |
| 1 | 3 | 3 | 0 |
| 0 | 1 | 0 | 1 |

regardless of the value of $q$. Therefore, the queries performed after Stage 1 will benefit greatly from a reduction in the number of Stage 1 queries. Please note that since no inconsistencies have been observed, there is no way to properly estimate $q$ at this point.

After a deterministic monotone Boolean function is obtained in Stage 1, the inference process enters Stage 2. At this point it is unclear as to how to select queries for Stage 2, so a random selection procedure will be used for this stage. After the first error occurs in Stage 2, the inference process enters Stage 3, in which it will remain until termination. Stage 3 is the focus of this section, because it is the only stage in which the likelihood ratio can be properly evaluated and $q$ can be estimated based on the observed vectors.

Recall that the likelihood function is given by

$$L(f) = q^{e(f)}(1-q)^{m-e(f)},$$

and the likelihood ratio is given by

$$\lambda(f^*) = \frac{L(f^*)}{\sum_{f \in F(V)} L(f)}.$$

As an example of the likelihood ratio computations consider the sample data given in Table 10.2. The likelihood values for all the possible monotone Boolean functions are given in Table 10.3. The function $f^* = v_1v_3 \vee v_2v_3$ produces 16 errors. Its associated estimated misclassification probability $q^*$ is $16/36 = 4/9$, since the total number of observations is $m = 36$. Therefore, the likelihood value of this function $L(f^*)$ is $(4/9)^{16}(1-4/9)^{36-16} = 1.818 \times 10^{-11}$. Notice how small this value is after only 36 observations. Adding up the likelihood values the monotone Boolean functions yields $(13 \times 1.455 + 2 \times 1.536 + 5 \times 1.818) \times 10^{-11} = 3.107 \times 10^{-10}$. Then the maximum likelihood ratio is computed as follows: $\lambda(f^*) = 1.818 \times 10^{-11}/3.107 \times 10^{-10} = 0.0585$.

Now let us return to the vector selection (or guided inference/learning) problem. As shown above, the probability that the correct function is inferred during Stage 1 decreases rapidly with the number of queries used during that stage. Therefore, the

**Table 10.3.** Example Likelihood Values for All Functions in $M_3$.

| *F* | *e(f)* | *q(f)* | *L(f)* | *λ(f)* |
|---|---|---|---|---|
| $F$ | 20 | 1/2 | $1.455 \times 10^{-11}$ | 0.0468 |
| $v_1v_2v_3$ | 21 | 1/2 | $1.455 \times 10^{-11}$ | 0.0468 |
| $v_1v_2$ | 23 | 1/2 | $1.455 \times 10^{-11}$ | 0.0468 |
| $v_1v_3$ | 18 | 1/2 | $1.455 \times 10^{-11}$ | 0.0468 |
| $v_1v_2 \vee v_1v_3$ | 20 | 1/2 | $1.455 \times 10^{-11}$ | 0.0468 |
| $v_1$ | 21 | 1/2 | $1.455 \times 10^{-11}$ | 0.0468 |
| $v_2v_3$ | 19 | 1/2 | $1.455 \times 10^{-11}$ | 0.0468 |
| $v_1 \vee v_2v_3$ | 19 | 1/2 | $1.455 \times 10^{-11}$ | 0.0468 |
| $v_1v_3 \vee v_2v_3$ | 16 | 36,989 | $1.818 \times 10^{-11}$ | 0.0585 |
| $v_1v_2 \vee v_1v_3 \vee v_2v_3$ | 18 | 1/2 | $1.455 \times 10^{-11}$ | 0.0468 |
| $v_1v_2 \vee v_2v_3$ | 21 | 1/2 | $1.455 \times 10^{-11}$ | 0.0468 |
| $v_2$ | 19 | 1/2 | $1.455 \times 10^{-11}$ | 0.0468 |
| $v_1 \vee v_2$ | 17 | 17/36 | $1.536 \times 10^{-11}$ | 0.0495 |
| $v_2 \vee v_1v_3$ | 16 | 36,989 | $1.818 \times 10^{-11}$ | 0.0585 |
| $v_3$ | 16 | 36,989 | $1.818 \times 10^{-11}$ | 0.0585 |
| $v_2 \vee v_3$ | 16 | 36,989 | $1.818 \times 10^{-11}$ | 0.0585 |
| $v_1 \vee v_2 \vee v_3$ | 17 | 17/36 | $1.536 \times 10^{-11}$ | 0.0495 |
| $v_1 \vee v_3$ | 19 | 1/2 | $1.455 \times 10^{-11}$ | 0.0468 |
| $v_3 \vee v_1v_2$ | 18 | 1/2 | $1.455 \times 10^{-11}$ | 0.0468 |
| $T$ | 16 | 36,989 | $1.818 \times 10^{-11}$ | 0.0585 |

selection criterion $\min |K_0(v) - K_1(v)|$ will be used as a standard for Stage 1, when comparing different approaches for the following Stage 3. This avoids bias in the sense that all Stage 3 approaches will benefit from using $\min |K_0(v) - K_1(v)|$ during Stage 1.

One important property of the selection criterion for Stage 3 is that the maximum likelihood ratio converges to 1. It is possible to define selection criteria that do not converge. If, for instance, the same vector is invariably selected, the estimated value of $q$ will converge to its true value. In this case, the likelihood values may remain equal for several monotone Boolean functions and hence the maximum likelihood ratio will never converge to 1.

Intuition may lead to an inefficient selection criterion. For example, let $E_z(v)$ be defined by the number of errors associated with assigning the function value $f(v)$ to $z$, as follows:

$$E_0(v) = \sum_{w \le v} m_1(w) - m_0(w) \quad \text{and} \quad E_1(v) = \sum_{w \le v} m_0(w) - m_1(w).$$

Then, consider defining the vector $v$ which "contributes the most errors" by $\max(E_0(v) + E_1(v))$. This vector selection criterion may lead to the same vector

**Table 10.4.** Updated Likelihood Ratios for $m_z(001) = m_z(001) + 1$.

| $f$ | $\lambda(f)$ | $e_1(001, f)$ | $\lambda_1(001, f)$ | $e_0(001, f)$ | $\lambda_0(001, f)$ |
|---|---|---|---|---|---|
| $F$ | 0.0468 | 21 | 0.0462 | 20 | 0.0468 |
| $v_1v_2v_3$ | 0.0468 | 22 | 0.0462 | 21 | 0.0468 |
| $v_1v_2$ | 0.0468 | 24 | 0.0462 | 23 | 0.0468 |
| $v_1v_3$ | 0.0468 | 19 | 0.0462 | 18 | 0.0474 |
| $v_1v_2 \vee v_1v_3$ | 0.0468 | 21 | 0.0462 | 20 | 0.0468 |
| $v_1$ | 0.0468 | 22 | 0.0462 | 21 | 0.0468 |
| $v_2v_3$ | 0.0468 | 20 | 0.0462 | 19 | 0.0468 |
| $v_1 \vee v_2v_3$ | 0.0468 | 20 | 0.0462 | 19 | 0.0468 |
| $v_1v_3 \vee v_2v_3$ | 0.0585 | 17 | 0.0522 | 16 | 0.0657 |
| $v_1v_2 \vee v_1v_3 \vee v_2v_3$ | 0.0468 | 19 | 0.0462 | 18 | 0.0474 |
| $v_1v_2 \vee v_2v_3$ | 0.0468 | 22 | 0.0462 | 21 | 0.0468 |
| $v_2$ | 0.0468 | 20 | 0.0462 | 19 | 0.0468 |
| $v_1 \vee v_2$ | 0.0495 | 18 | 0.0469 | 17 | 0.0529 |
| $v_2 \vee v_1v_3$ | 0.0585 | 17 | 0.0522 | 16 | 0.0657 |
| $v_3$ | 0.0585 | 16 | 0.0649 | 17 | 0.0529 |
| $v_2 \vee v_3$ | 0.0585 | 16 | 0.0649 | 17 | 0.0529 |
| $v_1 \vee v_2 \vee v_3$ | 0.0495 | 17 | 0.0522 | 18 | 0.0474 |
| $v_1 \vee v_3$ | 0.0468 | 19 | 0.0462 | 20 | 0.0468 |
| $v_3 \vee v_1v_2$ | 0.0468 | 18 | 0.0469 | 19 | 0.0468 |
| $T$ | 0.0585 | 16 | 0.0649 | 17 | 0.0529 |

being invariably queried and hence it might suffer from convergence problems, as will be demonstrated empirically in Section 10.4.

The likelihood framework seems to form a great basis for defining a Stage 3 vector selection criterion. Since the goal is to make the likelihood ratio converge to 1 as fast as possible, a reasonable approach would be to select the vector that maximizes the expected maximum likelihood ratio (denoted as $\Delta\lambda(v)$) at each inference step. To do this, the expected maximum likelihood ratio $\Delta\lambda(v) = p(v)\lambda_1(v) + (1 - p(v))\lambda_0(v)$ has to be estimated for each vector $v$. Here $\lambda_z(v)$ denotes the resulting maximum likelihood ratio when $f(v) = z$ is observed. Recall that $p(v)$ is the probability of observing $f(v) = 1$. That is, it can be estimated by $p^*(v) = q^*(1 - f^*(v)) + (1 - q^*)f^*(v)$.

As an example consider observing the vector (001). Table 10.4 gives the updated likelihood ratios for each monotone Boolean function when $m_z(001) = m_z(001)+1$, for $z = 0$ or 1. For a monotone Boolean function $f$, and a classification $z$, $e_z(001, f)$ and $\lambda_z(001, f)$ denote here the updated number of errors and the likelihood ratio, respectively. The updated maximum likelihood ratios are $\lambda_1(001) = \lambda_1(001, T) = 0.0649$ and $\lambda_0(001) = \lambda_0(001, v_1v_3 \vee v_2v_3) = 0.0657$. Since the optimal function assigns the vector (001) to 0 (i.e., $f^*(001) = 0$), the estimated probability of observing $f(001) = 1$ is given by $p^*(001) = q^* = 4/9$. Therefore, the expected maximum likelihood ratio when querying vector 001 is given by $\Delta\lambda(001) = p^*(001)\lambda_1(001) + (1 - p^*(001))\lambda_0(001) = 4/9 \times 0.0649 + 5/9 \times 0.0657 = 0.0653$.

Similar computations for the other vectors yield $\Delta\lambda(000) = 0.0651$, $\Delta\lambda(010) = 0.0654$, $\Delta\lambda(011) = 0.0592$, $\Delta\lambda(100) = 0.0652$, $\Delta\lambda(101) = 0.0592$, $\Delta\lambda(110) = 0.0654$, and finally $\Delta\lambda(111) = 0.0592$. The vectors with the largest expected likelihood ratio value are (010) and (110). Since no further improvement of the selection criterion is obvious, ties are broken arbitrarily.

The simulations in Section 10.4 reveal the efficiency of the selection criterion $\max \Delta\lambda(v)$ in terms of the query complexity. In terms of computational complexity it may take an exponential time (in the size of $V$) to compute $\max \Delta\lambda(v)$. Since the computational time for incrementally finding the inferred function is of $O(V^2)$ complexity, it would be nice to find a selection criterion that does not take more time than this and still makes the likelihood converge to 1 at a faster rate than randomly selecting vectors.

One such possibility may be based on the inferred border vectors. For the sake of argument suppose that the underlying monotone Boolean function $f$ to be inferred is known. Then randomly selecting vectors from its corresponding border vectors will make the maximum likelihood ratio converge to 1. As the number of queries $m$ goes to infinity, the ratios $m_0(v)/(m_0(v) + m_1(v)) \forall v \in \mathrm{LU}(f)$ and $m_1(w)/(m_0(w) + m_1(w)) \forall w \in \mathrm{UZ}(f)$ all converge to $q$. Recall that $\mathrm{LU}(f)$ and $\mathrm{UZ}(f)$ denote the lower units and upper zero vectors of a monotone Boolean function $f$, respectively (see also Section 10.2.3). The number of errors performed by any other monotone Boolean function $g$ is at least $x = \min\{\min\{m_1(v) - m_0(v), v \in \mathrm{LU}(f)\}, \min\{m_0(w) - m_1(w), w \in \mathrm{UZ}(f)\}\}$ greater than the number of errors performed by function $f$. Furthermore, $x \approx qm - (1 - q)m = m(2q - 1)$ for large $m$. That is, the number of additional errors increases at least linearly with $m$. Then, as $m$ goes to infinity, so does the number of additional errors performed by each of the other monotone Boolean functions. In other words, the relative likelihoods $L(f)/L(g) > (q/(1 - q))^x$ converge to 0 as $m$ goes to infinity. Since the number of other monotone Boolean functions is a finite number that does not depend on $m$, the likelihood ratio $\lambda(f) = L(f)/(L(f) + \Sigma L(g))$ converges to 1 as $m$ goes to infinity.

Focusing the queries at the border vectors of the underlying function probably allows this convergence to occur at a faster rate than randomly selecting from all the vectors. In situations where the underlying function is unknown, it may be that focusing the queries on the border vectors of the inferred function (i.e., $v \in \mathrm{LU}(f^*) \cup \mathrm{UZ}(f^*)$) is better than completely random selection. In the long run, an inferred border vector will not prevail if it is not an underlying border vector. Since the misclassification rate is less than 2, the rate at which the incorrectly classified inferred border vectors become correctly classified is greater than the rate at which correctly classified inferred border vectors become incorrectly classified. Therefore, in the long run all the classifications become correct when the queries are selected from the set of border vectors of the inferred function.

Notice that this convergence holds even if the misclassification probability is different for each vector, as long as they are all less than 2. Another added benefit is that finding the border vectors is easy, since they are readily available from the inferred function $f^*$. In fact, a simple modification of the incremental maximum flow algorithm can store each of these vectors as they are found. For each monotone Boolean function there are at most $O(V)$ border vectors in a set of vectors $V$. During the inference process the inferred function may take on any of these monotone Boolean functions. Therefore, randomly selecting one of the border vectors takes $O(V)$ time.

## 10.4 Experimental Results

### 10.4.1 Experimental Results for Problem 1

The preexisting inference algorithms described in Section 10.2.4 do not specify which vector to select when there are ties. In particular, the Sokolov and Hansel algorithms may have to choose between two vectors that make up the middle of a particular chain. Furthermore, the subroutine FIND-BORDER needs to be fed unclassified vectors, of which there may be many. Even the selection criterion $\min |K_1 - K_0|$ may result in ties. For the purpose of comparing the algorithms on the same ground and without introducing another aspect of randomness, ties were broken by selecting the first vector in the list of tied vectors.

The results in Figure 10.5 are based on an exhaustive analysis (i.e., all the monotone functions were generated) for $n$ up to and including 5. Random samples of 2,000 functions were generated for $n = 6, 7$, and 8; for $n = 9, 10$, and 11 they were composed of 200 functions. These functions were generated using the algorithm described in [Torvik and Triantaphyllou, 2002].

The Horvitz–Thompson [1952] estimator was used to compute the averages for $n$ greater than 5. The average number of queries was normalized by the maximum possible number of queries $2^n$ so that the magnitudes of the averages in Figure 10.5 were not overshadowed by the large values obtained for $n$ equal to 11. As a consequence, two algorithms that result in parallel curves in such a plot, have an exponential (in terms of $n$) difference in the average number of queries. Also, the gap between the curves in Figure 10.5 and the horizontal line *Average Number of Queries*/$2^n = 1$ (not shown in the figure) can be thought of as the benefit of the monotone assumption. This is due to the fact that $2^n$ is the number of required queries when the underlying function is not necessarily monotone.

The curve titled "Teacher" represents the lower bound on the number of queries for every single function. Therefore, it is expected that a few extra queries are required on the average. Since the heuristic based on the selection criterion $\min |K_1 - K_0|$ achieves the minimum average number of queries for $n$ up to 4, it can be thought of as a lower bound on the average, and its gap between "Teacher" quantifies the benefits of knowing the actual function beforehand.

Figure 10.5 paints a clear picture of how the preexisting inference algorithms fare against each other. Hansel's algorithm was the best performer by far, Sokolov's came in second, and an algorithm using the subroutine FIND-BORDER (which is also used by Gainanov [1984]; Valiant [1984]; Makino and Ibaraki [1995]; Boros, *et al.*, [1997]) was a distant third. In fact, since the curve differences between Hansel

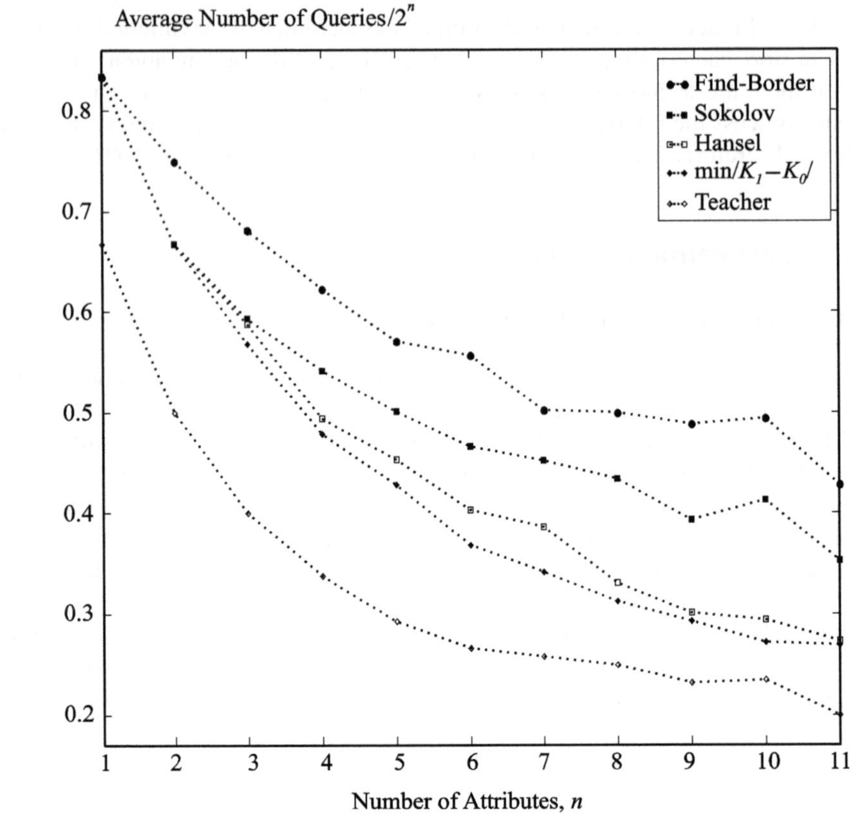


**Figure 10.5.** The Average Query Complexities for Problem 1.

and Sokolov, and Sokolov and the subroutine FIND-BORDER implementation, seem to increase with $n$, the corresponding difference in the average number of queries increases at a rate greater than exponentially with $n$.

The difference between the curves for Hansel and "Teacher" decreases as $n$ increases. The algorithm based on the criterion $\min|K_1 - K_0|$ has a curve that is almost parallel to Hansel's curve, indicating that this selection criterion performs about 2% better than Hansel's algorithm. This decrease is especially clear in Figure 10.5 for $n$ up to and including 8. For larger values of $n$, the high variance of our estimates makes it hard to distinguish the two curves, but the overall decreasing trends remain intact. It might seem that a 2% decrease is insignificant, but writing it as $2^n \times 0.02$ shows its real magnitude.

Another nice characteristic of this selection criterion is that it is the most consistent of all the algorithms. For example, it performs between 10 and 18 queries for 99.6% of the monotone Boolean functions in $M_5$. In contrast, the algorithm based on the subroutine FIND-BORDER is the least consistent with between 8 and 25 queries for 99.6% of the same monotone Boolean functions.

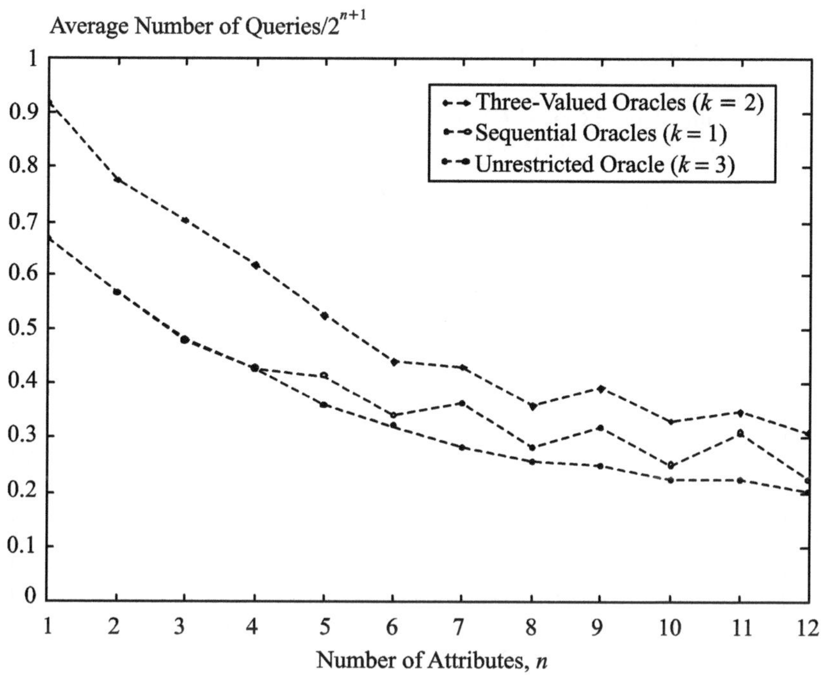


**Figure 10.6.** The Average Query Complexities for Problem 2.

### 10.4.2 Experimental Results for Problem 2

The results in Figures 10.6, 10.7, and 10.8 are based on an exhaustive analysis (i.e., all the monotone functions were generated) for $n$ up to and including 4. For $n = 4, 5, \ldots, 12$ random samples of functions were generated and the Horvitz–Thompson [1952] estimator was used to compute the averages for $n$ greater than 4. The number of pairs of nested monotone Boolean functions generated were 2,000 for $n = 5, 6, 7$, and 200 for $n = 8, 9, 10$, and 100 for $n = 11$ and 12.

Figure 10.6 shows the average number of queries for Problem 2 when using the selection criteria. The lower curve corresponds to the unrestricted case (Problem 2.3), which achieves the fewest number of queries on the average. The sequential case (Problem 2.1), corresponding to the middle curve, is not as efficient as the unrestricted oracles in general, although they are very close for $n = 1, 2, 3$, and 4. The least efficient of the three types of oracles is the three-valued (Problem 2.2) corresponding to the upper curve.

The gap between the curves in Figure 10.6 and the horizontal line *Average Number of Queries/*$2^{n+1} = 1$ (the uppermost line of the box around the curves) can be thought of as the benefit of the monotone and nestedness assumptions put together. This is due to the fact that $2^{n+1}$ is the number of required queries when the underlying pair of functions are neither nested nor monotone. For example, when $n = 12$ in the unrestricted problem ($k = 3$) the average number of queries is reduced

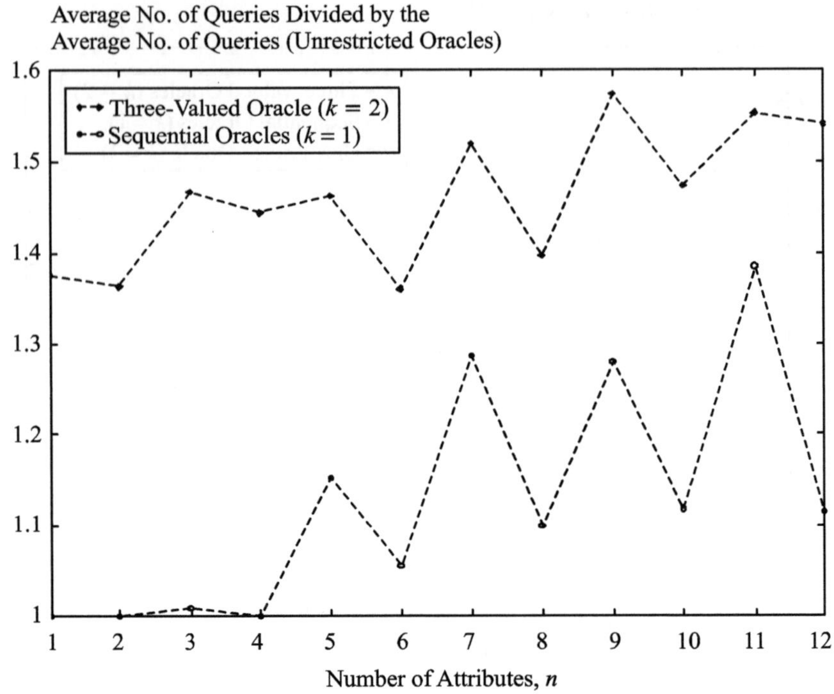


**Figure 10.7.** Increase in Query Complexities Due to Restricted Access to the Oracles.

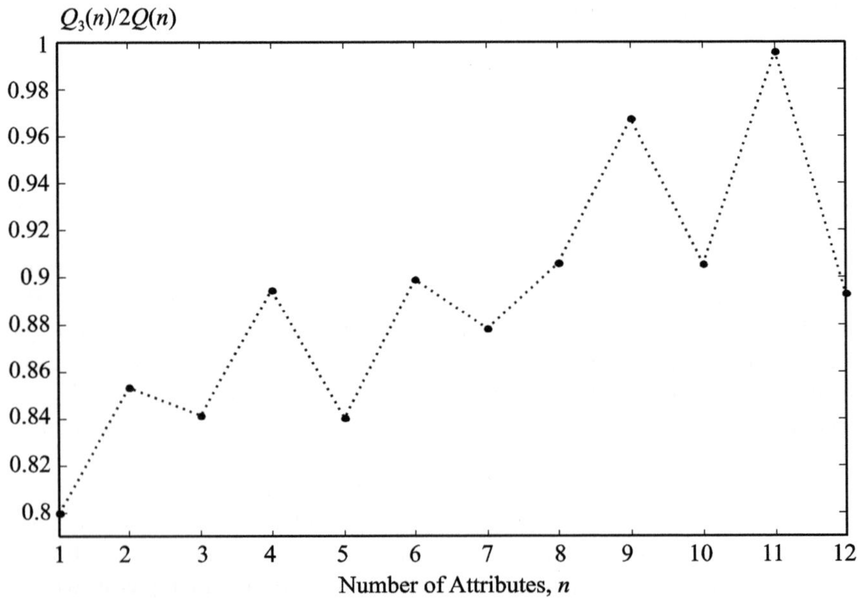


**Figure 10.8.** Reduction in Query Complexity Due to the Nestedness Assumption.

to about 20% of the maximum number of queries $2^{13} = 8{,}192$ due to the monotone and nestedness assumptions.

Figure 10.7 quantifies the increase in the average number of queries due to the two restrictions on the oracles for $n = 1, 2, 3, \ldots, 12$. As mentioned earlier, the sequential oracles are practically unrestrictive for $n = 1, 2, 3$, and 4. For $n$ greater than 4, the increase in average query complexity oscillates between 12% and 33% due to odd and even $n$, being much greater for odd $n$. In contrast, the three-valued oracle is much more restrictive across all the observed $n$, where the increase in the average number of queries oscillates between 35% and 55%, again due to odd and even $n$, being greater for odd $n$. In summary, the increases in the average number of queries for the sequential and three-valued cases are dramatic. This is probably due to the fact that the average number of queries increases exponentially with the number of attributes.

If the nested property of the two functions defined on $\{0, 1\}^n$ is ignored, the minimum total number of questions is, on the average, equal to $2Q(n)$. The benefit from the nestedness assumption for Problem 2 is quantified by the ratio of $Q_3(n)/2Q(n)$ which is given in Figure 10.8 for $n = 1, 2, \ldots, 12$. Therefore, the curves given in Figure 10.8 show the reduction in the average number of queries due to the nestedness assumption. This reduction decreases with the number of attributes. It starts out at 20% for $n = 1$, and oscillates between 1% and 10% for $n$ greater than 7.

### 10.4.3 Experimental Results for Problem 3

For the purpose of comparing the efficiency of the different selection criteria for Stage 3 on the same basis, ties resulting from the selection criteria (min $|K_0(v) - K_1(v)|$ for Stage 1, and $\max(E_0(v) + E_1(v))$, $\max \Delta\lambda(v)$, and $v \in \mathrm{LU}(f^*) \cup \mathrm{UZ}(f^*)$; the set of border vectors for Stage 3) were broken randomly. The four different inference processes using $\max \Delta\lambda(v)$, $v \in \mathrm{LU}(f^*) \cup \mathrm{UZ}(f^*)$, $\max(E_0(v) + E_1(v))$, or random selection for Stage 3 were simulated on the set of vertices $\{0, 1\}^n$. For all three Stage 3 selection criteria, the selection criterion min $|K_0(v) - K_1(v)|$ was used for Stage 1 and random selection was used for Stage 2. The resulting simulations were repeated 100, 50, 25, and 10 times for each of 6 representative functions of $M_n$, with misclassification probabilities equal to 0.1, 0.2, 0.3, and 0.4, for $n = 2, 3, 4$, and 5, respectively.

The representative functions are given in Table 10.5. For $n = 4$ and 5, these representative functions were randomly generated from a uniform distribution with individual probabilities of $1/\Psi(n) = 1/168$ and $1/7581$, respectively. For $n = 3$, the representative functions consist of nonsimilar functions (one from each similar subset of $M_3$). These functions represent all the functions in $M_3$, since the average case behavior is the same for a pair of similar monotone Boolean functions.

To compute the overall average for a given $q$, the individual curves were weighted by the number of similar functions the representative function has (including itself) in $M_3$. The individual curves for the monotone Boolean functions $F$, $v_1v_2v_3$, $v_1v_2$, $v_1v_2 \vee v_1v_3$, $v_1$, and $v_1v_2 \vee v_1v_3 \vee v_2v_3$, were therefore weighted by 2, 2, 6, 6, 3, and 1, respectively. For $n = 2, 4$, and 5, the overall averages were computed without

**Table 10.5.** The Representative Functions Used in the Simulations of Problem 3.

| $n = 2$ | $n = 3$ | $n = 4$ | $n = 5$ |
|---|---|---|---|
| $F$ | $F$ | $v_1v_2 \vee v_2v_4 \vee v_1v_3v_4$ | $v_1v_4 \vee v_1v_5 \vee v_2v_4 \vee v_2v_5$ |
| $v_1v_2$ | $v_1v_2v_3$ | $v_1v_2 \vee v_1v_3 \vee v_2v_3 \vee v_2v_4 \vee v_3v_4$ | $v_1v_3 \vee v_2v_3 \vee v_2v_4 \vee v_1v_2v_5$ |
| $v_1$ | $v_1v_2$ | $v_2v_3 \vee v_2v_4$ | $v_2 \vee v_1v_3v_4 \vee v_1v_4v_5$ |
| $v_2$ | $v_1v_2 \vee v_1v_3$ | $v_1v_2v_3 \vee v_1v_3v_4 \vee v_2v_3v_4$ | $v_1v_3 \vee v_2v_4 \vee v_3v_5 \vee v_1v_4v_5$ |
| $v_1 \vee v_2$ | $v_1$ | $v_1v_2 \vee v_2v_4 \vee v_3v_4$ | $v_2v_4 \vee v_2v_5 \vee v_3v_5 \vee v_4v_5$ |
| $T$ | $v_1v_2 \vee v_1v_3 \vee v_2v_3$ | $v_3 \vee v_1v_2 \vee v_1v_4$ | $v_2v_5 \vee v_1v_2v_3 \vee v_1v_3v_4 \vee v_1v_4v_5 \vee v_3v_4v_5$ |

weights. The overall averages for $n = 2$ and 3 benefit from a reduced variance, since no additional errors are added due to the sampling of functions as done for $n = 4$ and 5.

Figure 10.9 shows the resulting average maximum likelihood curves for the inference problem defined on $n = 2, 3, 4,$ and 5, and $q = 0.1, 0.2, 0.3,$ and 0.4. Each curve is the average of 600, 300, 150, and 60 simulated inference processes observed for $n = 2, 3, 4,$ and 5, respectively. In each plot, the horizontal axis corresponds to the number of Stage 3 queries, and the vertical axis corresponds to the maximum likelihood ratio. The curves are shown for the range of Stage 3 queries where the curve corresponding to the selection criterion $\max \Delta\lambda(v)$ has a maximum likelihood ratio that is less than 0.99.

Not only do the curves corresponding to the guided selection criteria $\max \Delta\lambda(v)$ and $v \in \mathrm{LU}(f^*) \cup \mathrm{UZ}(f^*)$ converge to 1 but they do so at a much faster rate than the curves corresponding to unguided random selection. In fact, the random selection achieves a maximum likelihood ratio of only about 0.7 after the same number of queries as the criterion $\max \Delta\lambda(v)$ uses to reach 0.99, and the criterion $v \in \mathrm{LU}(f^*) \cup \mathrm{UZ}(f^*)$ uses to reach about 0.9, for $n = 4$.

The difference between the curves for unguided selection and these two guided selections grows with the misclassification probability $q$ and with the dimension $n$. That is, the benefits from actively selecting vectors over passively receiving observations are greater when the values of $q$ and $n$ are large. In other words, the higher the misclassification probability and the dimension of the problem are, the greater become the benefits of guiding the inference process.

The curves associated with the criterion $\max(E_0(v) + E_1(v))$ seem to converge to a value significantly less than 1. For example, when $n = 3$ and $q = 0.3$, the maximum likelihood ratio converges to about 0.4, and this value decreases as the values of $q$ and $n$ increase. Therefore, the larger error rate and the vector domain are, the more important it becomes to define an appropriate vector selection criterion.

Table 10.6 gives the average number of queries needed by the selection criterion $\max \Delta\lambda(v)$ to converge to a maximum likelihood ratio of 0.99 for $n = 2, 3, 4,$ and 5, and for $q = 0.1, 0.2, 0.3,$ and 0.4. For a given $n$, these numbers increase dramatically as $q$ increases. In fact, there seems to be more than a doubling in the numbers

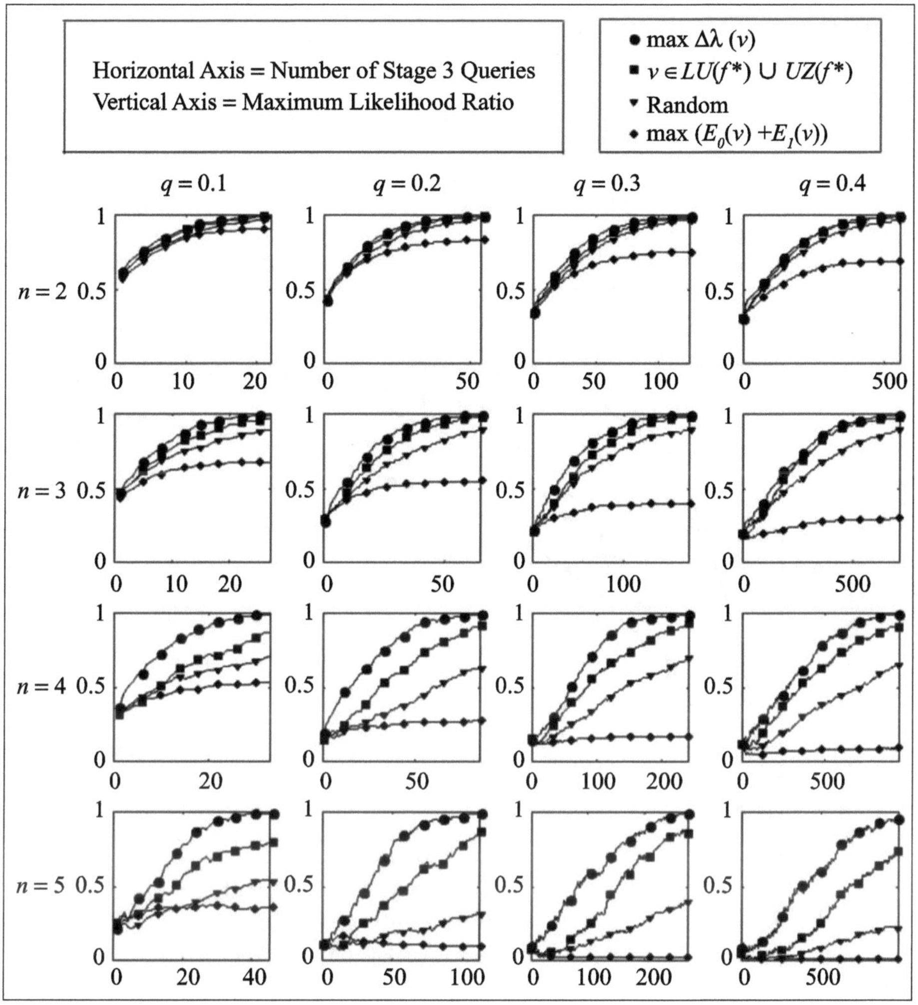


**Figure 10.9.** Average Case Behavior of Various Selection Criteria for Problem 3.

for fixed increments of $q$. For a given $q$, these numbers do not increase in such a dramatic fashion when $n$ increases. However, they do increase faster than linearly with $n$.

Randomly selecting the inferred border vectors (i.e., $v \in \mathrm{LU}(f^*) \cup Z(f^*)$) makes the maximum likelihood ratio converge to 1, as long as the misclassification probabilities are all less than $1/2$. That is, the misclassification probabilities do not necessarily have to be fixed. To see whether this holds for the selection criterion max $\Delta\lambda(v)$, consider an unrestricted model where the misclassification probability $q(v)$ is a random variable distributed uniformly on the interval $[q(1-\delta), q(1+\delta)]$, where $\delta \in [0, 1]$, for each vector $v \in \{0, 1\}^n$.

**Table 10.6.** The Average Number of Stage 3 Queries Used by the Selection Criterion $\max \Delta\lambda(v)$ to Reach $\lambda > 0.99$ in Problem 3 Defined on $\{0, 1\}^n$ with Fixed Misclassification Probability $q$.

| | $q = 0.1$ | $q = 0.2$ | $q = 0.3$ | $q = 0.4$ |
|---|---|---|---|---|
| $n = 2$ | 22 | 54 | 125 | 560 |
| $n = 3$ | 27 | 65 | 170 | 710 |
| $n = 4$ | 33 | 85 | 241 | 951 |
| $n = 5$ | 45 | 111 | 277 | 1,167 |

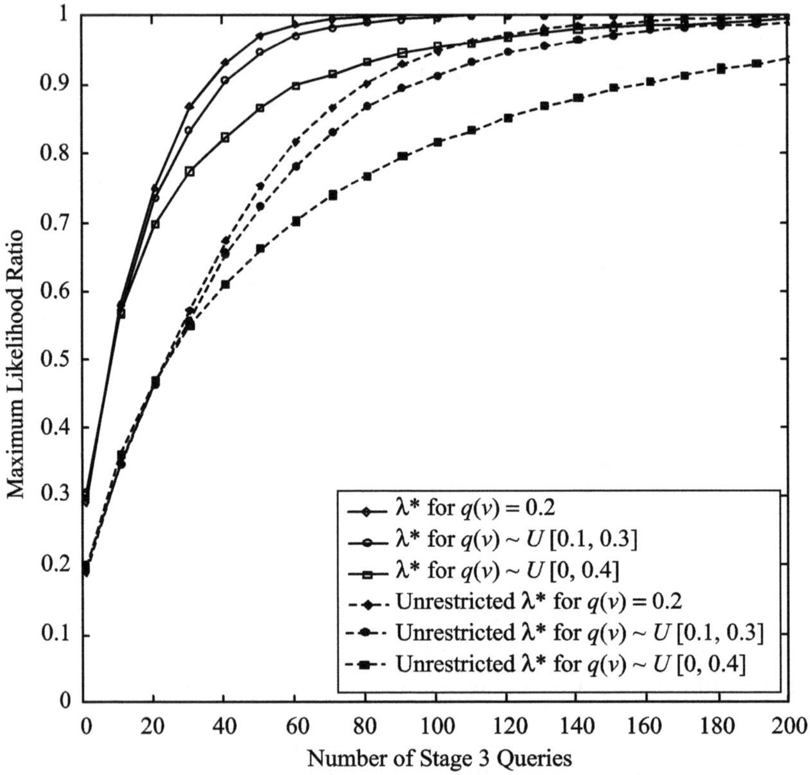


**Figure 10.10.** The Restricted and Regular Maximum Likelihood Ratios Simulated with Expected $q = 0.2$ and $n = 3$.

The case when $\delta = 0$ corresponds to the fixed misclassification probability model, that is, when $q(v)$ is equal to $q$ for all vectors $v \in \{0, 1\}^n$. The range of values for $q(v)$ increases with $\delta$, but the expected value of $q(v)$ is always equal to $q$. Therefore, the estimate of the maximum likelihood ratio based on the fixed $q$ model is worse for larger values of $\delta$. To compare this estimate to an unrestricted estimate, the inference process was simulated 200 times for each $\delta = 0, 0.5$, and 1, holding constant $n = 3$ and the expected $q = 0.2$. Figure 10.10 shows the average maximum likelihood ratio curves for the unrestricted model (dotted curves) and the fixed model (solid curves) when using the selection criterion max $\Delta\lambda(v)$.

The regular and the unrestricted maximum likelihood ratios both converge to 1, though at slower rates, as $\delta$ increases. In other words, the selection criterion max $\Delta\lambda(v)$ is appropriate in situations where the misclassification probability is not necessarily fixed. In general, the unrestricted maximum likelihood ratio is much smaller than the regular one. For the case when $q(v)$ is fixed at 0.2 (i.e., $\delta = 0$), the regular maximum likelihood ratio should be used. When $\delta > 0$, it is an overestimate of the true maximum likelihood ratio. For the case when $\delta = 1$, the unrestricted maximum likelihood ratio should be used, and when $\delta < 1$, it may be an underestimate. The true likelihood ratio lies somewhere in between the two.

## 10.5 Summary and Discussion

### 10.5.1 Summary of the Research Findings

The recent focus on the computational complexity has come at the expense of a dramatic increase in the query complexity for Problem 1. In fact, before the latest contributions described in this chapter, the more recent the inference algorithm is, the worse it performs in terms of the average query complexity! The subroutine, here referred to as FIND-BORDER, is the most commonly used in the recent literature (e.g., [Gainanov, 1984], [Valiant, 1984], [Makino and Ibaraki, 1995], [Boros, *et al.*, 1997]), and its performance was by far the worst. Therefore, the framework for unbiased empirical comparison of inference algorithms described in this chapter seems to be long overdue.

Even though guaranteeing the minimum average number of queries is currently only computationally feasible for relatively few attributes (i.e., up to 5 or 6), the recursive algorithm used for Problem 1 revealed the nonintuitive nature of the optimal solutions. These solutions paved the way for the new selection criterion min $|K_1 - K_0|$. This criterion would probably not have been developed (due to its nonintuitive nature) without the consultation of the optimal solutions.

The inference algorithm based on this selection criterion extends the feasible problem sizes to up to about 20 attributes (which involves about 1 million vectors) for Problem 1. When the number of attributes exceeds 20, computing the selection criterion might become intractable, while Hansel's algorithm will most likely still perform the best on the average. When creating the chain partition used in [Hansel, 1966] and [Sokolov, 1982] becomes intractable, perhaps finding border vectors one at a time by using the subroutine FIND-BORDER is still computationally feasible.

Problem 2 focused on the extension of the single monotone Boolean function inference problem to the inference of a pair of nested monotone Boolean functions. The benefits of the research results discussed in this chapter are manyfold. First, this chapter shows how the optimal and selection criterion approach to minimizing the average query complexity is extended to three different inference applications using a pair of nested monotone Boolean functions. The selection criteria seem to be good choices for the nested inference problem. They result in a slight increase in the average query complexity for the chain poset. For the poset $\{0, 1\}^n$, they are optimal for $n = 1, 2, 3$ and are probably very close to optimal for $n$ greater than 3.

Second, this chapter demonstrates how the nested monotone Boolean function model often is sufficient (i.e., a more complex model is not needed) and necessary (i.e., simpler models are not sufficient) for a wide variety of real-life applications. Suppose a simpler model, such as a single monotone Boolean function, is used for these applications. At best, the simpler model will provide a poor approximation of the phenomenon under study. At worst, it will be unable to model the phenomenon. Suppose a more complex model, such as a pair of independent monotone Boolean functions, is used for these applications. Then, at the very least, the query complexity will increase. In addition, the inferred functions may lead to conflicting knowledge and are more likely to contain errors.

Third, the developments in this chapter quantify the improvement (i.e., the reduction) in query complexity due to the nestedness assumption. The improvement due to the nestedness assumption is between 6% and 8% for larger chain posets ($h > 50$). This improvement is greater for smaller chain posets, reaching its maximum of 20% for $h = 2$. In general, the average query complexity on the chain poset is $O(\log(h))$, so this improvement is not very significant. For the poset $\{0, 1\}^n$, this improvement is a few percent points for $n > 8$. This improvement decreases with the number of attributes, reaching its maximum of 20% for $n = 1$. The average query complexity on the poset $\{0, 1\}^n$ is exponential in $n$. This fact makes this improvement far more dramatic than for the chain poset.

Fourth, this chapter compares the efficiency of the three major types of oracles. The three-valued (ternary) oracle provides the most significant restriction on the oracles. It causes up to 84% and 55% increase in the average number of queries for the chain poset and the poset $\{0, 1\}^n$, respectively. It is interesting to observe that the sequential oracles are just as efficient as the unrestricted oracles on the chain poset and for the poset $\{0, 1\}^n$ for $n$ up to 4. This implies that the pair of nested monotone Boolean functions defined on these posets can be inferred sequentially without losing optimality. For the poset $\{0, 1\}^n$ with $n > 7$, the sequential oracle causes a significant increase in the average query complexity of 12–33%.

The maximum likelihood ratio approach to modeling the inference process of Problem 3 yielded a number of benefits. It was demonstrated that an appropriately defined guided learner, such as maximizing the expected maximum likelihood ratio ($\max \Delta\lambda(v)$) or randomly selecting inferred border vectors ($v \in \mathrm{LU}(f^*) \cup \mathrm{UZ}(f^*)$), allowed the maximum likelihood ratio to converge to 1, even when the misclassification probability was not fixed. This avoids the bias problems associated with the variance approach reported in Cohn [1996], and also observed with the selection criterion $\max(E_0(v) + E_1(v))$ which is based on the number of errors.

For complete reconstruction of monotone Boolean functions, the guided approach showed a dramatic reduction in the average number of queries over a passive learner. The simulations also indicated that this improvement grows at least exponentially as the number of attributes $n$ and the error rate $q$ increase. Thus, defining an appropriate and efficient selection criterion is even more beneficial for large problems and applications with a high error rate.

For large problems (i.e., when $n > 5$), it may not be possible to compute the selection criterion $\max \Delta\lambda(v)$ since it takes exponential time (in the size of the query domain $V$) to do so. For such problems, queries can be selected randomly from the border vectors ($v \in \mathrm{LU}(f^*) \cup \mathrm{UZ}(f^*)$). This only takes $O(V)$ time, and results in much fewer queries than completely random selection on the average.

Hierarchical decomposition provides a way to address a large inference problem as a set of smaller independent inference problems. Even though it was not mentioned earlier, this decomposition is applicable to all three Problems 1, 2, and 3 where it can dramatically reduce the query complexity. Perhaps the greatest benefit of this decomposition is its simplified queries. This fact may not only improve the efficiency but also reduce the number of human errors, and hence increase the likelihood of inferring the correct function.

### 10.5.2 Significance of the Research Findings

The single most important discovery described in this chapter is the near-optimal selection criteria which take polynomial time to evaluate. This leads to the efficient inference of monotone Boolean functions. The significance of these criteria is further strengthened by the scope of real-life problems that can be modeled by using monotone Boolean functions. Even though only one (or a pair of nested) monotone Boolean function(s) defined on the set of Boolean vectors $\{0, 1\}^n$ was (were) studied here, the selection criterion approach to guiding the learner is appropriate for any monotone mapping $V \to F$, where the sets $V \subset R^n$ and $F \subset R^r$ are both finite. The query domain can be viewed as a finite poset by using the monotonicity constraints: $f_i(v) \leq f_i(w)$ iff $v \preceq w$, for $i = 1, 2, \ldots, r$, and whatever the relationships between the functions are, such as the nestedness constraints: $f_1(v) \geq f_2(v) \forall v \in V$. The selection criteria can be evaluated for any such poset in order to pinpoint "smart" queries.

Once the border vectors have been established for each monotone function, they can be used to classify new observations. In addition, they can be represented by a single monotone Boolean function, or a set of monotone Boolean functions, defined on a set of Boolean attributes. Representing the inferred knowledge in this intuitive manner is perhaps the most important aspect of this problem when human interaction is involved since people tend to make better use of knowledge they can easily interpret, understand, validate, and remember.

The use of Boolean functions for analyzing fixed data sets has recently gained a momentum due to their simple representation of intuitive knowledge. See, for example, [Triantaphyllou and Soyster, 1996], [Boros, *et al.*, 1995], [Torvik *et al.*, 1999], and [Yilmaz, *et al.*, 2003]. Boolean models are also becoming more popular because methods for solving their related hard logical optimization problems are emerging (e.g., [Triantaphyllou, 1994], [Chandru and Hooker, 1999], [Hooker, 2000], [Felici and Truemper, 2002], [Truemper, 2004], [Naidenova, 2006], and [Zakrevskij, 2006]). Some studies on guided inference of general Boolean functions from fixed data sets are provided in [Triantaphyllou and Soyster, 1995] and [Nieto Sanchez, *et al.*, 2002b] and are also described in Chapter 5 of this book.

The *narrow vicinity* hypothesis proposed by Kovalerchuk, *et al.* [2000] (and described in Chapter 9) suggests that the use of the monotonicity assumption is often necessary and sufficient. As such, it can greatly improve upon knowledge representations that are too simple or too complex. This chapter demonstrated that the problem of guided inference in the presence of monotonicity could be of great benefit in a wide variety of important real-life applications.

### 10.5.3 Future Research Directions

As mentioned in a previous section the selection criterion approach to learning monotone Boolean functions defined on $\{0, 1\}^n$ is applicable in the much more general monotone setting $V \to F$, where the sets $V \subset R^n$ and $F \subset R^r$ are both finite. The monotone mapping $V \to F$, where the set $V \subset R^n$ is infinite and the set $F \subset R^r$ is finite, forms another intriguing problem. It is well known that binary search is optimal when the query domain $V$ is a bounded subset of the real line, and $F = \{0, 1\}$. However, when the set $V$ is multidimensional and infinite (e.g., $V = [a, b]^2$), pinpointing the optimal queries is a much more complex problem. The selection criterion $\min |K_1 - K_0|$ can be modified to accommodate this case too. Let $U$ denote the unclassified set (i.e., a subset of $V$) and let the parameters $K_0(v)$ and $K_1(v)$ now denote the size of the subsets $\{w \in U : w \prec v\}$ and $\{w \in U : v \prec w\}$, respectively. For example, $K_z(v)$ is measured in terms of distance, area, volume, etc., when $n = 1, 2, 3$, etc., respectively. The selection criterion $\min |K_1 - K_0|$ is then optimal for $n = 1$. How well this criterion performs when $n > 1$ is an open question.

For the problems considered in this chapter, the selection criteria attempt to minimize the average query costs. This objective is based on certain assumptions of the query costs (fixed cost of querying an oracle in Problems 1, 2, and 3, and highly disproportionate or equal query costs for the two oracles in Problems 2.1 and 2.3, respectively). It would be interesting to see how the dialogue with the oracle(s) changes as these assumptions are modified.

When dealing with two oracles, it may be that the cost of querying the first oracle may be less than, yet of similar magnitude as, the cost of querying the second oracle. In this case, the first few queries should be directed at the first oracle. After a few queries it may be more cost beneficial to begin alternating between the two oracles. It could also be that the order of the queries has an effect on the total inference cost. In some applications, additional properties may be known about the underlying function. Some applications may put a limit on the number of lower units, shifting the focus of the optimal vertices from the vertical center to the vertical edge of the poset. It may be that the underlying function belongs to a subclass of monotone Boolean functions, such as threshold functions, 2-monotonic functions, etc.

## 10.6 Concluding Remarks

The methodologies presented in this chapter provide a framework for solving diverse and potentially very important real-life problems that can be modeled as guided inference problems in the presence of monotonicity. The benefits of these methodologies were shown to be dramatic for the specific problems studied here. However, these research findings are just the tip of the iceberg. The interested reader is referred to Torvik and Triantaphyllou [2002; 2004; 2006] for additional details on the solution methodologies for Problems 1, 2, and 3, respectively.

# Chapter 11

# Some Application Issues of Monotone Boolean Functions

## 11.1 Some Background Information

The property of monotonicity has many applications. Its attractive mathematical advantages in inferring a model of the system of interest with high accuracy make the search for this property in data and its consecutive algorithmic exploitation, to be of high potential in data mining and knowledge discovery applications. The following developments are based on the work described in [Kovalerchuk, Vityaev, and Triantaphyllou, 1996] and [Kovalerchuk, Triantaphyllou, and Vityaev, 1995].

This chapter discusses some general application issues of monotone Boolean functions to data mining and knowledge discovery problems. It also presents a rather simple design problem in the car industry and uses the same context to discuss some key issues on the accuracy of diagnostic systems.

## 11.2 Expressing Any Boolean Function in Terms of Monotone Ones

In order to help motivate the main development in this section, consider the arbitrary function depicted in Figure 11.1. Clearly, this function is *not* monotone. However, one may observe that this function is comprised by segments that *are* monotone. Actually, it is comprised by alternating increasing and decreasing monotone functions, not necessarily Boolean.

For instance, the segment between the points A and B in Figure 11.1 is an increasing monotone function, while the segment between the points B and C is a decreasing monotone function. From the same figure it is also suggested that one may decompose a general function in more than one way as a sequence of alternating increasing and decreasing monotone functions. For instance, between the two points D and E, one may identify the increasing monotone function defined between points D and $D_1$ and another between the points $D_1$ and E.

It turns out that any general Boolean function can also be viewed as a sequence of increasing and decreasing monotone Boolean functions. This was established

E. Triantaphyllou, *Data Mining and Knowledge Discovery via Logic-Based Methods*,
Springer Optimization and Its Applications 43, DOI 10.1007/978-1-4419-1630-3_11,

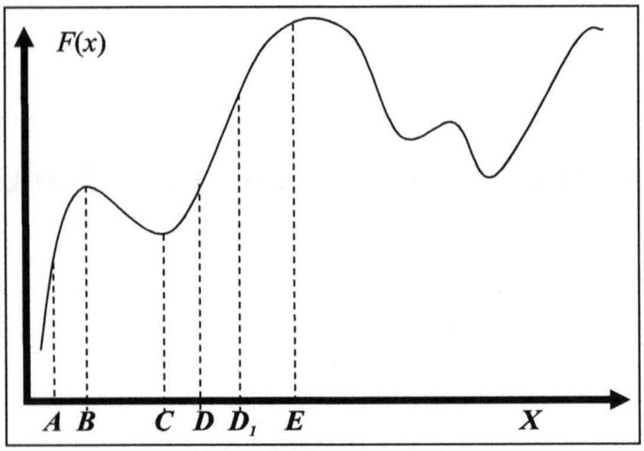


**Figure 11.1.** A Visualization of a Decomposition of a General Function into General Increasing and Decreasing Functions.

in [Kovalerchuk, Triantaphyllou, and Vityaev, 1995] and it is presented next as Theorem 11.1.

**Theorem 11.1.** *Any Boolean function can be described in terms of several increasing and decreasing monotone Boolean functions.*

*Proof.* It is well known (e.g., [Peysakh, 1987]) that any Boolean function can be presented by its DNF. Also, each conjunction from a DNF representation can be presented by a pair of monotone Boolean functions. One of them is an increasing and the other one is a decreasing function.

Let $x_1 \wedge \cdots \wedge x_i \wedge \bar{x}_{i+1} \wedge \cdots \wedge \bar{x}_k$ be a conjunction from the DNF representation of a given Boolean function, where for simplicity the first $i$ components (attributes) are positive while the next $k - i$ components are negations. Then, we can form the two Boolean functions

$$g(x_1, \ldots, x_n) = x_1 \wedge \cdots \wedge x_i$$

and

$$h(x_1, \ldots, x_n) = \bar{x}_{i+1} \wedge \cdots \wedge \bar{x}_k.$$

The function $g(x)$ is an increasing monotone Boolean function and the function $h(x)$ is a decreasing one [Radeanu, 1974]. Hence, the conjunction $x_1 \wedge \cdots \wedge x_i \wedge \bar{x}_{i+1} \wedge \cdots \wedge \bar{x}_k$ is equal to the conjunction of the two monotone (increasing and decreasing, respectively) functions $g$ and $h$:

$$g(x_1, \ldots, x_n) \wedge h(x_1, \ldots, x_n).$$

Therefore, any arbitrary Boolean function $q(x)$ can be presented in the form of the increasing and decreasing functions $g_j(x)$ and $h_j(x)$, respectively:

$$q(x) = \bigvee_{j=1}^{m} (g_j(x) \wedge h_j(x)), \tag{11.1}$$

where $x = (x_1, \ldots, x_n)$ and $m$ is some integer number. ■

Next, let us consider the special case in which a general Boolean function $q(x)$ can be written as follows:

$$q(x) = g(x) \wedge h(x),$$

where $q(x)$ and $h(x)$ are two monotone Boolean functions. For this case the following important property is true:

$$q^+ = g^+ \cap h^+,$$

where $q^+ = \{x : q(x) = 1\}$, $g^+ = \{x : g(x) = 1\}$, and $h^+ = \{x : h(x) = 1\}$ (i.e., these are the sets of the positive examples of these functions). The above observation follows easily since the function $q(x)$ is defined as the logical addition of the two functions $q(x)$ and $h(x)$.

Therefore, one can obtain the set of all positive examples for the *nonmonotone* function $q$ as the intersection of the sets of all positive examples for the two *monotone* functions $g$ and $h$.

For a general function $q(x)$, represented as in (11.1), the union of all these intersections gives the full set of positive examples:

$$q^+ = \cup q_j^+ = \cup (g_j^+ \cap h_j^+).$$

Oftentimes, we do not need so many separate monotone functions. It is noticeable that the union of all conjunctions, which do not include negations $\bar{x}_i$ forms a single increasing monotone Boolean function (see, for instance, [Yablonskii, 1986] and [Alekseev, 1988]).

## 11.3 Formulations of Diagnostic Problems as the Inference of Nested Monotone Boolean Functions

The idea of a pair of nested monotone Boolean functions discussed in the previous chapter may lead to some powerful modeling strategies. Recall that the visual interpretation of a pair of nested monotone Boolean functions is provided in Figure 10.4. The following two sections provide some typical examples of how such nested monotone Boolean functions can be applied. More applications can be found in a diverse spectrum of domains. The inference algorithms for such problems were discussed in Chapter 10.

### 11.3.1 An Application to a Reliability Engineering Problem

For illustrative purposes consider the problem of classifying the states of some mechanical system by a reliability expert. This expert is assumed to have worked with this particular system for a long time and thus can serve as an oracle (i.e., an operator who can correctly classify new states of the system). States of the system are represented by binary vectors from $E^n$ (the space defined on $n$ binary attributes or $\{0, 1\}^n$). This oracle is assumed so that he/she can answer questions such as:

"Is the reliability of a given state guaranteed?" (Yes/No) or "Is an accident for a given state guaranteed?" (Yes/No). In accordance with these questions, **two** interrelated nested classification tasks can be defined. In this way one can view the original complex system as comprised by two interrelated subsystems. Next, we define the four possible outcome classes (or patterns of operation) which are possible in this situation.

**Task 1:** (first subsystem)
**Pattern 1.1:** "Guaranteed reliable states of the system" (denoted as $E_1^+$).
**Pattern 1.2:** "Reliability of the states of the system is not guaranteed" (denoted as $E_1^-$).

**Task 2:** (second subsystem)
**Pattern 2.1:** "States of the system with some possibility for normal operation" (denoted as $E_2^+$).
**Pattern 2.2:** "States of the system which guarantee an accident" (denoted as $E_2^-$).

The goal here is to infer two monotone Boolean functions $f_1$ and $f_2$. The first function is related to task 1, while the second one is related to task 2. It can also be observed from the way all possible outcomes have been defined above that the following relations must be true: $E_2^+ \supset E_1^+$ and $f_2(x) \geq f_1(x)$ for all $v \in \{0, 1\}^n$ (see also Figure 10.4).

### 11.3.2 An Application to the Breast Cancer Diagnosis Problem

The diagnos problem considered in this application is also a *nested* one. That is, it is comprised of two interrelated subproblems. The **first subproblem** is related to the clinical question of whether a biopsy or short-term follow-up is necessary or not. The **second subproblem** is related to the question whether the radiologist believes that the current case is highly suspicious for malignancy or not. It is assumed that if the radiologist believes that the case is malignant, then he/she will also definitely recommend a biopsy. Formally, these two subproblems are defined as follows:

**The Clinical Management Subproblem:** (first subsystem)
One and only one of the following two disjoint outcomes is possible:
1) "Biopsy/short-term follow-up is necessary";
or: 2) "Biopsy/short-term follow-up is not necessary."

**The Diagnosis Subproblem:** (second subsystem)
Similarly as above, one and only one of the following two disjoint outcomes is possible. That is, a given case is
1) "Highly suspicious for malignancy";
or: 2) "Not highly suspicious for malignancy."

**Table 11.1.** Ratings of Midsize Cars that Cost Under $25,000 [*Consumer Reports*, 1994, page 160].

| | Attributes | | | | | | | |
|---|---|---|---|---|---|---|---|---|
| **Car Type** | $X_1$ | $X_2$ | $X_3$ | $X_4$ | $X_5$ | $X_6$ | $X_7$ | $X_8$ |
| Camry 6 | 4 | 5 | 21 | 1 | 0 | P | 1 | P |
| Camry 4 | 4 | 5 | 24 | 1 | 0 | P | P | P |
| Ford Taurus | 4 | 3 | 20 | 1 | P | P | 1 | P |
| Mercury Sable | 4 | 3 | 20 | 1 | 1 | P | 1 | 1 |
| Maxima | 4 | 5 | 21 | 1 | 0 | P | P | 1 |
| Chrysler NY | 3 | 3 | 21 | 1 | 0 | P | 1 | 1 |
| Buick Regal | 2 | 3 | 20 | 0 | 0 | P | 1 | 1 |
| Chevrolet Lumina | 1 | 2 | 22 | 0 | 0 | P | 1 | P |

It can be easily seen that the corresponding states satisfy the nesting conditions as in the previous example.

## 11.4 Design Problems

In the design domain our interest is focused on developing procedures to assist in the formulation of design criteria as formal requirements for a product of interest to some designer. This process can be very complicated and also be a multilevel and multiattribute task. An efficient solution to this problem may significantly speed up the design process.

We will present the key ideas in terms of a rather simple illustrative example. Suppose that one wishes to design a midsize car under $25,000 which will be better than the average one currently available in the market. Table 11.1 presents the key features (attributes) of cars on the market, as were taken from Consumer Reports in 1994 (on page 160). In this table we use the following notation:

$X_1$ : the overall score (the higher the better, the values are 1, 2, 3, and 4);
$X_2$ : the predicted reliability;
$X_3$ : its overall mileage per gallon (mpg);
$X_4$ : presence or not of an airbag on the driver's side (note that today this is a required feature for all cars);
$X_5$ : presence or not of an airbag on the passenger's side (note that today this is a required feature for all cars);
$X_6$ : presence or not of antilock brakes;
$X_7$ : presence of automatic transmission; and
$X_8$ : presence of air-conditioning.

For the above attributes in Table 11.1 "P" stands for "optional," while "1" and "0" stand for available and unavailable, respectively.

Next, the designer needs to identify a combination of car attributes which should be the design requirements for the new car. Such a task requires one to analyze a large amount of data (combinations of attribute values) with different overall scores (value of $X_1$ in Table 11.1).

In particular, we note that from the above data there are 5 different values for attribute $X_2$, 5 different values for attribute $X_3$, 2 for $X_4$, and 3 for each of $X_5$, $X_6$, $X_7$, and $X_8$. That is, there is a total of $5 \times 5 \times 2 \times 3 \times 3 \times 3 \times 3 = 4{,}050$ combinations. A designer should choose some of them. This number defines the size of the problem. Table 11.1 brings information of 5 unique cases (note that another 3 cases are identical to some of these 5 cases) from this population of 4,050 cases. Thus, the research problem is how to assist the designer to analyze these 4,050 cases and discriminate between acceptable and unacceptable ones.

Some of the cases can be excluded from further consideration rather easily. For instance, if all existing cars have antilock brakes, automatic transmission, and air-conditioning options, then probably it will not be wise to design a new car without these options. The same is true for the airbag on the driver's side. We can also restrict an acceptable range of reliability with values 3, 4, and 5. By using these observations, the actual number of all possible cases becomes $3 \times 5 \times 2 \times 2 \times 2 \times 2 \times 2 = 480$.

Each of these 480 cases needs to be classified into one of the following two categories (classes): (1) acceptable for design and (2) unacceptable for design. At this point, one may observe that this design problem is governed by the monotonicity property. If some car (i.e., a combination of values of the previous seven key attributes $X_2, X_3, X_4, \ldots, X_8$) is acceptable for design, then a better car (i.e., a combination of stronger values but still under the \$25,000 price level) should also be acceptable for design. The key property of monotonicity, along with the previous design attributes, may allow one to study the solution space effectively as is demonstrated in Section 11.7.

## 11.5 Process Diagnosis Problems

Many computer-aided process diagnosis systems are based on neural networks, decision trees, linear discriminant analysis, similarity-based classification, and various statistical models. Such systems are used in many application areas such as machinery monitoring, military target recognition, detection of radioactive contamination, non-destructive detection of damage in composite materials, drug design, and medical diagnosis just to name a few. Usually, for such systems an accuracy of 90%, 95%, or 99% is considered as satisfactory. However, as the following section illustrates, the above numbers may be grossly misleading.

## 11.6 Three Major Illusions in the Evaluation of the Accuracy of Data Mining Models

In this section we discuss the role of various measures of performance of diagnostic systems and data mining systems in general. In [Kovalerchuk, Vityaev, and Triantaphyllou, 1996] we defined three types of misleadings (we will call them here *illusions*) when one considers the accuracy rate of a system. Related to this subject is also the discussion at the end of Section 1.3.4, and most of Chapter 9.

### 11.6.1 First Illusion: The Single Index Accuracy Rate

Consider the following actual data: About 0.2% of 15,000 screened women have breast cancer. This information comes from unpublished data in 1995 from Woman's Hospital in Baton Rouge, Louisiana. Suppose that we have an "ultraoptimistic" diagnostic system which *indiscriminately* diagnoses any case as "noncancer." Such a system would have an accuracy rate of 99.8%, as it would miss only the 0.2% of the cases which in reality are the only ones with breast cancer.

At a first glance, a system with an accuracy of 99.8% may appear to be highly effective. However, in reality the above "ultraoptimistic" system is just useless if not terribly dangerous. That is, one needs to examine its accuracy in terms of the *false-positive* and *false-negative* error rates separately. In these terms the "ultraoptimistic" system makes no errors in diagnosing the noncancer cases (100% of them are diagnosed accurately), while it makes 100% errors in misdiagnosing all cancer cases. In light of these two indices, one may immediately reject such a system as profoundly inaccurate. A similar situation occurs with many other diagnostic/forecasting systems in a wide spectrum of applications.

Next, we will consider how to evaluate the performance of a diagnostic system in terms of these two indices. If both of these indices are equal to 99%, then the overall accuracy rate will also be equal to 99%. In such case, the overall performance of the system might be considered as satisfactory. However, if the system has 30% false-positive and 80% false-negative rates, then how can one evaluate it? Most likely, one needs to undertake a detailed study that considers the relative impact of each of these two types of errors.

In the previous discussion we highlighted the need to consider both rates, that is, the false-positive and the false-negative accuracy rates. This task requires one to determine the real border between the two diagnostic classes and compare it with the formal border which is determined by a particular model. The real border can be of any size, ranging from very narrow to very wide (see also Chapter 9 and Chapter 10 on monotonicity).

Another critical issue is the rate of cases that the system characterizes new observations as *"do not know"* or *undecidable* cases (not to be confused with the unclassified ones). This was discussed in more detail in Section 1.3.4. Such situations take place when the inferred knowledge from the training data is limited. That is, a system may have very low false-positive and false-negative rates but still be of dubious quality if it results in too many "undecidable" cases when it is called to classify new observations.

### 11.6.2 Second Illusion: Accurate Diagnosis without Hard Cases

Suppose that the border area of the previous example on breast cancer consists of only 10% of all possible cases and a system diagnoses incorrectly all of them. That is, it is wrong about 100% of the cases that belong to this border area. Furthermore, suppose that the same system is 100% accurate on the rest of the cases (i.e., on the nonborder cases which take 90% of all possible cases). Overall, this system is 90% accurate with a number of false-positive and false-negative errors. How effective is such a system? Clearly, this situation is different than the one described in the previous section.

Such a hypothetical system is highly (actually, perfectly) accurate on simple cases (as they are nonborder ones) and highly inaccurate (actually, perfectly inaccurate) on the challenging (i.e., "hard") cases as they are the ones which belong to the border area.

Should such a system be considered as effective? That is doubtful. Intuition calls that an effective diagnostic system (and, in general, any predictive system) should be highly accurate, and in particular, on the most challenging of the cases. This is exactly when one needs the aid of such a system! Many systems and human experts may diagnose cases that are well beyond the border area. The real value of a system is in assisting in accurately diagnosing border or marginal cases. These are the "hard" cases. In a situation like the previous one, the system appears to be highly effective (the overall accuracy is 90%), it is not of the extreme "ultraoptimistic" category, but its practical effectiveness is under serious doubt. The above issues are the basis of what we call the second illusion, or highly accurate on the nonborder (i.e., the "easy") cases only.

### 11.6.3 Third Illusion: High Accuracy on Random Test Data Only

The third illusion is related to a random choice of the testing data. The standard approach in evaluating a diagnostic system is to test it against a set of randomly chosen data for which the actual classification is known but it is kept unknown to the system.

In terms of the previous hypothetical example of having 10% of the cases in the border area and the other 90% in the nonborder area, such a testing would generate the previous performance measures and the system would be 90% accurate. This situation, however, would have disguised the severe deficiency and bias as described in the previous section. In other words, we would be subject to the second illusion but the real cause would be the random choice of the testing data. Thus, we call this type of weakness the *random choice illusion.*

## 11.7 Identification of the Monotonicity Property

How can the previous three accuracy illusions be prevented and the effectiveness of such systems be improved? The answer seems to be by developing methods that better define the border area between the diagnostic classes. In the past, different methods provided approximations of the real border by employing various **a priori** assumptions regarding these borders. Such assumptions are related to the metrics of the attribute space, type of distribution, class of discriminant functions,

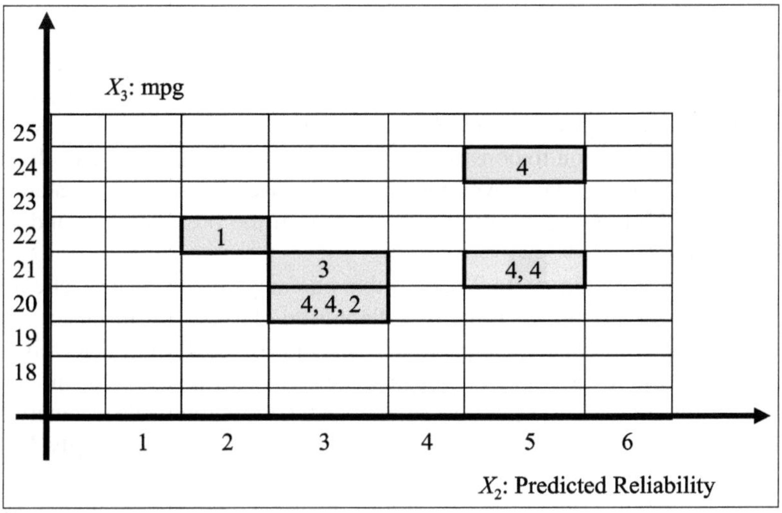


**Figure 11.2.** The Data Points in Terms of Attributes $X_2$ and $X_3$ Only.

type of rules to extract, etc. This a priori assumptions approach may often lead to confusing diagnostic solutions and/or to some of the illusions discussed in the previous sections.

Thus, it is important to have an approach that determines such borders directly from the data. Such a method is illustrated next on the car evaluation data presented in Section 11.4 of this chapter. This approach is based on the monotonicity property studied in Chapter 10 of this book.

In the small data set depicted in Table 11.1 observations (descriptions of cars) can be classified in either of two classes: acceptable and unacceptable car designs. In order to help fix ideas we consider two attributes only (although the proposed method can be generalized to include any number of attributes). These attributes are the predicted reliability (attribute $X_2$) and the miles per gallon or mpg (attribute $X_3$).

The data described in these two attributes are next plotted as in Figure 11.2. For each data point we also plot its overall rating score (attribute $X_1$). For instance, the data point with attribute values $(X_2, X_3) = (2, 22)$ corresponds to Chevrolet Lumina which has an overall rating value equal to 1 (hence the square with coordinates (2, 22) has a "**1**"). Similarly, the point $(X_2, X_3) = (3, 20)$ corresponds to the three cars Ford Taurus, Mercury Sable, and Buick Regal with overall scores of 4, 4, and 2, respectively. Thus, the square with coordinates (3, 20) has the "4,4,2" label. A similar interpretation follows for the rest of the data points in Figure 11.2.

Next, we use attribute $X_1$ (i.e., the overall rating score) to split this data set into two disjoint subsets. If $X_1 = 4$, then the design requirements lead to an *acceptable* solution (car), while values of $X_1 < 4$ lead to *unacceptable* solutions. That is, the acceptable and unacceptable characterization corresponds to the two diagnostic classes (positive and negative, respectively) for this data set. Note that the assignment of the acceptable (unacceptable) class to the positive (negative) class is arbitrary.

If one considers all possible values (i.e., the 5 different values) for attribute $X_2$ and all possible values for attribute $X_3$ (i.e., the 8 different values in Figure 11.2), one gets a total of 40 different design combinations. From Table 11.1 one can see that we have complete information for only 8 cars (design combinations). Thus, we do not have adequate information to construct a precise border between the two diagnostic classes as defined above. Moreover, the classes are overlapping as indicated in data point (3, 20) in Figure 11.2 by three cars; two from the positive class and one car from the negative class.

An examination of Figure 11.2 reveals that one may determine a practically infinite number of borders between the two diagnostic classes. For instance, a simple border is to have $X_2 > 4$ for the positive class and $X_2 < 4$ for the negative class. Another one is to have $X_2 \geq 3$ for the positive class and $X_2 < 3$ for the negative class and so on. Both of the previous borders (discrimination rules) generalize all available 8 cases (data points) and classify the remaining cases accordingly. Next one may observe that both of them classify point (5, 19) as "acceptable." However, what is the basic argument for doing so? Traditional methods such as neural networks and linear discriminant functions do not provide a good justification of such decisions other than to claim that such a decision is based on the generalization implied by the available training data points.

The above dilemma may be addressed by resorting to the monotone property embedded in the attribute definitions. A closer look at the *semantics* of the two attributes $X_2$ and $X_3$ reveals some interesting properties. To see this consider two hypothetical cars $a$ and $b$ with the following two *hypothetical* relationships:

(R1) The forecasted reliability $X_2$ of car $b$ is more than the forecasted reliability $X_2$ of car $a$. That is, we assume to have $X_2(b) > X_2(a)$.
(R2) Car $a$ covers less miles per gallon than car $b$. That is, we assume to have $X_3(b) > X_3(a)$.

Now we are ready to design car $a$ with properties $X_2$ and $X_3$ defined as above. That is, we have decided that car $a$ belongs to the acceptable (positive) class. From relations (R1) and (R2) one can determine that car $b$ is *better* than car $a$ (again, we assume that we have the two attributes $X_2$ and $X_3$ only). Then, in order to be logically consistent we should also accept that car $b$ belongs to the acceptable (positive) class as well. Let us denote that car $a$ (or car $b$) belongs to the acceptable class by saying that the predicate $D(a)$ (or $D(b)$) has true value. Then the above analysis can be formalized as follows:

$$\begin{array}{ll} \text{IF} & X_2(b) > X_2(a), \text{ and } X_3(b) > X_3(a), \text{ and } (D(a) = \text{true}), \\ \text{THEN} & (D(b) = \text{true}). \end{array}$$

This is exactly the meaning of the **property of monotonicity**. For the two attributes $X_2$ and $X_3$ the property of monotonicity is a direct consequence of the analysis of the semantics of these attributes. For other attributes, the property of monotonicity may not be as apparent.

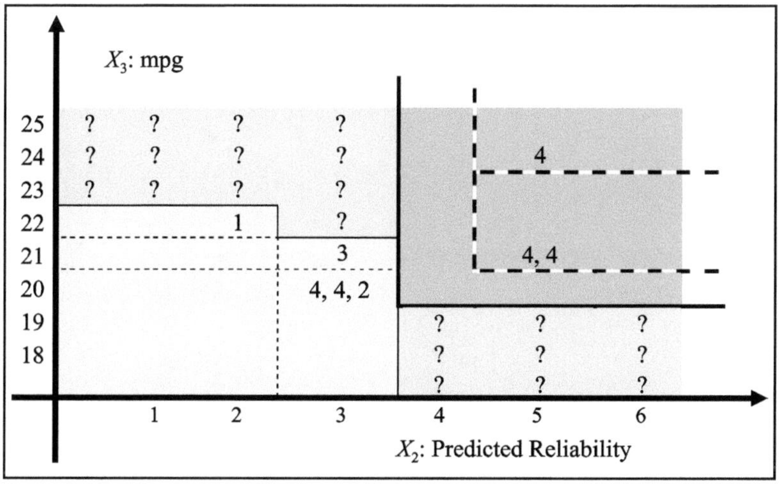


**Figure 11.3.** Monotone Discrimination of the Positive (Acceptable) and Negative (Unacceptable) Design Classes.

The main advantage of this semantic or empirical approach to discovering the property of monotonicity is that one can obtain *interpretable* properties of the diagnostic classes, instead of interpolating the data and defining the borders by using some a priori assumptions.

Next, we use the discovered property of monotonicity to define the borders for the current illustrative data set. Figure 11.3 shows the border between the positive class (dark region in the upper right area of Figure 11.3) and the negative class (the light region in the lower left area in Figure 11.3). The same figure also shows the "undecidable" region (designated by the "?" labels) for this simplified hypothetical example.

## 11.8 Concluding Remarks

The procedures discussed in this chapter are applicable to any situation in which we wish to infer the structure of a system of interest from observations of its behavior that can be grouped into two disjoint sets (i.e., the positive and negative groups) of training examples.

This chapter discussed some application issues of how to analyze any Boolean function (i.e., not necessarily monotone ones) in terms of a series of monotone Boolean functions. It also discussed some modeling techniques that rely on nested monotone Boolean functions.

Furthermore, this chapter provides a procedure for identifying the presence of monotonicity and then it combines it with a semantic analysis of the attributes to better identify the border between the two sets of training examples. It also discussed, in a comprehensive manner, three critical types of performance illusions which may occur when one evaluates the accuracy of a diagnostic system.

# Chapter 12

# Mining of Association Rules

## 12.1 Some Background Information

Mining of association rules from databases has attracted great interest because of its potentially very useful applications. Association rules are derived from a type of analysis that extracts information from coincidence [Blaxton and Westphal, 1998]. Sometimes called *market basket analysis*, this methodology allows a data analyst to discover correlations, or co-occurrences of transactional events. In the classic example, consider the items contained in a customer's shopping cart on any one trip to a grocery store. Chances are that the customer's own shopping patterns tend to be internally consistent, and that he/she tends to buy certain items on certain days. There might be many examples of pairs of items that are likely to be purchased together. This is the kind of information the store manager could use to make decisions about where to place items in the store so as to increase sales. This information can be expressed in the form of association rules. Such information may have tremendous potential on the marketing of new or existing products. This is the kind of approach used by many enterprises (such as Amazon.com for instance) to recommend new or existing products to their customers. Mining of association rules is applicable to many more domains [Bayardo, *et al.*, 1999]. This chapter is based on the results discussed in [Yilmaz, *et al.*, 2003].

Purchase records can be captured by using the bar codes on products. The technology to read them has enabled businesses to efficiently collect vast amounts of data, commonly known as *market basket data* [Agrawal and Srikant, 1994]. Typically, a purchase record contains the items bought in a single transaction, and a database may contain many such transactions. Analyzing such databases by extracting association rules may offer some unique opportunities for businesses to increase their sales, since association rules can be used in designing effective marketing strategies. The sizes of the databases involved can be very large. Thus, fast and effective algorithms are needed to mine association rules out of them.

For a more formal definition of association rules, some notation and definitions are introduced as follows. Let $I = \{A_1, A_2, A_3, \ldots, A_n\}$ be the set with the names of the items (also called attributes, hence the notation $A_i$) among which association

E. Triantaphyllou, *Data Mining and Knowledge Discovery via Logic-Based Methods*,
Springer Optimization and Its Applications 43, DOI 10.1007/978-1-4419-1630-3_12,

rules will be searched. This set is often called the *item domain* [Agrawal and Srikant, 1994], [Bayardo, *et al.*, 1999]. Then, a *transaction* is a set of one or more items obtained from the set $I$. This means that for each transaction $T$, the relation $T \subseteq I$ holds. Let $D$ be the set of all transactions. Also, let $X$ be defined as a set of some of the items in $I$. The set $X$ is contained in a transaction $T$ if the relation $X \subseteq T$ holds.

Using these definitions, an *association rule* is a relationship of the form $X \Rightarrow Y$, where $X \subset I$, $Y \subset I$, and $X \cap Y = \emptyset$. The set $X$ is the *antecedent part*, while the set $Y$ is the *consequent part* of the rule. An association rule holds true with some *confidence level* denoted as $CL$. The confidence level is the conditional probability (as it can be inferred from the available transactions in the target database) of having the consequent part $Y$ given that we already have the antecedent part $X$. Moreover, an association rule has *support* $S$, where $S$ is the number of transactions in $D$ that contain $X$ and $Y$ simultaneously. A *frequent item set* is a set of items that occur frequently in the database. That is, their support is above a predetermined minimum support level. A *candidate item set* is a set of items, possibly frequent, but not yet checked whether they meet the minimum support criterion. The association rule analysis in our approach will be restricted to those association rules which have only one item in the consequent part of the rule. However, a generalization can be made easily.

*Example 1.* Consider the following illustrative database:

$$D = \begin{bmatrix} 1 & 0 & 1 & 1 & 0 \\ 0 & 0 & 1 & 1 & 0 \\ 1 & 1 & 0 & 0 & 0 \\ 0 & 1 & 1 & 0 & 0 \\ 0 & 1 & 0 & 1 & 1 \\ 1 & 1 & 1 & 0 & 1 \\ 0 & 0 & 0 & 0 & 1 \\ 1 & 1 & 1 & 1 & 1 \\ 0 & 1 & 1 & 1 & 1 \\ 0 & 0 & 0 & 1 & 1 \end{bmatrix}.$$

This database is defined on five items, so $I = \{A_1, A_2, A_3, A_4, A_5\}$. Each row represents a transaction. For instance, the second row represents a transaction in which only items $A_3$ and $A_4$ were bought. The support of the rule $A_2A_4 \rightarrow A_5$ is equal to 3. This is true because the items $A_2$, $A_4$, and $A_5$ occur simultaneously in 3 transactions (i.e., the fifth, eighth, and ninth transactions). The confidence level of the rule $A_2A_4 \rightarrow A_5$ is 100% because the number of transactions in which $A_2$ and $A_4$ appear together is equal to the number of transactions that $A_2$, $A_4$, and $A_5$ appear (both are equal to three), giving a confidence level of 100%. ■

Given the previous definitions, the problem of interest is how to mine association rules out of a database $D$, that meet some preestablished minimum support and confidence level requirements. Mining of association rules was first introduced by Agrawal, Imielinski and Swami in [1993]. Their algorithm is called AIS (for *Agrawal, Imielinski, and Swami*). Another study used a different approach to solve the problem of mining association rules [Houtsma and Swami, 1993]. That study presented a new algorithm called SETM (for *Set-Oriented Mining*). The new algorithm was proposed to mine association rules by using relational operations in a relational database environment. This was motivated by the desire to use the SQL system to compute frequent item sets.

The next study [Agrawal and Srikant, 1994] received a lot more recognition than the previous ones. Three new algorithms were presented; the *Apriori*, the *AprioriTid*, and the *AprioriHybrid*. The Apriori and AprioriTid algorithms are fundamentally different from the AIS and SETM ones. As the name AprioriHybrid suggests, this approach is a hybrid between the Apriori and the AprioriTid algorithms.

Another major study in the field of mining of association rules is described in [Savasere, *et al.*, 1995]. These authors presented an algorithm called *Partition*. Their approach reduces the search by first computing all frequent item sets in two passes over the database. Another major study on association rules takes a sampling approach [Toivonen, 1996]. These algorithms make only one full pass over the database. The main idea is to select a random sample, and use it to determine representative association rules that are very likely to also occur in the whole database. These association rules are in turn validated in the entire database.

This chapter is organized as follows. The next section presents a formal description of the research problem under consideration. The third section describes the new approach which is based on the OCAT approach and the RA1 heuristic (as described in Chapters 2, 3, and 4). The fourth section presents an extensive computational study that compared the proposed approach for the mining of association rules with some existing ones. Finally, the chapter ends with a conclusions section.

## 12.2 Problem Description

Previous work on mining of association rules focused on extracting all conjunctive rules, provided that these rules meet the criteria set by the analyst. Such criteria can be the minimum support and confidence levels. Although previous algorithms mainly considered databases from the domain of market basket analysis, they have been applied to the fields of telecommunication data analysis, census data analysis, and to classification and predictive modeling tasks in general [Bayardo, *et al.*, 1999]. These applications differ from market basket analysis in the sense that they contain dense data. That is, such data sets may possess all or some of the following properties:

(i) Have many frequently occurring items;
(ii) Have strong correlations between several items;
(iii) Have many items in each record.

When standard association rule mining techniques are used (such as the Apriori approach [Agrawal and Srikant, 1994] and its variants), they may cause exponential resource consumption in the worst case. Thus, it may take too much CPU time for these algorithms to mine the association rules. The combinatorial explosion is a natural result of these algorithms, because they mine *exhaustively* all the rules that satisfy the minimum support constraint as specified by the analyst. Furthermore, this characteristic may lead to the generation of an excessive number of rules. Then, the end user will have to determine which rules are worthwhile. Therefore, the higher the number of the derived association rules is, the more difficult it is to review them. In addition, if the target database contains dense data, then the previous situation may become even worse.

The size of the database also plays a vital role in data mining algorithms [Toivonen, 1996]. Large databases are desired for obtaining accurate results, but unfortunately, the efficiency of the algorithms depends heavily on the size of the database. The core of today's algorithms is the Apriori algorithm [Agrawal and Srikant, 1994] and this algorithm will be the one to be compared with in this chapter. Therefore, it is highly desirable to develop an algorithm that has polynomial complexity and still being able to find a few rules of good quality.

## 12.3 Methodology

### 12.3.1 Some Related Algorithmic Developments

The proposed approach for mining of association rules is based on the OCAT approach and the RA1 heuristic. Recall that this heuristic infers a set of clauses (i.e., a Boolean function in CNF or DNF) from two mutually exclusive collections of binary examples.

Below are some definitions that are used in these approaches and are going to be used in the new approach as well.

| | |
|---|---|
| $C$ | is the set of attributes in the current clause (a disjunction for the CNF case); |
| $a_k$ | an attribute such that $a_k \in A$, where $A$ is the set of the attributes $A_1, A_2, \ldots, A_n$ and their negations; |
| $POS(a_k)$ | the number of all positive examples in $E^+$ which would be accepted if attribute $a_k$ were included in the current CNF clause; |
| $NEG(a_k)$ | the number of all negative examples in $E^-$ which would be accepted if attribute $a_k$ were included in the current clause; |
| $\ell$ | the size of the candidate list; |
| $ITRS$ | the number of times the clause forming procedure is repeated. |

The RA1 algorithm is described again in Figure 12.1. It is of polynomial time complexity as shown in Chapter 4. By examining the previous definitions, some key observations can be made at this point. When an attribute of high *POS* function value is chosen to be included in the CNF clause currently being formed, it is very likely that this will cause it to accept some additional positive examples.

```
DO for ITRS number of iterations
    BEGIN {Reset the E+ and E- sets};
        DO WHILE (E- ≠ Ø)
          C = Ø; {initialization}
            DO WHILE (E+ ≠ Ø)
                Step 1:  Rank in descending order all attributes a_i ∈ A
                         (where a_i is either A_i or Ā_i) according to their
                         POS(a_i)/NEG(a_i) value. If NEG(a_i) = 0, then
                         use as an alternative scoring function the product
                         of an arbitrarily large number times POS(a_i).
                         We call this the ALT(a_i) value;
                Step 2:  Form a candidate list of the attributes which have
                         the l highest POS(a_i)/NEG(a_i) ratios or ALT(a_i)
                         values (when NEG(a_i) = 0);
                Step 3:  Randomly choose an attribute a_k from the candi-
                         date list;
                Step 4:  Let the partial current clause be:
                                   C ← C ∨ a_k;
                Step 5:  Let E+(a_k) be the set of members of E+
                         accepted when a_k is included in the current CNF
                         clause;
                Step 6:  Let E+ ← E+ − E+(a_k);
                Step 7:  Let A ← A − a_k;
                Step 8:  Calculate the new POS(a_i) values for all a_i ∈ a;
            REPEAT
                Step 9:  Let E-(C) be the set of members of E- which
                         are rejected by C;
                Step 10: Let E- ← E- − E-(C);
                Step 11: Reset E+;
        REPEAT
    END;
CHOOSE   the final Boolean system among the previous ITRS systems which has the
         smallest number of clauses.
```

**Figure 12.1.** The RA1 Heuristic for the CNF Case (see also Chapter 4).

The reverse is true for attributes with a small *NEG* function value in terms of the negative examples. Therefore, attributes that have high *POS* function values and low *NEG* function values are a good choice for inclusion in the current CNF clause. In [Deshpande and Triantaphyllou, 1998] it was shown, through some empirical experiments, that the *POS/NEG* ratio is an effective evaluative criterion, since it is very likely to lead to Boolean functions with few clauses.

### 12.3.2 Alterations to the RA1 Algorithm

For a Boolean expression to reveal actionable information about associations in a database, it is more convenient to be expressed in DNF. The first step is to select an attribute about which associations will be sought. This attribute will form the consequent part of the desired association rules. By selecting such an attribute, the database can be partitioned into two mutual sets of records (binary vectors). Vectors that have value equal to "1" in terms of the selected attribute can be seen as the positive examples. A similar interpretation holds true for records that have value of "0" for that attribute. The latter vectors will be the set of the negative examples.

Given the above way for partitioning (dichotomizing) a database of transactions, it follows that each conjunction (clause in a DNF expression) of the target Boolean function will reject all the negative examples, while on the other hand, it will accept some of the positive examples. Of course, when all the conjunctions are considered together, they will accept all the positive examples.

In terms of association rules, each clause in the Boolean expression (which now is expressed in DNF) can be thought of as a set of frequent item sets. That is, such a clause forms a frequent item set. Thus, this clause can be checked further whether it meets the preset minimum support and confidence level criteria.

The requirement of having Boolean expressions in DNF does not mean that the RA1 algorithm has to be altered (although it can easily be done) to produce Boolean expressions in DNF. As was shown in Chapter 7, if one forms the complements of the positive and negative sets and then swaps their roles, then a CNF producing algorithm will produce a DNF expression (and vice versa). The last alteration is to swap the logical operators ($\wedge$) "AND" and ($\vee$) "OR" in the CNF (or DNF) expression.

Another interesting issue is to observe that the confidence level of the association rules produced by processing frequent item sets (i.e., clauses of a Boolean expression in DNF when the RA1 approach is used) will always be equal to 100%. This happens because each DNF clause rejects all the negative examples while it accepts some of the positive examples when a database with transactions is partitioned as described earlier.

A critical change in the RA1 heuristic is that for deriving association rules, it should only consider the attributes themselves and not their negations. This is not always the case, since some authors have also proposed to use association rules with negations [Savasere, *et al.*, 1998]. However, association rules are usually defined on the attributes themselves and not on their negations. If one considers only the attributes themselves and excludes their negations, this requirement may cause certain problems due to certain degenerative situations that could occur. These situations may occur as follows:

*Degenerative Case # 1.* If only one item is bought in a transaction, and if that particular item is selected to be the consequent part of the association rules sought, then the $E^+$ set will have an example (i.e., the one that corresponds to that transaction) with only zero elements. Thus, the RA1 heuristic (or any variant of it) will never terminate. Hence, for simplicity it will be assumed that such degenerative transactions do not occur in our databases.

*Degenerative Case # 2.* After forming a clause, and after the $E^-$ set is updated (Step 10 in Figure 12.1), the new *POS/NEG* values may be such that the new clause may be one of those that have been already produced earlier (i.e., it is possible to have "cycling").

*Degenerative Case # 3.* A newly generated clause may not be able to reject any of the negative examples.

The previous is an exhaustive list of all possible degenerative situations when the original RA1 algorithm is used on this particular type of problem (i.e., in the mining of association rules). Thus, the original RA1 algorithm needs to be altered in order to avoid them. Degenerative case #1 can be easily avoided by simply discarding all one-item transactions (which are very rare to occur in reality any way). Degenerative cases #2 and #3 can be avoided by establishing some upper limits on the number of iterations a single clause is generated (recall the randomized characteristic of the RA1 heuristic). Such cases are more likely to occur with clauses generated towards the end of the clause generation process. If such a limit is reached, then the entire Boolean function is generated again in a randomized fashion.

In order to mine association rules that have different consequents, the altered RA1 should be run for each of the attributes $A_1, A_2, A_3, \ldots, A_n$. After determining the frequent item sets for each of these attributes, one needs to calculate the support level for each frequent item set, and check whether the (preset) minimum support criterion is met. If it is, then the current association rule is reported. The proposed altered RA1 (to be denoted as ARA1) heuristic is summarized in Figure 12.2. It captures the greedy aspect of the OCAT approach (which is embedded into the algorithm). Please note that this version does not implement randomization for simplicity of the illustration of the main ideas. Randomization of the ARA1 heuristic can be done easily in a way analogous to the RA1 case (as shown in Figure 4.1). Finally, it should be stated here that the new heuristic is also of polynomial time complexity as was the case with the original RA1 algorithm. This follows easily from a complexity analysis similar to the one described in Chapter 4 for the case of the RA1 algorithm.

## 12.4 Computational Experiments

In order to compare the altered RA1 (i.e., the ARA1) heuristic with some of the existing mining of association rule methods, we applied them on several synthetic databases that were generated by using the data generation programs described in [Agrawal and Srikant, 1994]. The URL for these codes was: **http://www.almaden.ibm.com/cs/quest/syndata.html** (note: when this URL was tested recently it was not active but the user was directed to a general site in this domain). These databases contain transactions that would reflect the real world, where people tend to buy sets of certain items together. Several databases were used in these comparisons. The sizes of these databases are as follows:

**DO** for each consequent $A_1, A_2, A_3, \ldots, A_n$
**BEGIN**
Form the $E^+$ and $E^-$ sets according to the presence or absence of the current $A_i$ attribute.
Calculate the initial *POS* and *NEG* values.
Let $A = \{A_1, A_2, A_3, \ldots, A_n\}$.
**DO WHILE** $(E^- \neq \varnothing)$
$C = \varnothing$; {initializations}
**START1: DO WHILE** $(E^+ \neq \varnothing)$

**Step 1:** Rank in descending order all attributes $a_i \in A$ (where $a_i$ is the attribute currently under consideration) according to their $POS(a_i)/NEG(a_i)$ value;
If $NEG(a_i) = 0$, then the $ALT(a_i)$ value is used instead (see also Figure 12.1);
**Step 2:** Evaluate the *POS/NEG* ratios or the *ALT* values;
**Step 3:** Choose an attribute $a_k$ accordingly;
**Step 4:** Let the set of attributes in the current clause be $C \leftarrow C \cup \{a_k\}$;
**Step 5:** Let $E^+(a_k)$ be the set of members of $E^+$ accepted when $a_k$ is included in the current CNF clause;
**Step 6:** Let $E^+ \leftarrow E^+ - E^+(a_k)$;
**Step 7:** Let $A \leftarrow A - \{a_k\}$;
**Step 8:** Calculate the new $POS(a_k)$ values for all $a_k \in A$;
**Step 9:** If $A = \varnothing$ (i.e., checking for degenerative case #1), then goto START1;

**REPEAT**

**Step 10:** Let $E^-(C)$ be the set of members of $E^-$ which are rejected by $C$;
**Step 11:** If $E^-(C) = \varnothing$, then determine the appropriate degenerative case (i.e., case #2, or #3).
Check whether the corresponding counter has hit the preset limit.
If yes, then go to START1;
**Step 12:** Let $E^- \leftarrow E^-(C)$;
**Step 13:** Calculate the new *NEG* values;
**Step 14:** Let $C$ be the antecedent and $A_i$ the consequent of the rule.
Check the candidate rule $C \rightarrow A_i$ for minimum support.
If it meets the minimum support level criterion, then output the rule;
**Step 15:** Reset the $E^+$ set (i.e., select the examples which have $A_i$ equal to "1" and store them in set $E^+$);

**REPEAT**
**END**

**Figure 12.2.** The Proposed Altered Randomized Algorithm 1 (ARA1) for the Mining of Association Rules (for the CNF Case).

**Database #1:** 1,000 items with 100,000 transactions (the min support was set to 250).

**Database #2:** 1,000 items with 10,000 transactions (the min support was set to 25).

**Database #3:** 500 items with 5,000 transactions (the min support was set to 12).

**Database #4:** 500 items with 4,500 transactions (the min support was set to 11).

**Database #5:** 500 items with 4,000 transactions (the min support was set to 10).

The first results are from the densest databases used in [Agrawal and Srikant, 1994], that is, database #1. The Apriori algorithm was still in the process of generating the frequent item sets of length 2 after 80 hours 22 minutes and 8 seconds when database #1 was used. Therefore, the experiment with the Apriori algorithm was aborted. However, the ARA1 algorithm completed mining the very same database in only 44 hours 22 minutes and 1 second. The ARA1 algorithm mined a single rule with each of the following support levels: 259, 263, 308, 441, 535, 623, 624, 756, 784, 984, and 1,093. All the experiments were run on an IBM 9672/R53 mainframe computer. This processor is a 10-engine box with each engine being rated at 26 MIPS (millions of instructions per second).

For the experiments with Database #2, however, some parallel computing techniques were utilized for the Apriori algorithm. The frequent item sets were gathered into smaller groups, making it possible to build the next frequent item sets in shorter time. As a result, each group was analyzed separately, and the CPU times for each of these jobs were added together at the end. The Apriori algorithm completed mining this database in 59 hours 15 minutes and 3 seconds. Figure 12.3 illustrates the number of rules for this case. By "rules" we mean clauses in DNF (i.e., conjunctions). On the other hand, the ARA1 algorithm mined Database #2 in only 2 hours 54 minutes and 57 seconds. These results are depicted in Figure 12.4.

It should be noted here that the CPU times recorded for the Apriori experiments for this research were higher than the similar results reported in [Agrawal and Srikant, 1994]. For instance, it was reported in [Agrawal and Srikant, 1994] that the Apriori algorithm took approximately 500 seconds to mine Database #1. That result was obtained on an IBM RS/6000 530H workstation with a main memory of 64 MB and running AIX 3.2. On the other hand, for Database #1, the Apriori program written for this research was in the process of generating item sets of length 2 after 80 hours 22 minutes and 8 seconds. The only difference between the approach taken in [Agrawal and Srikant, 1994] and the one in [Yilmaz, *et al.*, 2003] is that the candidate item sets in [Agrawal and Srikant, 1994] were stored in a hash tree. Hashing is a data storage technique that provides fast direct access to a specific stored record on the basis of a given value for some field [Savasere, *et al.*, 1998].

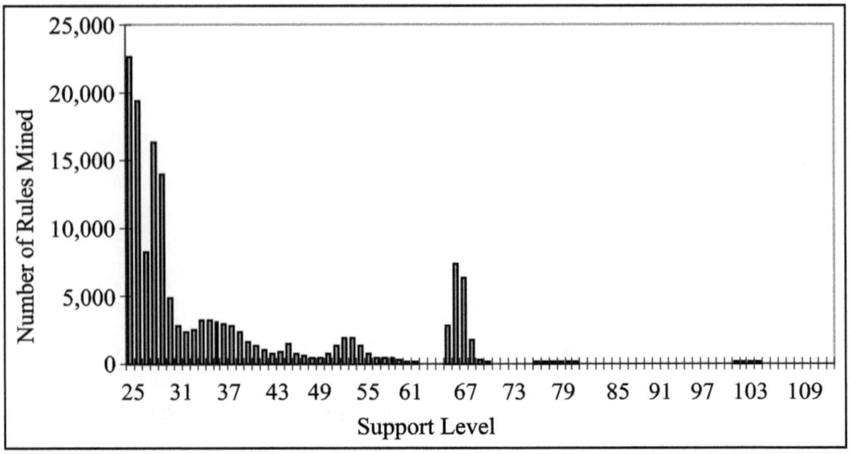


**Figure 12.3.** Histogram of the Results When the Apriori Approach Was Used on Database #2.

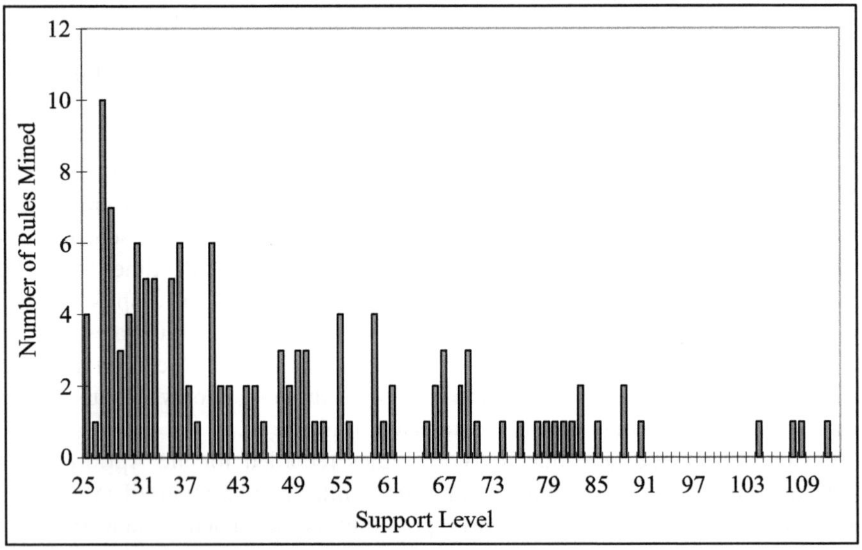


**Figure 12.4.** Histogram of the Results When the ARA1 Approach Was Used on Database #2.

In the work described in [Yilmaz, *et al.*, 2003], hash trees were not used in storing candidate item sets; instead they were kept in the main memory of the computer. This made it faster to access candidate item sets because direct access is generally very expensive CPU-wise. It is believed that the programming techniques and the type of the computers used in [Agrawal and Srikant, 1994] are causing the CPU time difference. In addition, the Apriori code in this research was run under a time-sharing option, which again could make a difference. As was mentioned earlier, the computer codes for the Apriori and the ARA1 algorithms were run on an IBM 9672/R53 computer. The results obtained by using Database #2 suggest that ARA1 produced a reasonable number of rules quickly. Also, these rules were of high quality, since by construction, all had 100% confidence level.

After obtaining these results, it was decided to mine the remaining databases by also using a commercial software package, namely, MineSet by Silicon Graphics. MineSet is one of the most commonly used data mining computer packages. Unfortunately, MineSet used in those tests worked with transactions of a fixed length. Therefore, the transactions were coded as zeros and ones, zeros representing that the corresponding item was not bought, and ones representing that the corresponding item was bought. However, this causes MineSet to also mine negative association rules. Negative association rules are rules based on the absence of items in the transactions too, rather having only their presence. Now negations of attributes may also appear in a rule's structure. Another drawback of MineSet is that only a single item is supported in both the left- and the right-hand sides of the rules to be mined. Also, the version of MineSet used in [Yilmaz, *et al.*, 2003] allowed for a maximum of 512 items in each transaction. The MineSet software used for this study was installed on a Silicon Graphics workstation, which had a CPU clock rate of 500 MHz and a RAM of 512 MB.

As stated above, MineSet supports only a single item in both the left and the right hand sides of the association rules. This suggests that MineSet uses a search procedure of also polynomial time complexity. Such an approach would have first to count the support of each item when it is compared with every other item, and store these supports in a triangular matrix of dimension $n$ (i.e., equal to the number of attributes).

During the pass over the database, the supports of the individual items could be counted, and the rest will only be a matter of checking whether the result is above the preset minimum confidence level. For instance, when checking the candidate association rule $A_2 \rightarrow A_6$, the confidence level would be the support of $A_2$ divided by the support of $A_2A_6$. On the other hand, when doing the same for rule $A_6 \rightarrow A_2$, then the confidence level would be the support of $A_6$ divided by the support of $A_2A_6$. Therefore, such an approach requires $n(n-1)/2$ operations (where $n$ is the number of attributes or items). If $|D|$ is the number of transactions (records) in the database, then the time complexity is $O(|D|n^2)$.

This is almost of the same time complexity that the ARA1 approach has. However, for the ARA1 case, this complexity is for the worst-case scenario. The ARA1 algorithm will stop as soon as it has produced a Boolean function (i.e., a set of clauses or rules) that accepts all the positive and rejects all the negative examples. In addition, the ARA1 approach is able to mine rules with multiple items in the antecedent part of an association rule. The ARA1 approach can also be easily adapted to mine association rules with multiple items in the consequent part. The only change that has to be made is in the partitioning (dichotomization) of the original database into the sets of the positive and the negative examples. On the other hand, the Apriori approach has an exponential time complexity because it follows a combinatorial search approach.

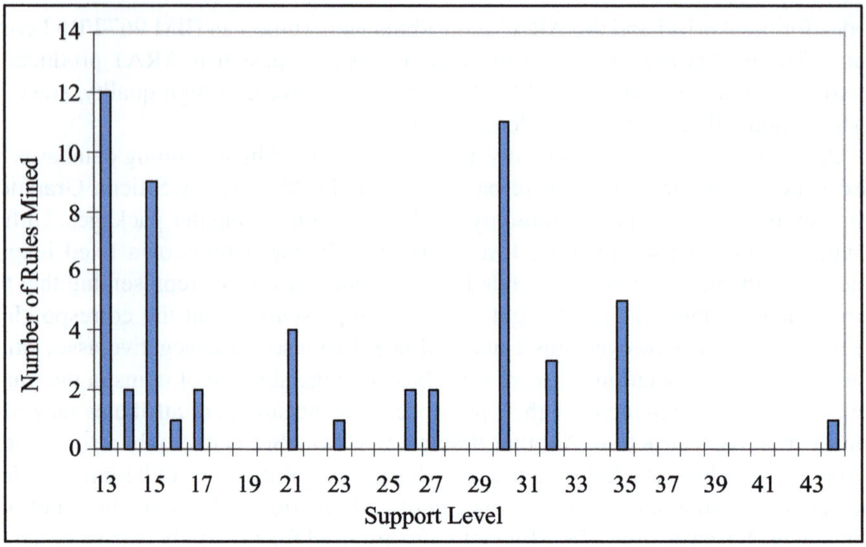


**Figure 12.5.** Histogram of the Results When the MineSet Software Was Used on Database #3.

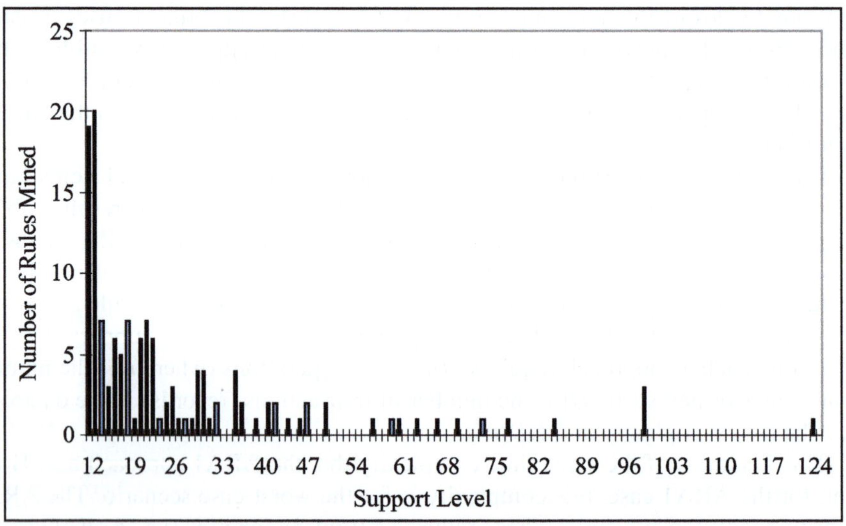


**Figure 12.6.** Histogram of the Results When the ARA1 Approach Was Used on Database #3.

When Database #3 was used, it took MineSet 31 minutes and 40 seconds to mine the association rules. On the other hand, it took ARA1 just 6 minutes and 5 seconds to mine the same database. Figures 12.5 and 12.6 provide the number of the mined rules from database #3. When Database #4 was used, it took MineSet 28 minutes and 30 seconds to mine the association rules. For the ARA1 approach, the required

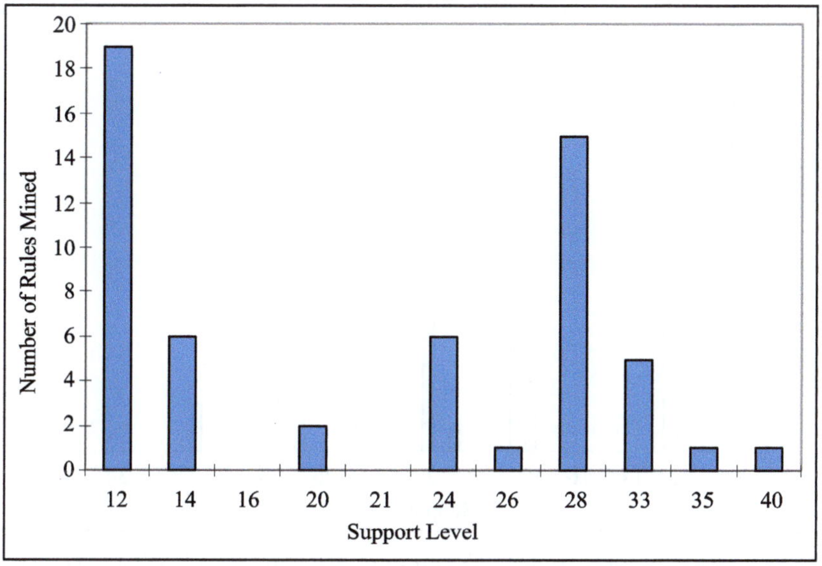


**Figure 12.7.** Histogram of the Results When the MineSet Software Was Used on Database #4.

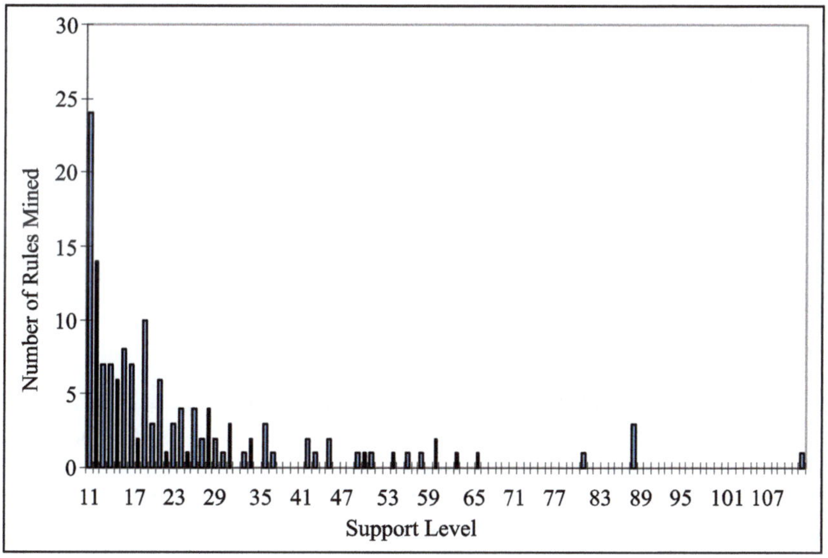


**Figure 12.8.** Histogram of the Results When the ARA1 Approach Was Used on Database #4.

time was 5 minutes and 26 seconds only. These results are depicted in Figures 12.7 and 12.8. For Database #5, it took MineSet 25 minutes and 20 seconds to mine the association rules. On the other hand, it took only 4 minutes and 23 seconds when

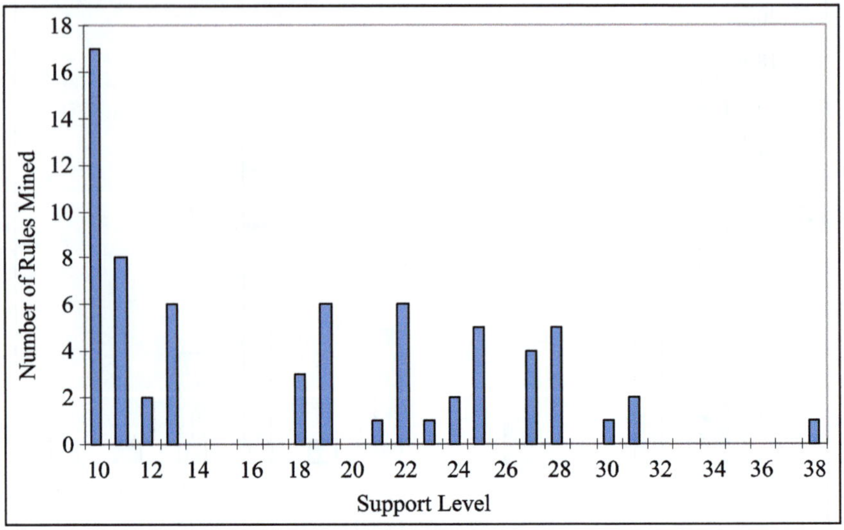


**Figure 12.9.** Histogram of the Results When the MineSet Software Was Used on Database #5.

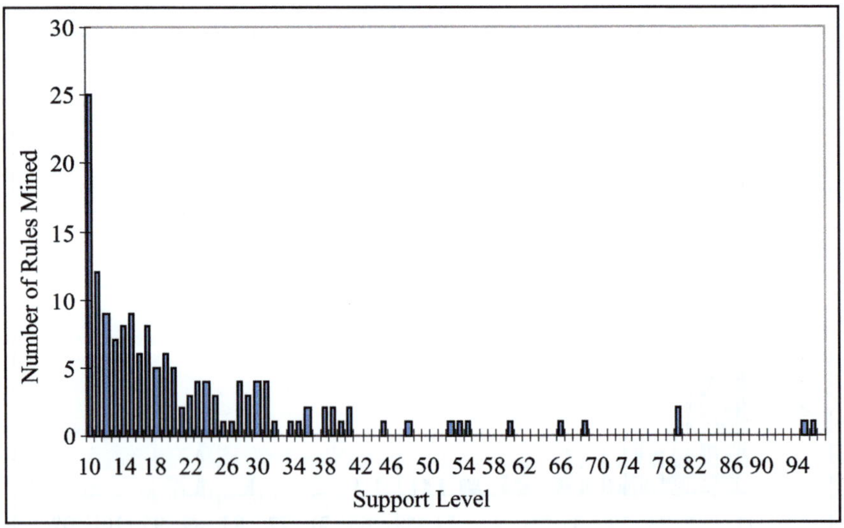


**Figure 12.10.** Histogram of the Results When the ARA1 Approach Was Used on Database #5.

the ARA1 approach was used on the same database. The corresponding results are depicted in Figures 12.9 and 12.10. Table 12.1 presents a summary of all the above CPU times. From these results it becomes evident that the ARA1 approach derives association rules faster and also these rules have much higher support levels.

**Table 12.1.** Summary of the Required CPU Times Under Each Method.

| | **Apriori Method** (*hh:mm:ss*) | **ARA1 Method** (*hh:mm:ss*) | **MineSet Method** (*hh:mm:ss*) |
|---|---|---|---|
| Database #1 | Not completed | 44:22:01 | N/A |
| Database #2 | 59:15:03 | 02:54:57 | N/A |
| Database #3 | N/A | 00:06:05 | 00:31:40 |
| Database #4 | N/A | 00:05:26 | 00:28:30 |
| Database #5 | N/A | 00:04:23 | 00:25:20 |

## 12.5 Concluding Remarks

This chapter presented the developments of a new approach for deriving association rules from databases. The new approach is called ARA1 and it is based on a previous logic-based algorithm (i.e., the RA1 approach as was explained in Chapter 4). This algorithm has the OCAT approach embedded into it. The ARA1 heuristic can easily be randomized.

The proposed ARA1 approach produces a small set of association rules in polynomial time. Furthermore, these rules are of high quality with 100% support levels. The 100% support level of the derived rules is a characteristic of the way the ARA1 approach constructs (i.e., mines) association rules. The ARA1 approach can be further extended to handle cases with less than 100% support levels. This can be done by introducing stopping criteria that terminate the appropriate loops in Figure 12.2, that is, to have a predetermined lower limit (i.e., a percentage less than 100%) of the positive examples to be accepted by each clause (in the CNF case) and also a predetermined percentage of the negative examples which are rejected. The current version of the ARA1 algorithm builds clauses which accept all the positive examples while the final Boolean function it builds rejects all the negative examples (and accepts all the positive ones).

An extensive empirical study was also undertaken. The Apriori approach and the MineSet software (a year 2003 version of it) by Silicon Graphics were compared with the proposed ARA1 algorithm. The computational results demonstrated that the RA1 approach could be both highly efficient and effective. The above observations strongly suggest that the proposed ARA1 approach is very promising for mining association rules in today's world with the always-increasing-in-size and diverse-in-nature databases.

# Chapter 13

# Data Mining of Text Documents

## 13.1 Some Background Information

This chapter investigates the problem of classifying text documents into two disjoint classes. It does so by employing a data mining approach based on the OCAT algorithm. This chapter is based on the work discussed in [Nieto Sanchez, Triantaphyllou, and Kraft, 2002]. In the present setting two sample sets of training examples (text documents) are assumed to be available. An approach is developed that uses indexing terms to form patterns of logical expressions (Boolean functions) that next are used to classify new text documents (which are of unknown class). This is a typical case of supervised "crisp" classification.

A typical application of this type of classification problem with text documents occurs in the declassification process of vast amounts of documents originally produced by the U.S. Federal Government. For reasons of national security, today there are huge numbers of documents that remain classified as secret. These documents are being kept in secured places because they were considered to be important to national security. However, high maintenance costs and new laws dictate that these documents should be reevaluated, and the ones that are not critical any more should become public.

Thus, such documents need to be (roughly speaking) classified into the two disjoint categories of "secret" and "nonsecret." In this context, when a document becomes "nonsecret" after being "secret," it is termed *declassified*. In reality, documents have been classified into multiple levels of criticality to the national security, but in this chapter we will consider only two classes as described above. It should also be stated here that once a document becomes public (i.e., it has been declassified), there is no way to make it secret again (especially now with the proliferation of the Internet). Other similar applications can be found in analyzing vast amounts of text data for the detection of potentially illegal activities or for identifying trends when using the Web (i.e., for Web mining).

In order to highlight the complexity of this kind of classification problem, consider the evaluation of the following three illustrative sentences: "An atomic test is to be conducted at Site $X$," "An atomic test is to be conducted at 1:00 p.m.," and

E. Triantaphyllou, *Data Mining and Knowledge Discovery via Logic-Based Methods*,
Springer Optimization and Its Applications 43, DOI 10.1007/978-1-4419-1630-3_13,

"An atomic test is to be conducted at Site *X* at 1:00 p.m." which come from three hypothetical documents, *A*, *B*, and *C*, respectively. According to [DynMeridian, 1996] and [DOE, 1995], only document *C* is both specific and sensitive and should not be declassified and instead should continue to be kept secret. The reason for this DOE (U.S. Department of Energy) classification rule is because document *C* includes a sentence with specific reference to the "place and time" of an "atomic test." On the other hand, documents *A* and *B* can be declassified (assuming that the rest of their contents is not critical) and become available to the general public. In this illustrative example some key text features that can be used to characterize the two classes are references to "place," "time," and "atomic test."

Traditionally, this declassification process is carried out by employing vast numbers of human experts. However, the sheer amount of documents under consideration can make this process extremely ineffective and inefficient. Although there are guidelines on how to declassify secret documents, directly computerizing the human effort would require developing sophisticated parsers. Such parsers would have to analyze syntactically a document and then determine which, if any, guideline is applicable. The poor quality of the documents (many of which are decades old) and the complexity of the declassification guidelines could make such an approach too risky to national security. Thus, a reasonable alternative is to seek the employment of data mining techniques. More specifically, techniques that use logic-based approaches might be appropriate. Thus, this chapter is centered on the following three interrelated data mining problems:

1) Employ data mining techniques for extracting (mining) from two sets of examples some text features which could be used to correctly group documents into two disjoint classes.
2) Use such features to form Boolean functions (patterns) which could explain how the training examples are grouped together, and accurately classify new documents.
3) When considering a guided learning strategy for extracting the Boolean functions, determine the next training document to include in the sets with the training examples so that accurate Boolean functions are inferred as quickly as possible.

Since being able to justify this kind of classification decisions is important (given the severity of erroneously releasing a sensitive document to the public), methods that do not clearly allow for an explanation of their decision-making process are not appealing here. Therefore, an impetus for this application domain is to seek the development of an approach that is based on mathematical logic, versus approaches that do not provide satisfactory explanation capabilities.

Traditional text classification and *information retrieval* (IR) techniques may have some limitations in solving this problem because they group documents that share a similar content. The prime example of such techniques is the *vector space model* (VSM) [Salton, 1989], which according to the literature ([Buckley and Salton, 1995] and [Shaw, 1995]) is the most effective methodology for this type of classification. The limitation of this technique is that it is based on similarity measures. Thus, it may not be able to distinguish between the three illustrative sentences given earlier in terms of the critical classification issues despite all of them having similar contents. Other techniques, such as fuzzy set approaches (FSA), neural networks (NN), nearest neighbor methods, and computational semantic analysis (SA), have limitations in addressing these data mining problems. This happens because either their time complexity or the resulting sizes of their outputs are still unacceptable and do not possess satisfactory explanatory capabilities (see, for instance, [Scholtes, 1993] and [Chen, 1996]).

An alternative approach to address the present problems is the "One Clause At a Time" (OCAT) algorithm (see also Chapters 2 and 3). As was stated earlier, the OCAT approach extracts (mines) key features from the training examples and next uses them to infer Boolean functions which can be used in classifying the training examples into the two original disjoint classes. These Boolean functions can also easily be transformed into IF-THEN type of classification rules. The OCAT approach applies to examples that can be represented by binary vectors (although attributes with continuous values can be transformed into ones with binary values as was described in Section 2.2). However, this is not a limitation because it is the mere presence or absence of certain key words that can cause a document to be grouped in one class or another. On the other hand, the typical document classification done by traditional IR systems uses term frequencies (which are continuous numbers usually normalized in the interval [0, 1]) of keywords to group together documents of seemingly similar context.

This chapter is organized as follows. Section 13.2 presents an overview of the document clustering process. Section 13.3 briefly describes the OCAT approach in the context of the classification of text documents. Section 13.4 presents an overview of the VSM algorithm. Section 13.5 presents a quick overview of the guided learning approach. Sections 13.6 to 13.9 describe the experimental data, testing methodology, and analyses of the derived computational results. Finally the chapter ends with some concluding remarks.

## 13.2 A Brief Description of the Document Clustering Process

The traditional process for automatic clustering of text documents results in a grouping of documents with similar content into meaningful groups in order to facilitate their storage and retrieval [Salton, 1989]. This is a four-step process as follows. In the first step a computerized system compiles a list of the unique words that co-occur in a sample of the documents from various classes (see, for example, [Salton, 1989] and [Cleveland and Cleveland, 1983]). In the second step, the co-occurring frequency of these words is analyzed and the best set of indexing terms is extracted. Usually, indexing terms (also known as *keywords* or *content descriptors*) are selected among the words with moderate frequency. The most *common* and the most *rare* words (i.e., the most frequent and infrequent words, respectively) are discarded as keywords because they convey little lexical meaning (see, for example, [Zipff, 1949], [Luhn, 1957; 1958], [Salton, 1968], [Cleveland and Cleveland, 1983], [Fox, 1990], and [Meadow, 1992]). Some examples of common words are "a," "an," "and," and "the" [Fox, 1990]. Rare words depend on a document's subject domain [Meadow, 1992].

In the third step, a document is indexed by affixing it with the set of keywords that only occur in its text. According to Cleveland and Cleveland [1983], "this assignment is correct because authors usually repeat words that conform with the document's subject." A list (vector) of keywords represents the content of a document and usually it is referred to as a *document representation* or *surrogate*. An example of such a surrogate is the list of the seven words or phrases: {"*Document classification*," "*document indexing*," "*vector space model*," "*data mining*," and "*OCAT algorithm*"}. This surrogate could be used to represent the content of this chapter, which is composed of thousands of words, symbols, and numbers. Hence, the goal of the third step is to construct a surrogate for representing the content of each document.

An advantage of using such surrogates is that they can be further simplified by expressing them as numerical vectors [Salton, 1989]. One way to construct such vectors is by expressing their elements as binary values to indicate the presence (denoted by 1) or absence (denoted by 0) of certain keywords in a document. For instance, the vector's element $w_{ij} = 1$ (or 0) expresses the presence (or absence) of keyword $T_i$ $(i = 1, 2, 3, \ldots, t)$ in document $D_j$ $(j = 1, 2, 3, \ldots, N)$. Thus, the surrogate $D_j = [0\ 1\ 1\ 1\ 1\ 0]$ of six binary elements indicates the presence of keywords (terms) $T_2$, $T_3$, $T_4$, and $T_5$ and the absence of keywords (terms) $T_1$ and $T_6$ in $D_j$.

Another way to construct these numerical vectors is by expressing (i.e., weighting) their elements using real values from the range [0, 1]. In this case, the value of an element $w_{ij}$ indicates the relative occurrence frequency of a keyword within a document. For instance, a hypothetical surrogate such as $D_j = [0.00\ 1.00\ 0.10\ 0.75\ 0.90\ 1.00]$ may indicate that term $T_3$ occurs a few times, terms $T_4$ and $T_5$ occur moderately frequently, and terms $T_2$ and $T_6$ occur with high frequency in $D_j$. In the remainder of this chapter, however, only binary surrogates will be considered. As stated earlier the reason for considering binary vectors as surrogates is because the mere presence or absence of a keyword (or some pattern of keywords) may be detrimental in assigning a document to one of the two disjoint classes considered in this chapter.

In the last step of the (traditional) classification process, documents sharing similar keywords (i.e., similar content) are grouped together. This classification follows from the pairwise comparison of all the surrogates [Salton, 1989].

## 13.3 Using the OACT Approach to Classify Text Documents

In order to help illustrate the main issues of this process consider the two sets of training examples depicted in Figure 13.1. These are the same data used in previous chapters. The set with the positive examples is comprised of four examples, while the set of the negative examples is comprised of six examples. These examples (document surrogates in our context) are defined on four binary features (i.e., index terms

$$E^+ = \begin{bmatrix} 0 & 1 & 0 & 0 \\ 1 & 1 & 0 & 0 \\ 0 & 0 & 1 & 1 \\ 1 & 0 & 0 & 1 \end{bmatrix} \quad E^- = \begin{bmatrix} 1 & 0 & 1 & 0 \\ 0 & 0 & 0 & 1 \\ 1 & 1 & 1 & 1 \\ 0 & 0 & 0 & 0 \\ 1 & 0 & 0 & 0 \\ 1 & 1 & 1 & 0 \end{bmatrix}$$

**Figure 13.1.** A Sample of Four Positive and Six Negative Examples.

$$E^+ = \begin{bmatrix} 1 & 0 & 1 & 0 \\ 0 & 0 & 0 & 1 \\ 1 & 1 & 1 & 1 \\ 0 & 0 & 0 & 0 \\ 1 & 0 & 0 & 0 \\ 1 & 1 & 1 & 0 \end{bmatrix} \quad E^- = \begin{bmatrix} 0 & 1 & 0 & 0 \\ 1 & 1 & 0 & 0 \\ 0 & 0 & 1 & 1 \\ 1 & 0 & 0 & 1 \end{bmatrix}$$

**Figure 13.2.** The Training Example Sets in Reverse Roles.

or attributes) or their negations. A value of 1 indicates the presence of the corresponding index term, while a value of 0 indicates the absence of the index term.

When the OACT approach, combined with the RA1 heuristic, is applied on the previous data, the following Boolean function is derived:

$$(A_2 \vee A_4) \wedge (\bar{A}_2 \vee \bar{A}_3) \wedge (A_1 \vee A_3 \vee \bar{A}_4). \tag{13.1}$$

Recall that a fundamental property of Boolean function (13.1) is that it accepts (i.e., evaluates to 1) all the examples in $E^+$, while it rejects (i.e., evaluates to 0) all the examples in $E^-$. However, since such Boolean function is usually constructed from a relatively small collection of training examples, it is possible for the Boolean function to be inaccurate when it classifies new examples. The error may occur either if the new example is positive, and the Boolean function rejects it, or if the new example is negative and the Boolean function accepts it.

Let us consider generating a second Boolean function for classifying new examples. The second Boolean function is derived by treating the second set of training examples as positive and the first as the negative training examples. For instance, Figure 13.2 depicts the same training examples as the ones in Figure 13.1, but now they have reverse roles.

When the OCAT approach, combined with the RA1 heuristic, is applied on the new inference problem, the following Boolean function (13.2) is derived:

$$(A_3 \vee \bar{A}_2) \wedge (\bar{A}_4 \vee A_2 \vee \bar{A}_1) \wedge (A_1 \vee \bar{A}_3). \tag{13.2}$$

As with Boolean function (13.1), a property of the corresponding Boolean function $(A_3 \vee \bar{A}_2) \wedge (\bar{A}_4 \vee A_2 \vee \bar{A}_1) \wedge (A_1 \vee \bar{A}_3)$ is to accept (i.e., to evaluate to 1) the former negative examples and to reject (i.e., to evaluate to 0) the former positive examples. For convenience, following the setting of the examples in Figures 13.1 and 13.2, Boolean function (13.1) will be called the *positive rule* (denoted as $R^+$) and Boolean function (13.2) the *negative rule* (denoted as $R^-$).

The disadvantage of using only one rule (Boolean function) can be overcome by considering the combined decisions of $R^+$ and $R^-$ when classifying a new example $e$. If $e$ is a positive example, it will be denoted as $e^+$, while if it is a negative example, it will be denoted as $e^-$. Under this setting, the classification of $e$ can be:

1. *Correct* if and only if:
   (a) $R^+(e^+) = 1$ and $R^-(e^+) = 0$;
   (b) $R^+(e^-) = 0$ and $R^-(e^-) = 1$;
2. *Incorrect* if and only if:
   (c) $R^+(e^+) = 0$ and $R^-(e^+) = 1$;
   (d) $R^+(e^-) = 1$ and $R^-(e^-) = 0$;
3. *Undecidable* if and only if:
   (e) $R^+(e^+) = 1$ and $R^-(e^+) = 1$;
   (f) $R^+(e^-) = 1$ and $R^-(e^-) = 1$;
   (g) $R^+(e^+) = 0$ and $R^-(e^+) = 0$;
   (h) $R^+(e^-) = 0$ and $R^-(e^-) = 0$.

Cases (a) and (b) correspond to "correct" classifications because both rules perform according to the desired properties described above. However, as indicated above it is possible that the rules could incorrectly classify an example (cases (c) and (d)). Or the rules could simultaneously accept (cases (e) and (f)) or simultaneously reject (cases (g) and (h)) the example. Cases (e) through (h) are called *undecidable* (or *unclassifiable*) because one of the rules does not possess enough classification knowledge, and thus such a rule must be reconstructed. Therefore, "undecidable" situations open the path to improve the accuracy of a classification system. This chapter exploits the presence of "undecidable" situations in order to guide the reconstruction of the rule (Boolean function) that triggered an erroneous classification decision. Recall that the above were also described in a nontechnical manner in Figure 5.1.

## 13.4 An Overview of the Vector Space Model

The *vector space model* (VSM) is a mathematical model of an *information retrieval* (IR) system that can also be used for the classification of text documents (see, for instance, [Salton and Wong, 1975] and [Salton, 1989]). It is often used as a benchmarking method when dealing with document retrieval and classification related problems. Figure 13.3 illustrates a typical three-step strategy of the VSM approach to clustering.

To address Step 1 Salton [1989] suggested that a suitable measure for comparing in pairwise manner any two surrogates $X$ and $Y$ is the cosine coefficient (CC) as

**Input**: A sample of surrogates.
**Output**: Clusters of documents and clusters' centroids.
**Step 1:** Compute the pairwise similarity coefficients among all surrogates in the sample.
**Step 2:** Cluster documents with sufficiently large pairwise similarities.
**Step 3:** Compute the centroids of the clusters.

**Figure 13.3.** The Vector Space Model (VSM) Approach.

defined in Equation (13.3) (other similarity measures are listed in [Salton, 1989], Chapter 10):

$$CC = \frac{|X \cap Y|}{|X|^{1/2} \cdot |Y|^{1/2}}. \tag{13.3}$$

In this formula, $X = (x_1, x_2, x_3, \ldots, x_n)$ and $Y = (y_1, y_2, y_3, \ldots, y_n)$, where $x_i$ indicates the presence (value 1) or absence (value 0) of the $i$-th indexing term in $X$ and similarly for $y_i$ with respect to $Y$. Moreover, $|X| = |Y| = n$ is the number of indexing terms, and $|X \cap Y|$ is the number of indexing terms appearing simultaneously in $X$ and $Y$. To be consistent with the utilization of binary surrogates, formula (13.3) provides the CC expression for the case of Boolean vectors. This coefficient measures the angle between two surrogates (Boolean vectors) on a Cartesian plane. Salton [1989] indicates that "the magnitude of this angle can be used to measure the *similarity* between any two documents." This statement is based on the observation that two surrogates are identical, if and only if the angle between them is equal to $0°$.

In Step 2 the VSM clusters together documents that share a similar content based on their surrogates. According to Salton [1989], any clustering technique can be used to group documents with similar surrogates. A collection of clustering techniques is given in [Anderberg, 1973], [Van Rijsbergen, 1979], [Aldenderfer, 1984], and [Späth, 1985]. However, it is important to mention here that with any of these techniques, the number of generated classes is always a function of some predefined parameters. This is in contrast to the requirements of our problem here in which the number of classes is exactly equal to two. When the VSM works under a predefined number of classes, it is said to perform a pseudo-classification. In this chapter the training examples are already grouped into two (disjoint) classes. Thus, the VSM is applied on the examples (documents) in each class and the corresponding centroids are derived. Hence, we will continue to use this kind of pseudo-classification.

To address Step 3 [Salton, 1989], [Salton and Wong, 1975], and [Van Rijsbergen, 1979] suggest the computation of a class centroid to be done as follows. Let $w_{rj}$ $(j = 1, 2, 3, \ldots, n)$ be the $j$-th element of the centroid for class $C_r$ which contains $q$ documents. Also, the surrogate for document $D_i$ is defined as $\{D_{ij}\}$. Then, $w_{rj}$ is computed as follows:

$$w_{rj} = (1/q) \sum_{i=1}^{q} D_{ij}, \quad \text{for } j = 1, 2, 3, \ldots, n. \tag{13.4}$$

That is, the centroid for class $C_r$ is also a surrogate (also known as the "average" document) defined on $t$ keywords.

Finally, the VSM classifies a new document by comparing (i.e., computing the CC value) its surrogate against the centroids that were created in Step 3. A new document will be placed in the class for which the CC value is maximum.

In the tests to be described later in this chapter, the VSM is applied on the documents (training examples) available for each class. In this way, the centroid of each of the two classes is derived. For instance, consider the training examples depicted in Figures 13.1 and 13.2. The VSM is now applied on these data. The centroids in expression (13.5) have been computed from the data in Figure 13.1 and the centroids in expression (13.6) from the data in Figure 13.2. Obviously, the centroids for the second set are in reverse order of those for the first set of data.

$$C_{+} = [1/2, 1/2, 1/4, 1/2] \tag{13.5}$$
$$C_{-} = [2/3, 1/3, 1/2, 1/3]$$

$$C'_{+} = [2/3, 1/3, 1/2, 1/3] \tag{13.6}$$
$$C'_{-} = [1/2, 1/2, 1/4, 1/2]$$

The variables $C_{+}$ and $C_{-}$ stand for the centroids for the data in Figure 13.1, and the variables $C'_{+}$ and $C'_{-}$ stand for the centroids for the data in Figure 13.2. In order to match the names of the positive and negative rules described for the OCAT algorithm, the two centroids for the data in Figure 13.1 will be called the *positive centroids* while the two centroids for the data in Figure 13.2 will be called the *negative centroids*. As with the OCAT algorithm, the utilization of two sets of centroids has been investigated in order to tackle the new classification problem by using the VSM as new examples become available.

## 13.5 A Guided Learning Approach for the Classification of Text Documents

The central idea of the guided learning approach (GLA) in the context of this chapter can be illustrated as follows. Suppose that the collection to be classified contains millions of documents. Also, suppose that an oracle (i.e., an expert classifier) is queried in order to classify a small sample of examples (documents) into the two groups $E^{+}$ and $E^{-}$. Next, suppose that the OCAT algorithm is used to construct the positive and negative rules, such as was the case with the illustrative samples in Figures 13.1 and 13.2. As indicated earlier, these rules may be inaccurate when classifying examples not included in the training set, and therefore they will result in one of the classification outputs provided in cases (a) through (h), as described in Section 13.3. One way to improve the classification accuracy of these rules is to add more documents, one at a time, to the training set (either in $E^{+}$ or $E^{-}$) and have them reconstructed. Therefore, the question GLA attempts to answer is: Which next document should be inspected by the expert classifier so that the classification performance of the derived rules could improve as fast as possible?

One way to provide the expert with a new document is to randomly select one from the remaining unclassified documents. We call this the random input learning strategy (to be denoted as RANDOM). A drawback of this strategy may occur if the oracle and *incumbent* rules frequently classify a new document in the same class. If this occurs frequently, then the utilization of the oracle and the addition of the new example to the training set is of no benefit. An alternative and more efficient way is to provide the expert with a document selected from the "undecidable" (i.e., the undecidable/unclassifiable) group. This strategy (in a general form) was first introduced in [Triantaphyllou and Soyster, 1995] and was explained in Chapter 5. This approach appears to be a more efficient way for selecting the next document because an "undecidable" situation implies that one of the two rules misclassified the document. Therefore, the expert's verdict will not only guide the reconstruction of the rule that triggered the misclassification, but it may also improve the accuracy of the two rules. We call this the guided learning strategy (to be denoted as GUIDED). It should also be stated here that an incremental learning version of the OCAT approach was given in Chapter 6.

## 13.6 Experimental Data

In order to better understand the classification performance of the OCAT approach in addressing this new problem, OCAT's classification accuracy was compared with that of the VSM. Both approaches were tested under three experimental settings: (i) a leave-one-out cross validation (or CV) (also known as the Round-Robin test); (ii) a 30/30 cross validation (or 30CV), where 30 stands for the number of training documents in each class; and (iii) in an experimental setting in which the OCAT algorithm was studied under a random and a guided learning strategy. These will be defined below. This multiple testing strategy was selected in order to gain a more comprehensive assessment of the effectiveness of the various methods.

For these tests, a sample of 2,897 documents was randomly selected from four document classes of the TIPSTER collection ([Harman, 1995] and [Voorhees, 1998]). The previous number of documents and those below for each class were determined based on memory limitations of the computing platform used (an IBM Pentium II PC running Windows 95). The TIPSTER collection is a standard data set for experimentations with information retrieval systems. The four document classes were as follows:

1) U.S. Department of Energy (DOE) documents,
2) Wall Street Journal (WSJ) documents,
3) Associated Press (AP) documents, and
4) ZIPFF class documents.

**Table 13.1.** Number of Documents Randomly Selected from Each Class.

| Class | DOE | AP | WSJ | ZIPFF | Total |
|---|---|---|---|---|---|
| **Number of Documents** | 1,407 | 336 | 624 | 530 | 2,897 |

NOTES: DOE, AP, WSJ, and ZIPFF stand for the U.S. Department of Energy, the Associated Press, and the Wall Street Journal, respectively; ZIPFF is a collection of technical documents on various topics.

**Table 13.2.** Average Number of Indexing Words Used in Each Experiment.

| Type of Experiment | DOE vs. AP | DOE vs. WSJ | DOE vs. ZIPFF | AP vs. WSJ | WSJ vs. ZIPFF |
|---|---|---|---|---|---|
| CV | 511 | 605 | 479 | 448 | 501 |
| 30CV | 803 | 911 | 890 | 814 | 811 |

NOTES: In order to keep the size reasonable for our computing environment, only the first hundred words from each document were considered. Stop words were always removed.

We chose documents from the TIPSTER collection because for security reasons we did not have access to actual secret DOE documents. Table 13.1 shows the number of documents that were used in the computational experiments. These documents were randomly extracted from the four classes of the TIPSTER collection.

Two mutually exclusive classes were simulated by forming the following five class-pairs: (DOE vs. AP), (DOE vs. WSJ), (DOE vs. ZIPFF), (AP vs. WSJ), and (WSJ vs. ZIPFF). These five class-pairs were randomly selected from all possible combinations of class-pairs. To comply with the notation presented in the previous sections, the first class of each class-pair was denoted as $E^+$, while the second class was denoted as $E^-$. Thus, the problem now is to find a Boolean function which accurately classifies a document surrogate according to the appropriate TIPSTER class.

Table 13.2 shows the average number of keywords that were extracted from the five class-pairs mentioned above. The data in this table can be interpreted as follows. For the class-pair (DOE vs. AP), the average number of keywords used in all the experiments was 511 under the CV validation and 803 under the 30CV validation. A similar interpretation applies to the data in the other columns.

It should be stated at this point that a number of alternative indexing terms were used. Besides single words, sequences of two words at a time, sequences of three words at a time, and sequences of four words at a time were also used. However, some pilot studies indicated that the best results would be derivable by using as indexing terms single words only. Thus, the attributes (binary variables) in the derived Boolean functions are single keywords and not sequences of them.

## 13.7 Testing Methodology

This section first summarizes the methodology for the Leave-One-Out Cross Validation and the 30/30 Cross Validation. These two alternative testing methods have been employed in order to gain a better appreciation of the effectiveness of the various procedures proposed here to classify text documents. The same section also presents the statistical tests employed to determine the relative performance of the VSM and the OCAT/RA1 algorithm. This section ends with the methodology for the guided learning approach.

### 13.7.1 The Leave-One-Out Cross Validation

The cross validation (CV) testing was implemented on samples of 60 documents as follows. First, 30 documents from each class were randomly selected. Note that the size 60 was used due to storage limitations in our computing environment. Then, one document was removed from these sets of documents with its class noted.

After that, the *positive* and *negative rules* under the OCAT/RA1 approach and the *positive* and *negative centroids* under the VSM were constructed using the remaining 59 documents. In the third step, the class of the document left out was inferred by both algorithms. Then, the correctness of the classification was determined according to the cases (a) through (h), as defined in Section 13.3.

The previous second and third steps were repeated with different sets of training examples until all 60 documents had their class inferred one at a time. This experimental setting was replicated ten times with different subsets of the training data, at which point the results of the two algorithms were tested for statistical difference.

### 13.7.2 The 30/30 Cross Validation

The 30/30 cross validation (30CV) was implemented on samples of 254 documents as follows. The number of 254 documents was used to avoid excessive computational time. Initially, the *positive* and *negative rules* under the OCAT/RA1 approach and the *positive* and *negative centroids* under the VSM were constructed by using only 30 documents (randomly selected) from each class (i.e., a total of 60 documents were used in each run).

Next, the classification of the remaining 194 documents was inferred. As before, the correctness of this classification was determined according to the cases (a) through (h), as defined in Section 13.3. As with the first experimental setting, the 30CV validation was replicated ten times, at which point the results of the two algorithms were tested for statistical difference.

### 13.7.3 Statistical Performance of Both Algorithms

To determine the statistical performance of both algorithms, the following hypotheses were tested. The first test was needed to determine the relative dominance of the algorithms. The second test was implemented based on a sign test in order to determine the consistency of the dominance of the algorithms as follows:

1. $H_o : P_{\text{OCAT/RA1}} \leq P_{\text{VSM}}$

   $H_1 : P_{\text{OCAT/RA1}} > P_{\text{VSM}}$

2. $H_o : p = 0.50$

   $H_1 : p \neq 0.50$

where $P_{\text{OCAT/RA1}}$ and $P_{\text{VSM}}$ are the numbers of documents with "correct" classification under each algorithm, divided by the total number of documents in the experiment. In addition, $p$ is the probability of finding an equal number of positive and negative differences in a set of outcomes. More on how these tests were performed is provided in Section 13.8, which presents the computational results.

### 13.7.4 Experimental Setting for the Guided Learning Approach

Let us consider the following question: What is the best next document to be given to the oracle in order to improve the performance of the two classification rules? Three samples of 510 documents (255 from each class) from the three class-pairs (DOE vs. ZIPFF), (AP vs. DOE), and (WSJ vs. ZIPFF) were used in an investigation of this question. The number of 510 documents was determined by the available RAM memory on the Windows PC we used. The previous three class-pairs were processed by the OCAT/RA1 algorithm (only) under the RANDOM and the GUIDED learning approaches. Note that the VSM produces *symmetric systems* and thus it cannot be used with this guided learning strategy (see also Chapter 5).

These two learning approaches were implemented as follows. At first, 30 documents from each class in the experiment were randomly selected, and the positive and negative rules (Boolean functions) were constructed. Next, the class membership of all 510 documents in the experiment was inferred based on the two sets of classification rules. The criteria expressed as cases (a) through (h) in Section 13.3 were used to determine the number of "correct," "incorrect," and "undecidable" classifications. Next, a document was added to the initial training sample as follows. For the case of the RANDOM approach, this document was selected randomly from among the documents not included in the training sets yet (i.e., neither in $E^+$ nor in $E^-$).

In contrast, under the GUIDED approach this document was selected from the set of documents which the positive and negative rules had already termed as "undecidable." However, if the two rules did not detect an "undecidable" case, then the GUIDED approach was replaced by the RANDOM approach until a new "undecidable" case was identified (see also Chapter 5). This process for the RANDOM and GUIDED approaches was repeated until all 510 documents were included in the two training sets $E^+$ and $E^-$. The results of all these experiments are presented next.

## 13.8 Results for the Leave-One-Out and the 30/30 Cross Validations

Table 13.3 (parts (a) and (b)) summarizes the experimental results for the CV validation, while Table 13.4 (parts (a) and (b)) summarizes the results for the 30CV validation. The abbreviations "C," "I," and "U" in the first column of both tables correspond to the "correct," "incorrect," and "undecidable" classification outcomes, respectively. These outcomes can be obtained by using the positive and the negative rules (for the OCAT/RA1 case) or the positive and the negative centroids (for the VSM case). For instance, the data in Table 13.3 (part (a)), column 2 (i.e., class-pair (DOE vs. AP)) indicate that the VSM identified 334 "correct," 261 "incorrect," and 5 "undecidable" cases.

Similarly, the data in Table 13.3 (part (a)), column 3 (i.e., class-pair (DOE vs. AP)) indicate that the OCAT algorithm identified 410 "correct," 5 "incorrect," and 185 "undecidable" classifications. The data in the other columns can be interpreted in a similar manner. The last two columns of Table 13.3 (part (b)) and Table 13.4 (part (b)) summarize the results across all five class-pairs. Figure 13.4 compares the proportions of these results for both algorithms.

Two key observations can be derived from the size of the dark areas (or areas of "undecidable" classifications) in Figure 13.4 which was derived from the data in Tables 13.3 and 13.4. First, it can be observed that the proportion of "undecidable" cases detected by the VSM algorithm is almost 0% (but not equal to zero). These have occurred when the two positive and the two negative centroids accepted the same document and, therefore, the classes predicted by both sets of centroids have to be

**Table 13.3a.** Summary of the First Experimental Setting: Leave-One-Out Cross Validation (part a).

| | DOE vs. AP | | DOE vs. WSJ | | DOE vs. ZIPFF | |
|---|---|---|---|---|---|---|
| | **VSM** | **OCAT/RA1** | **VSM** | **OCAT/RA1** | **VSM** | **OCAT/RA1** |
| C | 334 | 410 | 286 | 296 | 280 | 358 |
| I | 261 | 5 | 314 | 66 | 320 | 25 |
| U | 5 | 185 | 0 | 238 | 0 | 217 |

**Table 13.3b.** Summary of the First Experimental Setting: Leave-One-Out Cross Validation (part b).

| | AP vs. WSJ | | WSJ vs. ZIPFF | | TOTAL | |
|---|---|---|---|---|---|---|
| | **VSM** | **OCAT/ RA1** | **VSM** | **OCAT/ RA1** | **VSM** | **OCAT/ RA1** |
| C | 316 | 365 | 286 | 303 | 1,502 (or 50.1%) | 1,732 (or 57.7%) |
| I | 284 | 47 | 314 | 76 | 1,493 (or 49.8%) | 219 (or 7.3%) |
| U | 0 | 188 | 0 | 221 | 5 (or 0.1%) | 1,049 (or 35.0%) |

**Table 13.4a.** Summary of the Second Experimental Setting: 30/30 Cross Validation (part a).

| | DOE vs. AP | | DOE vs. WSJ | | DOE vs. ZIPFF | |
|---|---|---|---|---|---|---|
| | VSM | OCAT | VSM | OCAT | VSM | OCAT/RA1 |
| C | 975 | 1,406 | 975 | 1,266 | 1,035 | 1,320 |
| I | 975 | 70 | 975 | 134 | 915 | 124 |
| U | 0 | 474 | 0 | 550 | 0 | 506 |

**Table 13.4b.** Summary of the Second Experimental Setting: 30/30 Cross Validation (part b).

| | AP vs. WSJ | | WSJ vs. ZIPFF | | TOTAL | |
|---|---|---|---|---|---|---|
| | VSM | OCAT/ RA1 | VSM | OCAT/ RA1 | VSM | OCAT/ R1 |
| C | 1,088 | 1,283 | 1,088 | 1,145 | 5,161 (or 52.9%) | 6,420 (or 65.9%) |
| I | 846 | 140 | 837 | 176 | 4,548 (or 46.7%) | 644 (or 6.6%) |
| U | 16 | 527 | 25 | 41 | 41 (or 0.4) | 2,686 (or 27.5%) |

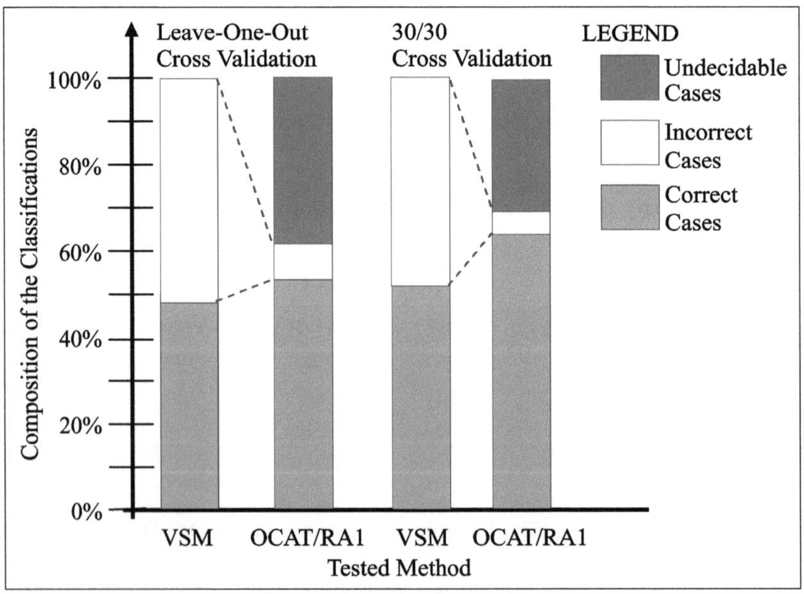


**Figure 13.4.** Comparison of the Classification Decisions Under the VSM and the OCAT/RA1 Approaches.

selected randomly. More specifically, these "undecidable" instances occurred even when these randomly predicted classes were identical. The VSM was implemented based on the CC value, following the suggestions in [Salton, 1989].

In contrast, as the second observation, one has large proportions of "undecidable" classifications with the OCAT/RA1 algorithm. Recall that this type of classification decision is preferred to making a wrong classification because such cases demonstrate that the positive, or the negative, or both rules are unable to correctly classify new documents. Therefore, in these results the large dark areas in the above figure show that both rules were unable to correctly classify a large proportion of the documents in the experiments. More importantly, the size of these areas indicates that positive or negative rules may be improved if they are modified when an "undecidable" situation is detected.

Consider the proportion of the "incorrect" classifications (i.e., the least gray areas in Figure 13.4). One can derive two conclusions. First, the number of "incorrect" classifications the VSM made amounts to 49.8% with the leave-one-out cross validation and to 46.7% with the 30/30 cross validation. These large proportions of "incorrect" classifications can be attributed to the inability of the positive and negative centroids to distinguish between "incorrect" and "undecidable" classifications. Second, the results show that the OCAT/RA1 algorithm made 7.3% and 6.6% of "incorrect" classifications with the two test settings. In this case, these rather small error rates can be attributed to the utilization of the positive and the negative rules (Boolean functions) of small size which enabled the OCAT/RA1 algorithm to distinguish between the "incorrect" and "undecidable" classifications.

Despite the disparate proportions of the "inaccurate" and "undecidable" classifications for both of these algorithms, their performances were statistically compared using *only* the proportions with the "correct" classifications. That is, the undecidable cases were not considered here. In this way the VSM approach was not placed in an unfair setting when it was compared with the OCAT/RA1 approach. This comparison was needed in order to determine which algorithm better addressed the classification problem studied in this chapter. For this comparison, it was assumed that no additional improvement of the two algorithms was possible under the CV and 30CV cross validations. Furthermore, the "incorrect" and "undecidable" outcomes were considered as incorrect classifications.

The results of these tests (as shown in Table 13.5) indicate that the OCAT/RA1 approach is more accurate than the VSM in both types of computational experiments. Furthermore, the very low $p$-values in Table 13.6 indicate that it is extremely unlikely to find a similar number of positive and negative differences in the proportions of the "correct" classifications under the two approaches [Barnes, 1994]. Therefore, the results of these two statistical tests profoundly indicate that the OCAT/RA1 approach is better suited to address the document classification problem studied here.

**Table 13.5.** Statistical Difference in the Classification Accuracy of the VSM and OCAT/RA1 Approaches.

| | | | | Binomial Test | |
|---|---|---|---|---|---|
| | $P_{OCAT}$[§] | $P_{VSM}$[£] | $P_{VSM} - P_{OCAT}$ | Half-length[¶] | Interval |
| CV | 0.577 | 0.501 | −0.0760 | 0.025 | (−0.035, −0.085) |
| 30CV | 0.658 | 0.529 | −0.1287 | 0.014 | (−11.47, −14.27) |

NOTES:
[§]1,732/$n$ and 6,420/$n$; where $n$ is 3,000 for CV and 9,750 for 30CV.
[£]1,502/$n$ and 5,161/$n$; where $n$ is 3,000 for CV and 9,750 for 30CV.
[¶]Denotes that both approaches performed statistically differently.

**Table 13.6.** Data for the Sign Test to Determine the Consistency in the Ranking of the VSM and OCAT/RA1 Approaches.

| | Type of Experiment | |
|---|---|---|
| | CV | 30CV |
| Number of "+" signs | 4 | 7 |
| Number of "−" signs | 46 | 43 |
| $p$-value $= \sum_{i=0}^{m} \binom{50}{i} \cdot p^i \cdot (1-p)^{50-i}$, where $m = 4$ for CV and 7 for 30CV and $p = 0.50$ | $p$-value $= 2.23 \times 10^{-10}$ | $p$-value $= 1.04 \times 10^{-7}$ |

## 13.9 Results for the Guided Learning Approach

Figures 13.5 through 13.7 show the results of the OCAT/RA1 algorithm under the RANDOM and GUIDED learning approaches. The horizontal axis indicates the percentage of training documents used during the experiment. For instance, at the beginning of the experiment there were 60 training documents or 11.76% of the 510 documents in the experiment. Next, when one more document was added to the training set, following the recommendation of the GUIDED and RANDOM approaches, there were 12.16% of the documents in the experiment.

The vertical axis shows the proportions of "correct," "incorrect," and "undecidable" classifications for the various percentages of training documents used in the experiment. The abbreviations *Rc*, *Ri*, *Ru*, *Gc*, *Gi*, and *Gu* stand for the proportions of "correct," "incorrect," and "undecidable" outcomes for the RANDOM and GUIDED approaches, respectively. For instance, *Rc* is the proportion of "correct" classifications under the RANDOM approach, and *Gu* is the proportion of "undecidable" classification under the GUIDED approach.

Table 13.7 shows the percentages of training documents the OCAT/RA1 algorithm needed before it classified all 510 documents in each class-pair correctly (i.e., until it became 100% accurate). The position of the vertical dotted line in the previous

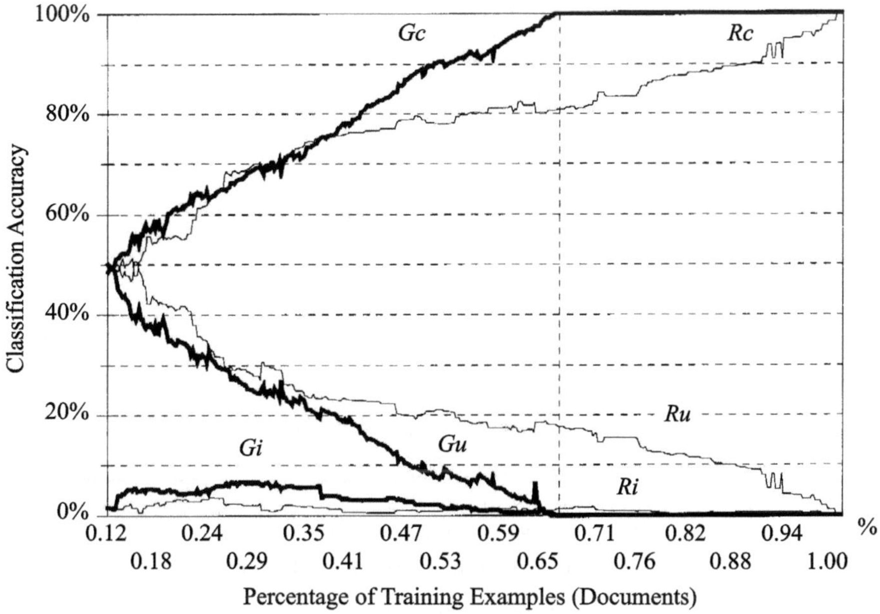


**Figure 13.5.** Results When the GUIDED and RANDOM Approaches Were Used on the (DOE vs. ZIPFF) Class-Pair.

three figures corresponds to the percentages shown in this table. For instance, in Figure 13.5 it is shown that for the class-pair (DOE vs. ZIPFF) this line is at the 65.69% mark on the horizontal axis.

Some important observations can be made regarding the proportions of "correct," "incorrect," and "undecidable" classifications in Figures 13.5 to 13.7. First, the rate of "correct" classifications under the GUIDED approach, denoted as *Gc*, was higher than the rate *Rc* under the RANDOM approach. Actually, the last row in Table 13.7 indicates that the OCAT/RA1 algorithm needed on the average about 34% *less* training documents to classify correctly all 510 documents under the GUIDED approach than under the RANDOM approach.

These results are very interesting for a number of reasons. They confirm the key assumption stated in Section 13.5 which indicated that inquiring about the class membership of new documents from the "undecidable" group could increase the accuracy of the OCAT/RA1 algorithm. This is the key aspect of the guided learning strategy presented in Chapter 5. These results are also encouraging because they help to answer the second question stated in the introduction of this section. That is, queries to the oracle could stop when about 66% of the 510 documents from the three class-pairs of the TIPSTER collection had been inspected and were included in the training sets. More importantly, these results are significant because they suggest that the OCAT/RA1 algorithm can be employed for the classification of large collections of text documents.

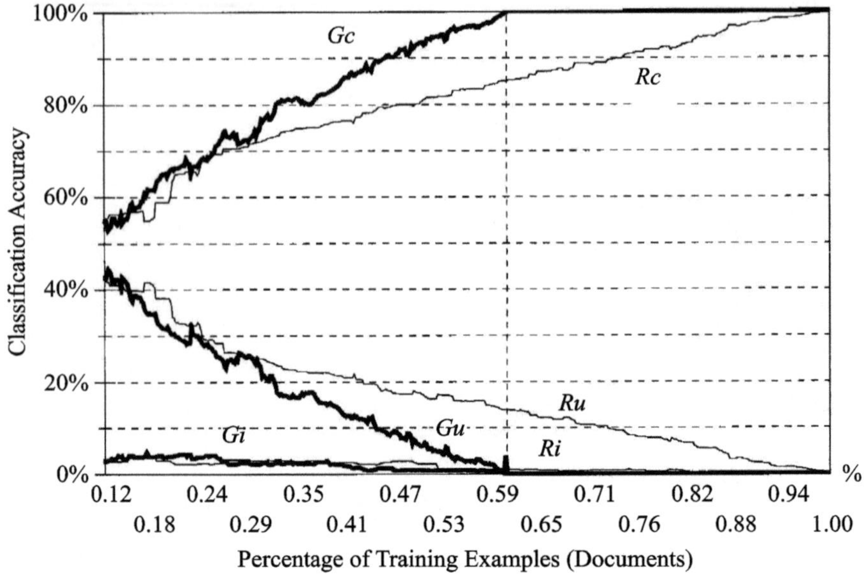


**Figure 13.6.** Results When the GUIDED and RANDOM Approaches Were Used on the (AP vs. DOE) Class-Pair.

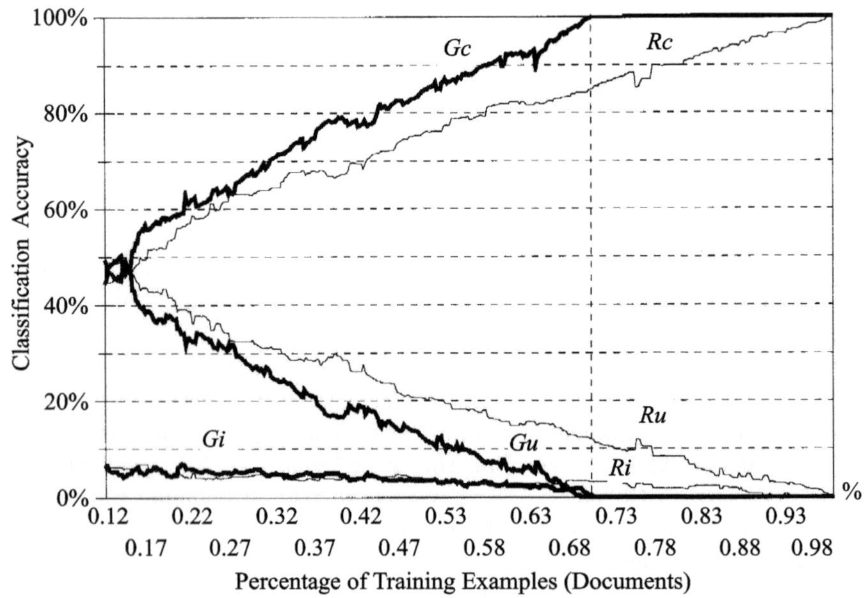


**Figure 13.7.** Results When the GUIDED and RANDOM Approaches Were Used on the (WSJ vs. ZIPFF) Class-Pair.

**Table 13.7.** Percentage of Documents from the Population that Were Inspected by the Oracle Before an Accuracy of 100% Was Reached.

| Class-pairs | % Under GUIDED | % Under RANDOM |
|---|---|---|
| (DOE vs. ZIPFF) | 65.69 | 100.00 |
| (AP vs. DOE) | 60.98 | 99.80 |
| (WSJ vs. ZIPFF) | 71.18 | 99.80 |
| Average | 65.95 | 99.87 |

NOTE: 100% accuracy was achieved when the number of "incorrect" and "undecidable" classifications were 0%.

Other observations are related to the rates at which "incorrect" and "undecided" classifications were eliminated. From the previous figures it can be seen that these rates were a direct consequence of improving the classification rules. The figures show that the rates $G_i$ and $G_u$ reach 0% when, on the average, about 66% (or 336 documents) of the 510 documents in the experiment were included in the $E^+$ and $E^-$ sets of training examples. On the other hand, it can be seen that under the RANDOM learning approach, the rates *Ri* and *Ru* reached 0% when 99.8% (or 509) of the 510 documents were processed.

## 13.10 Concluding Remarks

This chapter examined a classification problem in which a document must be classified into one of two disjoint classes. As an example of the importance of this type of classification, one can consider the possible release to the public of documents that may affect national security. The method proposed in this chapter (being an automatic method) is not infallible. This is also true because its performance depends on how representative the training examples (documents) are. The application of such an approach to a problem of critical importance can be seen as an important and useful automatic tool for a preliminary identification of the documents to be classified.

This chapter considered an approach to this problem based on the vector space model (VSM) algorithm and compared it with the OCAT approach, as it is embedded in the RA1 heuristic, for inferring the individual clauses. These two approaches were tested on almost 3,000 documents from the four document classes of the TIPSTER collection: the U.S. Department of Energy (DOE), the Wall Street Journal (WSJ), Associated Press (AP), and the ZIPFF class of documents. Furthermore, these documents were analyzed under two types of experimental settings: (i) a leave-one-out cross validation and (ii) a 30/30 cross validation (where 30 indicates the initial number of training documents from each document class). The experimental results suggest that the OCAT/RA1 algorithm performed significantly better in classifying documents into two disjoint classes than the VSM.

Moreover, the results of a third experiment suggested that the classification efficiency of the OCAT/RA1 algorithm can be improved substantially if the guided learning approach presented in Chapter 5 is implemented. Actually, experiments on samples of 510 documents from the previous four classes of the TIPSTER collection indicated that the OCAT/RA1 algorithm needed, on the average, only about 336 (i.e., 66% of the) training documents before it correctly classified all of the documents. The results presented here, although limited to a relatively small collection of almost 3,000 documents, are encouraging because they suggest that the OCAT approach can be used in the classification of large collections of documents.

The importance of text mining becomes even more profound when one considers the endless potential of the World Wide Web. Most of the information in the Web is available in the form of text (along with the usual hyperlinks, audio, video, and photo items). Another potential avenue for more applications is when analyzing (mining) communications data for intelligence purposes. This has already attracted the interest of many governments and agencies, especially after the 9/11 events in New York City. As more information becomes available in digital form, the significance of mining of text documents will only increase.

# Chapter 14

# First Case Study: Predicting Muscle Fatigue from EMG Signals

## 14.1 Introduction

Most of the previous chapters discussed some application issues on a number of areas. This chapter discusses a case study in detail. The emphasis is on some comparative issues with other data mining techniques that do not use logic-based approaches. This chapter also provides a link to the data used in this study.

This case study is based on the problem of predicting muscle fatigue from electromyographic (EMG) signals. The main results were published in [Torvik, *et al.*, 1999]. The data used in this study are easily downloadable from URL: *http://www.csc.lsu.edu/trianta* (the link is in the "Research on Data Mining" part of that webpage). The original data are continuous. Also, part of this data set is depicted in Table 14.1. These data were first described in [Garcia, *et al.*, 1997] and [Waly, *et al.*, 1997].

## 14.2 General Problem Description

This chapter presents the development, testing, and comparison of a number of models for the prediction of muscle fatigue associated with sustained muscle contraction. This study aimed at two goals. The first goal was to compare a number of predictive methods, especially a number of statistical techniques, with some data mining methods. The second goal was to develop an accurate model for the prediction of muscle fatigue via electromyography.

An experimental study was conducted in order to evaluate the effects of heavy isometric loading (maximum and at 80% of the maximum) on recorded electromyography (EMG) signals. Furthermore, this study investigated any possible effects of electrode orientation on the detection of muscle fatigue during heavy isometric loading.

E. Triantaphyllou, *Data Mining and Knowledge Discovery via Logic-Based Methods*,
Springer Optimization and Its Applications 43, DOI 10.1007/978-1-4419-1630-3_14,

**Table 14.1.** A Part of the EMG Data Used in This Study.

| | | | | | | | | | | |
|---|---|---|---|---|---|---|---|---|---|---|
| 66.02 | 13.21 | 25.53 | 32.07 | 41.1 | 61.37 | 70.16 | 96.33 | 114.68 | 207.43 | 0 |
| 60.02 | 5.33 | 25.59 | 36.21 | 51.52 | 65.13 | 84.82 | 118.49 | 137.15 | 216.01 | 0 |
| 58.02 | 9.03 | 18.56 | 23.75 | 35.27 | 57.91 | 73.09 | 96.83 | 116.73 | 192.19 | 0 |
| 36.01 | 0.00 | 2.21 | 5.11 | 30.68 | 49.84 | 71.24 | 97.59 | 114.1 | 164.80 | 0 |
| 68.02 | 3.15 | 23.10 | 30.23 | 42.71 | 63.67 | 70.31 | 88.67 | 96.99 | 180.73 | 0 |
| 68.02 | 3.99 | 25.59 | 35.47 | 54.51 | 69.18 | 97.22 | 128.02 | 157.79 | 204.10 | 0 |
| 50.02 | 2.94 | 17.01 | 32.37 | 48.41 | 56.56 | 77.71 | 106.13 | 128.73 | 215.56 | 0 |
| 0.00 | 0.00 | 0.00 | 0.00 | 0.00 | 48.59 | 66.56 | 76.85 | 78.94 | 89.73 | 0 |
| 36.01 | 0.00 | 12.57 | 17.80 | 36.09 | 55.22 | 67.16 | 81.78 | 94.10 | 142.01 | 0 |
| 0.00 | 0.00 | 0.00 | 0.00 | 0.00 | 37.26 | 49.24 | 67.41 | 73.88 | 79.64 | 0 |
| 52.02 | 12.31 | 20.53 | 31.20 | 41.96 | 52.07 | 58.74 | 69.61 | 86.83 | 179.73 | 0 |
| 0.00 | 0.00 | 0.00 | 0.00 | 35.01 | 45.14 | 54.14 | 64.84 | 69.74 | 84.64 | 0 |
| 72.02 | 12.50 | 17.02 | 27.69 | 42.63 | 59.35 | 78.89 | 92.56 | 101.10 | 171.97 | 0 |
| 0.00 | 0.00 | 0.00 | 0.00 | 30.43 | 56.83 | 76.82 | 88.58 | 94.46 | 98.90 | 0 |
| 70.02 | 11.24 | 17.45 | 26.07 | 41.12 | 61.50 | 78.80 | 99.25 | 112.34 | 165.26 | 0 |
| 0.00 | 0.00 | 0.00 | 0.00 | 21.20 | 42.73 | 69.70 | 83.66 | 91.30 | 106.18 | 0 |
| 42.01 | 14.38 | 25.99 | 36.54 | 50.59 | 67.99 | 90.05 | 107.34 | 124.49 | 195.60 | 0 |
| 34.01 | 0.00 | 8.53 | 29.75 | 54.78 | 92.96 | 116.91 | 139.69 | 169.89 | 191.70 | 0 |
| 96.03 | 17.68 | 31.33 | 40.41 | 61.41 | 83.80 | 122.28 | 151.86 | 169.50 | 209.69 | 0 |
| 0.00 | 0.00 | 0.00 | 2.58 | 44.48 | 67.07 | 93.96 | 113.64 | 126.87 | 149.93 | 0 |
| 14.00 | 12.98 | 14.86 | 18.22 | 45.09 | 61.55 | 92.91 | 112.47 | 127.46 | 189.91 | 0 |
| 56.02 | 0.00 | 0.00 | 25.37 | 50.46 | 73.28 | 97.66 | 150.58 | 178.15 | 207.24 | 0 |
| 64.02 | 12.77 | 25.41 | 30.87 | 52.20 | 65.88 | 95.25 | 114.49 | 129.43 | 211.07 | 0 |
| 64.02 | 0.00 | 0.00 | 12.22 | 39.89 | 64.99 | 93.46 | 105.06 | 121.97 | 148.81 | 0 |
| 62.02 | 16.40 | 30.80 | 34.83 | 42.49 | 61.87 | 78.84 | 88.66 | 99.58 | 156.85 | 0 |
| 34.01 | 22.04 | 34.13 | 35.53 | 48.59 | 68.55 | 85.48 | 113.44 | 136.44 | 195.92 | 0 |
| 36.01 | 9.17 | 21.82 | 28.85 | 37.18 | 51.90 | 68.09 | 83.86 | 101.27 | 182.72 | 0 |
| 58.02 | 0.00 | 23.75 | 30.20 | 38.5 | 56.59 | 75.53 | 109.93 | 129.01 | 147.16 | 0 |
| 60.02 | 13.89 | 22.34 | 24.61 | 36.35 | 61.11 | 78.34 | 101.32 | 123.83 | 180.96 | 0 |
| 68.02 | 0.00 | 0.00 | 22.61 | 41.61 | 68.98 | 90.79 | 107.28 | 125.61 | 151.00 | 0 |
| 26.01 | 7.79 | 13.75 | 24.30 | 31.12 | 46.78 | 67.19 | 91.28 | 96.08 | 163.42 | 0 |
| 26.01 | 0.00 | 0.00 | 20.50 | 32.32 | 49.61 | 76.35 | 95.98 | 111.42 | 125.44 | 0 |
| 52.02 | 17.81 | 27.16 | 36.91 | 51.53 | 70.55 | 89.59 | 105.71 | 119.33 | 201.05 | 0 |
| 74.02 | 12.05 | 27.10 | 32.80 | 46.91 | 67.18 | 84.79 | 106.85 | 166.76 | 226.97 | 0 |

## 14.3 Experimental Data

The EMG data for this study were derived from 18 healthy male subjects with no history of musculoskeletal injuries (see also Table 14.1). All subjects were selected on a voluntary basis from a student population. They represented a wide spectrum of body weights, heights, age, and muscle strengths. They ranged in age between 22 and 40 years old with a mean value of 27.2 years. Their weights ranged from 53.2 kg to 105.9 kg (117 lb to 233 lb) with a mean value of 75.86 kg (166.89 lb). Their heights ranged from 160 cm to 187.5 cm (5′4″ to 6′3″) with a mean value of 172.5 cm (5′9″).

The subjects were required to perform a static muscle effort corresponding to a predetermined load level. The load was applied to permit static contraction of the biceps brachii muscle. The load was placed in the dominant hand of each subject with the upper arm hanging freely in a neutral adducted position to the side of the body. The forearm was flexed at 90° at the elbow joint. The wrist was maintained in a straight neutral position with the hand supinated to support the load. The load consisted of a bar and two balanced weights attached to both sides of the bar. Two levels of loading were studied. These loads were set to the maximum amount of weight the individual could hold for a few (e.g., 3–5) seconds and at 80% of the maximum weight. The maximum weight was determined on a separate day prior to the experimental sessions. The subjects were instructed to hold the weight, as described earlier, as long as possible.

The EMG data were recorded from the biceps brachii muscle using two sets of electrodes simultaneously. One set was placed along the muscle fibers and the other across the muscle fibers. The electrodes used in this experiment were of the Beckman type, 11-mm silver/silver chloride surface electrodes. The EMG signals were recorded using an R611 Multichannel Sensormedics Dynograph via a type 9853A voltage/pulse/pressure coupler. The amplifier gain was adjusted to allow full utilization of the dynamic range of the A/D converter (+10 volts). A sampling rate of 512 Hz was used to digitize the EMG signals using a 12-bit A/D converter model DT 2801-A.

The EMG signals were recorded from the onset of the load under investigation until the subject could not hold the load anymore. A preprocessing of the EMG signals was conducted as described earlier. The window size used in this preprocessing was selected to be 512,000 per second based on the results of the first experiment. The EMG parameters were estimated for the first and last window in the recorded signal. Furthermore, the EMG parameters were calculated at fixed periods of time as a percentage of the total time an individual was able to maintain the task of holding the load. The center of the EMG window used in the analysis was set at the selected fixed periods of time. These periods were selected at 5% through 95% of the total time with an increment of 5%.

## 14.4 Analysis of the EMG Data

Several traditional statistical and data mining analyses were conducted to achieve the objectives of this study. The recorded EMG signals were analyzed using the MATLAB numeric computation software and its signal processing toolbox developed by the MathWorks, Inc. Both time and frequency domain analyses were conducted.

### 14.4.1 The Effects of Load and Electrode Orientation

The results obtained indicated that the time domain parameters did not change significantly for all the interactions and main effects of the three independent variables (load, electrode orientation, and muscle condition at rest or fatigue). The frequency domain parameters were significantly affected by the main effects of electrode orientation and muscle condition. The load had no significant effect on these parameters. The effect of electrode orientation on the characteristic frequencies used in this investigation was more pronounced for the lower frequencies of the spectrum. Electrodes placed across the muscle fibers showed lower fractile frequencies compared to electrodes along the muscle fibers. The effect of electrode orientation was only significant for the lower fractiles with the exception of the 99-th fractile (peak frequency, 1, 5, 10, 25, and 99 fractile).

In this study, the full wave rectified integral (FWRI) and the root mean square (RMS) values were estimated. For a discrete signal which consists of $N$ equally spaced samples $x(n)$, for $n = 1, 2, 3, \ldots, N$ these measures are given algebraically as follows:

$$\text{FWRI} = \frac{\sum_{n=1}^{N-1} \frac{|x(n)|+|x(n+1)|}{2}}{N-1}$$

and

$$\text{RMS} = \sqrt{\frac{\sum_{n=1}^{N} x(n)}{N}}.$$

### 14.4.2 The Effects of Muscle Condition, Load, and Electrode Orientation

The first and last window of the recorded EMG signals were used to represent the muscle at a resting condition and the fatigue state, respectively. The EMG indices were the dependent variables. The independent variables were the muscle condition (rest or fatigue), load, and electrode orientation. The effect of muscle fatigue was significant for all the characteristic frequencies used. A significant shift toward lower frequencies was observed. However, the amount of shift in these frequencies was higher for the submaximum load. Also, it is worth noting that the shift in these frequencies was not linear across the spectrum. Therefore, monitoring a single characteristic frequency may not be adequate for the quantification of the spectrum shift.

In the frequency domain analysis, the estimated power spectrum and its characteristic fractile frequencies were calculated. The estimated power spectrum, also known as the periodogram, was calculated as the modulus squared Fourier transform [Garcia, *et al.*, 1997], [Waly, *et al.*, 1997].

The characteristic frequencies used in this study are the 1, 5, 10, 25, 50, 75, 90, 95, 99 fractile frequency, and the peak frequency. The $p$-th fractile frequency $fp$, analogous to the statistical definition of fractile, is defined as the frequency for which the relation

$$\frac{\int_{f=0}^{fp} Gs(f)df}{S} = p$$

holds; where $Gs(f)$ is the one-sided power spectrum of $s(t)$ and $S$ is defined as

$$S = \int_{f=0}^{\infty} Gs(f)df.$$

## 14.5 A Comparative Analysis of the EMG Data

Since there are numerous prediction methods in statistics and AI (including the newer ones which are based on data mining techniques) an important goal of this study was to use the derived EMG data to compare the prediction accuracy of some of these methods. Of particular interest was to compare the OCAT approach (as it is embedded in the RA1 heuristic, see also Chapter 4) with other methods from the statistics and data mining fields. Besides the OCAT approach, other methods were Fisher's linear discriminant analysis, logistic regression, a neural network approach, fuzzy $c$-means, and fuzzy $k$-nearest neighbor approaches.

As was explained in Chapters 2 and 3, the OCAT (*One Clause At a Time*) approach considers two sets of rules: the positive and the negative sets of rules (inferred Boolean functions). As a result of this, classifying a new (i.e., unclassified) observation will result in one of the following three outcomes: the classification will be either correct, or incorrect, or will be an undecidable case. To provide a common ground to compare our results to those obtained using the other methods, we decided to fix the number of undecidable cases (in the testing data) to that obtained by the OCAT approach. As a result, we could find the accuracy, on the same number of actual classifications, for each method.

To determine which cases should be deemed undecidable for each of the various other methods, we expanded an interval symmetric about the respective cutoff value. This led to a unique set of undecidable cases and hence a unique classification accuracy on the remaining cases. A symmetric interval is reasonable when the underlying classification function is monotone and symmetric. Note that a classification function is monotone in the sense that once the function has been determined, each variable has a monotone effect on the classification. That is, each variable will either have a nonnegative effect or a nonpositive effect (but not both) on the outcome.

In all of the methods except the neural network, symmetry and monotonicity are reasonable assumptions. This is probably the most apparent in the logistic regression model and the fuzzy models with their respective intuitive probability and membership value interpretations. The hyperplane that separates the two groups in linear discriminant analysis is obviously monotone. The sigmoid function (used in logistic regression as well as the transfer function in the following neural network) is also monotone in each variable.

In addition, Fisher's linear classification rule assumes equal variance–covariance matrices, which results in symmetric probabilities of misclassification. Even though the sigmoid transfer function, used in individual neurons, is monotone, the overall network of neurons is not monotone when it has hidden layers. Despite this fact, our neural network had a single hidden layer, and the symmetric interval was computed on the single output neuron.

We split the dataset into a training and a testing set, and the classification functions were derived for each method on the training set. Due to the unsupervised nature of the fuzzy $c$-means algorithm, it was trained and tested on the testing data. The OCAT/RA1 approach achieved 100% accuracy on the training data (due to the way it builds a Boolean function) and, for this particular data split, labeled 20 of the test cases as undecidable.

Since logistic regression, linear discriminant analysis, and neural networks use monotone classification functions, a fixed number of undecidable cases corresponds to a single one-dimensional interval. Note that this interval may be of many different sizes as long as it captures the given number of cases. Also, recall that each function is monotone. In the logistic regression, neural network, and Fisher's linear discriminant cases, these intervals are also symmetric, which further simplifies this process. As was stated earlier, the sigmoid function is symmetric in itself, while Fisher's linear classification rule assumes equal variance–covariance matrices, which results in symmetric probabilities of misclassification. As a result of this symmetry, for a fixed number of undecidable observations (20 in this case), there corresponds an interval lying between the two groups of points. Therefore, the sum of the numbers of correctly and incorrectly classified points is also fixed.

In order to find the interval corresponding to a fixed number of undecidable cases, it is only necessary to find one of its borders (because of symmetry and the middle point is given). To find an appropriate upper border point we performed a binary search. Note that one may also start with 0 captured cases (i.e., at the cutoff point) and expand the interval, using its symmetry property, until one finds an interval that captures 20 points (so they can be compared meaningfully with the OCAT/RA1 approach which had 20 undecidable cases).

### 14.5.1 Results by the OCAT/RA1 Approach

The OCAT/RA1 sets of rules (i.e., the inferred positive and negative Boolean functions by this method) classified all of the 192 training points correctly (by construction), while for the 66 testing observations the results were

Number of correct classifications = 41
Number of incorrect classifications = 5
Number of undecidable cases = 20.

Therefore, the accuracy rate is 89.1% (41 correct out of $41+5$ cases) on the cases that were actually classified. The number of undecidable cases of 20 was fixed for the remaining approaches to provide the same number of actual classifications.

### 14.5.2 Results by Fisher's Linear Discriminant Analysis

#### A Quick Description of the Method

Fisher's approach finds the linear projection (multivariate to single values) that maximizes the standardized (by the sample variance) squared distances between the two group means. A new observation is classified based on where it is projected onto this line. If we want to place all possible observations into one of the two groups, then we can use the average of the two projected group means as the cutoff value. However, we may also wish to describe a class of uncertain points (as with the undecidable cases above) by an interval where the misclassification probability is high. If the variance–covariance matrices of the two groups are equal, this interval is symmetric about the previous cutoff point. Note that Fisher's linear function estimates the common variance–covariance matrix, and therefore it implicitly assumes that the matrices are equal. The interested reader may wish to consult with the work in [Fisher, 1936], [Fauset, 1994], [Johnson and Wichern, 1992], [Kleinbaum, Klein, and Pryor, 2005], and [Hosmer and Lemeshow, 2000].

#### Results

The linear discriminant function we obtained from the training data is

$$
\begin{aligned}
Y = & -0.0297X_1 - 0.0121X_2 + 0.0459X_3 - 0.0775X_4 \\
& + 0.0020X_5 + 0.0459X_6 + 0.0783X_7 + 0.0124X_8 \\
& - 0.0251X_9 + 0.0228X_{10},
\end{aligned}
$$

and it provided 86.0% accuracy on the training data by using a cutoff value of 7.1959.

For the testing data, we found that the interval [6.6264, 7.7654] determined that 20 cases were undecidable. That is, when fixing the number of undecidable cases to 20, the linear discriminant rule will result in the following classifications on the remaining 46 testing observations:

$$
\begin{aligned}
\text{Number of correct classifications} &= 39 \\
\text{Number of incorrect classifications} &= \ 7.
\end{aligned}
$$

Notice that even if we change this interval (while maintaining 20 points within it), the number of correctly and incorrectly classified points will not change due to the monotonicity and symmetry properties discussed in the introduction. Also, observe that the interval edges are symmetrically displaced by 0.5695 about the original cutoff value.

### 14.5.3 Results by Logistic Regression

#### A Quick Description of the Method

Logistic regression models the logit of the probability of an observation belonging to class 0 (or 1) as a linear function of the variables. To obtain the maximum likelihood estimates of its parameters, a nonlinear continuous optimization method has to be utilized. Once these estimates are obtained, we can easily find the classification probabilities corresponding to each observed point. If we want to place all possible observations into one of the two groups, then using 0.5 as the cutoff value on the probabilities, will minimize the misclassification probability.

We may also determine the interval of uncertainty (where the misclassification probability is high) directly by choosing a maximum misclassification probability (less than 0.5). That is, to fix the number of undecidable cases in the testing test, we need to find a corresponding misclassification probability cutoff point. The interested reader should consult the procedures presented in [Kleinbaum, Klein, and Pryor, 2005] and [Hosmer and Lemeshow, 2000] for further insight.

#### Results

The parameter estimates we obtained from the training data are

| **Theta** | **SE** |
|---|---|
| −9.2761 | 1.4505 |

| **Beta** | **SE** |
|---|---|
| 0.0375 | 0.0165 |
| 0.0244 | 0.0429 |
| −0.0447 | 0.0492 |
| 0.0720 | 0.0501 |
| −0.0101 | 0.0401 |
| −0.0435 | 0.0499 |
| −0.1318 | 0.0531 |
| 0.0023 | 0.0475 |
| 0.0266 | 0.0305 |
| −0.0223 | 0.0128 |

which provided 87.5% accuracy on the training data by using a cutoff value of 0.5 on the probability.

For the testing data, we found that the interval [0.3125, 0.6875] corresponds to 20 undecidable cases. That is, when fixing the number of undecidable cases to 20, the logistic regression rule will result in the following classifications on the remaining 46 testing observations:

Number of correct classifications = 39

Number of incorrect classifications = 7.

Even though the above accuracy rate is equivalent to that of Fisher's linear discriminant, the actual classifications are not equivalent. In fact, their classifications differed at four individual data points.

Notice that even if we change this interval (while maintaining 20 points within it), the number of correctly and incorrectly classified points will not change due to the monotonicity and symmetry properties discussed earlier. Also, observe that the interval edges are symmetrically displaced by about 0.5.

### 14.5.4 A Neural Network Approach

#### A Quick Description of the Method

A neural network (NN) consists of processing elements (PEs) and weighted connection between them (see, for instance, [Myers, 1990]). We used a feedforward network which consisted of an input layer, a hidden layer, and an output layer. The input layer corresponds to the 10 independent variables and a bias term. The PE in the output layer corresponds to the binary classification. The hidden layer is created to make the network capable of using more complex classification rules. The more processing elements in the hidden layer, the more complex the rules become. Note that if the hidden layer were eliminated, the resulting rule would be a single hyperplane as with Fisher's rule above. We experimented with between 2 and 10 hidden layer PEs and did not find much difference in performance, so we decided to go with 2 hidden PEs.

The connections between the three layers are represented by two matrices. These matrices are first randomly initialized with some small values before they are trained to accommodate the observed classifications. We did this using the well-known *Back Propagation Algorithm.* It feeds the inputs forward in the network, and the errors are fed backward. This enables one to update the weight matrices according to the errors of the gradient descent based on the transfer function. We used the bipolar sigmoid transfer function which not only provides a natural association ($-$infinity and $+$infinity map to $-1$ and $+1$, respectively) but it is also easily differentiable. Using bipolar variables seems more reasonable than binary since the 0 value has a different effect than 1 in multiplication and we do not want the classes to be treated differently.

As with any gradient-based optimization procedure, the Back Propagation Algorithm needs to determine the magnitude of the updates (training rate). The training rate is commonly set to some small value (less than 1). We experimented with different values and found that 0.1 worked well. Often, a momentum term is included to improve the convergence rate. It did not seem to have much of an effect in this case, so we excluded it. When running the training algorithm, we observed that the accuracy on the training data reached a plateau around 82–85%, so we decided to terminate the algorithm when an accuracy of 85% or better was achieved (accuracy when using 0 as the cutoff value in output PE).

Once the trained weight matrices are obtained, we can classify the observations in the testing set. If we want to place all possible observations into one of the two groups, then we use 0 as the cutoff value for the bipolar sigmoid transfer function. As with the logistic regression and the discriminant analyses performed above, we wanted to fix the number of undecidable cases in the testing data. That is, we needed to find a corresponding cutoff point. More details on neural networks can be found in [Arbib, 2002] and [Dayan and Abbot, 2001].

### Results

After running the training algorithm several times (different initial matrices give differing results), *the most accurate* classifications on the testing data were found with the following weight matrices:

| $w_1$ | $w_2$ | |
|---|---|---|
| −0.7761 | 0.6067 | 0.4334 |
| −0.6696 | 0.5701 | −2.4373 |
| −0.0742 | 0.0608 | 2.0423 |
| −0.0105 | 0.0230 | |
| −0.3816 | 0.3126 | |
| 0.4831 | −0.3956 | |
| 1.4933 | −1.2890 | |
| 1.4208 | −1.1971 | |
| 0.9058 | −0.7587 | |
| 0.5988 | −0.5306 | |
| 0.4624 | −0.4206 | |

After seven epochs the accuracy on the training data (denoted as "*acc_trn*") improved as follows:

$$acc\_trn = 0.4688 \quad 0.5938 \quad 0.7708 \quad 0.8073 \quad 0.8281 \quad 0.8438$$

and reached 85.4% accuracy on the training data as it was terminated.

For the testing data, we found the interval corresponded to 20 undecidable cases. That is, when fixing the number of undecidable cases to 20, we got the following classifications on the remaining 46 testing observations:

Number of correct classifications = 41
Number of incorrect classifications = 5.

Note that this is the best observed classification accuracy and that we observed as poor as 37 correct classification (and 9 incorrect), with the same accuracy on the training data. In a situation where the new observations are not given, we could have obtained a classification accuracy anywhere in between 37/46 and 41/46.

**Table 14.2.** Summary of the Prediction Results.

| | Accuracy on Data Set (in %) | |
|---|---|---|
| **Method Used** | **Training Data** | **Testing Data** |
| Linear Discriminant Analysis | 86.0 | 84.8 |
| Logistic Regression | 87.5 | 84.8 |
| Neural Network | 85.4 | 80.4 to 89.1 |
| Fuzzy $c$-Means | N/A | 69.6 |
| Fuzzy $k$-Nearest Neighbors | | |
| (when $k = 5$) | 100.0 | 82.6 |
| (when $k = 10$) | 100.0 | 82.6 |
| The OCAT/RA1 Approach | 100.0 | 89.1 |

## 14.6 Concluding Remarks

The analyses reported in this chapter, including those under the $c$-means and $k$-nearest neighbors methods, are summarized in Table 14.2. The results indicate that the model derived by using the OCAT/RA1 approach is the most accurate one. Furthermore, analyzing EMG data on muscle fatigue is important in its own right (e.g., [Garcia, *et al.*, 1997] and [Waly, *et al.*, 1997]).

This comparative study indicates that the OCAT/RA1 approach has an appealing potential when it is compared with some of the existing prediction approaches. More similar studies are required before one can fully assess the merits and the full potential of the OCAT/RA1 approach.

# Chapter 15

# Second Case Study: Inference of Diagnostic Rules for Breast Cancer

## 15.1 Introduction

For this case study we used a data set that described a number of clinical cases of breast cancer diagnoses. The data were divided into two disjoint sets of malignant and benign cases. We applied the OCAT approach, as it is embedded in the RA1 heuristic (see also Chapter 4), after the data were transformed into binary ones according to the method described in Section 2.2. The following sections describe the data and inferred diagnostic rules in more detail.

## 15.2 Description of the Data Set

The data were collected with the help of Dr. James F. Ruiz, a Radiologist at the Woman's Hospital in Baton Rouge, Louisiana. The data represent the characteristics of various cases of breast tumors, as taken from historic records from this hospital. It is to be noted that at the time of this study (in 1995) the Woman's Hospital possessed the second largest such collection in the United States (the largest collection is at the Pittsburgh Hospital in Pennsylvania). Annually, more than 30,000 new cases are added to this collection. Each case has been independently verified by at least two radiologists.

The set we used represents a miniscule sample of this collection. This data set was used in this study to demonstrate the flexibility and robustness of the OCAT/RA1 method as described in Chapters 2, 3, and 4. The data are defined on at most 24 attributes as shown in Tables 15.1 and 15.2.

Table 15.3 presents a sample of the training data. The entire data set used in this study can be easily downloaded from the personal webpage of the author (i.e., from *http://www.csc.lsu.edu/trianta*). For the interpretation of the data in Table 15.3 consider any row, say the third one (which has been boldfaced for easy reference). Then, the coding conventions presented in Tables 15.1 and 15.2 are used as follows.

E. Triantaphyllou, *Data Mining and Knowledge Discovery via Logic-Based Methods*, Springer Optimization and Its Applications 43, DOI 10.1007/978-1-4419-1630-3_15,

**Table 15.1a.** Attributes for the Breast Cancer Data Set from Woman's Hospital in Baton Rouge, LA (Part (a); Attributes 1 to 16).

| Attribute Number | Interpretation | Domain of Values |
|---|---|---|
| 1 | Case ID | An integer number |
| 2 | Number of calcifications per $cm^2$ | **A**, if less than 10;<br>**B**, if between 10 and 20;<br>**C**, if more than 20 |
| 3 | Approximate volume of lesion (in $cm^3$) | A continuous number |
| 4 | Total number of calcifications | An integer number |
| 5 | Irregularity of the shape of the calcifications | **A**, if *mild*;<br>**B**, if *moderate*;<br>**C**, if *marked* |
| 6 | Variation in the shape of the calcifications | **A**, if *mild*;<br>**B**, if *moderate*;<br>**C**, if *marked* |
| 7 | Irregularity in the size of the calcifications | **A**, if *mild*;<br>**B**, if *moderate*;<br>**C**, if *marked* |
| 8 | Variation in the density of the calcifications | **A**, if *mild*;<br>**B**, if *moderate*;<br>**C**, if *marked* |
| 9–13 | *Le Gal* type | **(1, 2, 3, 4, 5)**.<br>Note that a given lesion may include more than one *Le Gal* type |
| 14 | Ductal orientation | **A**, if *yes*;<br>**B**, if *no* |
| 15 | Density of the calcifications | **A**, if *low*;<br>**B**, if *moderate*;<br>**C**, if *high* |
| 16 | Density of the parenchyma | **A**, if *low*;<br>**B**, if *moderate*;<br>**C**, if *high* |

The first field denotes the case ID (= 3). The second field denotes that the number of calcifications per $cm^2$ is less than 10. The next field indicates that the approximate volume of the lesion is equal to $1.040\,cm^3$. The field which follows (and which is equal to B) indicates that the total number of calcifications is between 10 and 30. The fifth field (equal to B) indicates that the shape of the calcifications is moderate. A similar interpretation follows for the rest of the fields and their values. The last two fields indicate that there is only one (value of field #19 is equal to 1) diagnostic

**Table 15.1b.** Attributes for the Breast Cancer Data Set from Woman's Hospital in Baton Rouge, LA (Part (b); Attributes 17 to 26).

| Attribute Number | Interpretation | Domain of Values |
|---|---|---|
| 17 | Comparison with previous exam | **A**, if *change in the number or character of the calcifications*;<br>**B**, if *not defined*;<br>**C**, if *newly developed*;<br>**D**, if *no previous exam* |
| 18 | Associated findings | **A**, if *multifocality*;<br>**B**, if *architectural distortion*;<br>**C**, if *a mass*;<br>**D**, if *nothing* |
| 19 | Number of applicable diagnostic classes | An integer from 1 to 8 |
| 20–27 | Diagnostic classes | See Table 15.2 (parts (a) and (b)) for the details |

**Table 15.2a.** Interpretation of the Breast Cancer Diagnostic Classes (Part (a); Malignant Classes Only).

| Symbol Used for the Class Coding | Interpretation (Malignant Classes Only) |
|---|---|
| **A** | intraductal carcinoma |
| **B** | infiltrating ductal carcinoma |
| **C** | intraductal comedo type |
| **D** | tubular carcinoma |

**Table 15.2b.** Interpretation of the Breast Cancer Diagnostic Classes (Part (b); Benign Classes Only).

| Symbol Used for the Class Coding | Interpretation (Benign Classes Only) |
|---|---|
| **E** | fibrosis |
| **G** | cysts |
| **I** | mild hyperplasia |
| **K** | apocrine metaplasia |
| **M** | fibroadenoma |
| **O** | LCIS |
| **F** | adenosis |
| **H** | sclerosing adenosis |
| **J** | moderate hyperplasia |
| **L** | atypical hyperplasia |
| **N** | fibrocystic change – not specified |
| **P** | papillomatosis |

**Table 15.3.** A Part of the Data Set Used in the Breast Cancer Study.

| | | | | | | | | | | | | | | | | | | | | | |
|---|---|---|---|---|---|---|---|---|---|---|---|---|---|---|---|---|---|---|---|---|---|
| 1 | A | 1.000 | B | A | C | C | C | 0 | 2 | 3 | 4 | 0 | B | C | A | C | D | 1 | A | | |
| 2 | A | 211.000 | C | A | B | B | B | 0 | 2 | 3 | 0 | 0 | B | A | C | A | A | 1 | A | | |
| **3** | **A** | **1.040** | **B** | **B** | **C** | **C** | **A** | **0** | **2** | **0** | **4** | **0** | **B** | **C** | **A** | **A** | **D** | **1** | **A** | | |
| 4 | A | 0.180 | A | A | B | A | A | 0 | 0 | 3 | 4 | 0 | B | A | B | C | D | 2 | A | B | |
| 5 | A | 0.600 | A | B | A | A | A | 0 | 0 | 0 | 4 | 0 | B | B | C | C | D | 1 | A | | |
| 6 | C | 2.400 | C | C | C | C | C | 0 | 0 | 0 | 4 | 5 | B | C | A | A | D | 1 | C | | |
| 7 | C | 0.972 | C | A | C | A | A | 0 | 0 | 3 | 0 | 0 | B | B | C | C | D | 1 | A | | |
| 8 | B | 0.216 | B | B | C | C | C | 0 | 0 | 0 | 4 | 0 | B | B | A | C | C | 2 | A | D | |
| 10 | A | 0.336 | A | A | C | B | B | 0 | 0 | 0 | 0 | 5 | B | A | B | D | C | 1 | B | | |
| 13 | C | 15.000 | C | C | C | C | C | 0 | 0 | 0 | 4 | 5 | A | B | B | D | D | 1 | C | | |
| 14 | A | 0.072 | A | A | B | B | A | 0 | 0 | 0 | 4 | 0 | B | B | B | D | B | 2 | B | C | |
| 15 | B | 70.750 | C | A | C | C | C | 0 | 2 | 0 | 4 | 0 | B | C | B | C | C | 2 | A | C | |
| 17 | A | 0.018 | A | B | B | A | A | 0 | 2 | 0 | 0 | 0 | B | B | B | D | D | 3 | E | H | I |
| 18 | B | 0.024 | B | A | A | A | A | 0 | 2 | 0 | 0 | 0 | B | C | A | D | D | 2 | E | H | |
| 20 | B | 0.648 | B | C | B | B | A | 0 | 0 | 0 | 4 | 5 | A | A | B | A | D | 3 | E | F | K |
| 21 | B | 0.120 | B | A | C | B | B | 0 | 2 | 0 | 4 | 0 | B | A | B | D | D | 3 | E | G | H |
| 23 | A | 0.060 | A | B | A | B | A | 0 | 2 | 0 | 0 | 0 | B | A | C | A | D | 3 | F | G | K |
| 24 | B | 0.150 | B | A | C | C | B | 0 | 0 | 0 | 4 | 0 | B | C | C | A | D | 3 | E | H | L |
| 25 | A | 0.027 | A | B | A | A | A | 0 | 0 | 0 | 4 | 0 | B | C | B | D | D | 3 | E | G | K |
| 26 | A | 6.910 | B | B | C | B | A | 0 | 0 | 0 | 0 | 5 | A | C | A | A | D | 3 | E | K | L |
| 27 | A | 0.054 | A | B | B | A | A | 0 | 0 | 0 | 4 | 5 | B | A | C | A | D | 1 | E | | |
| 29 | B | 0.150 | A | A | A | A | A | 0 | 2 | 3 | 0 | 0 | B | A | B | C | D | 1 | I | | |
| 30 | A | 0.144 | C | C | C | B | A | 0 | 0 | 0 | 4 | 0 | B | C | A | A | D | 1 | N | | |

class which is “intraductal carcinoma” (the value is equal to A). This is a malignant class.

## 15.3 Description of the Inferred Rules

In this study we focused on diagnostic class A (i.e., “intraductal carcinoma”). Thus, the inferred rules are the ones related to the “intraductal carcinoma” class. That is, the training data set was split (dichotomized) into two disjoint groups. The first group had records related to the above particular malignant case, and the second group had the rest (which were benign and malignant as defined in the last attribute of the data).

We first transformed the previous data into their equivalent binary representation (by applying the procedure described in Section 2.2) and then we applied the RA1 heuristic (as described in Chapter 4), with value of IRS $=$ 1 (i.e., no randomized runs of the algorithm were executed). After that we transferred the extracted Boolean function into IF-THEN type of classification rules defined on binary attributes. Next we used the inverse of the previous transformation to express the binary attributes back into the original ones (i.e., on attributes with the values mentioned in Tables 15.1 and 15.2). These are only the “positive” rules.

The antecedent parts of these rules are presented in Table 15.4 (parts (a) to (c)) in the order they were generated. Each antecedent part (rule) is represented by a set

**Table 15.4a.** Sets of Conditions for the Inferred Rules for the "Intraductal Carcinoma" Diagnostic Class (Part (a); Rules #1 to #5).

| Rule Number | Condition Numbers | Interpretation |
|---|---|---|
| **#1** | Cond. 1.1 | The volume of the calcifications is more than 0.03 $cm^3$. |
| | Cond. 1.2 | The total number of calcifications is greater than 10. |
| | Cond. 1.3 | The variation in *shape* is moderate or marked. |
| | Cond. 1.4 | The irregularity in *size* of the calcifications is marked. |
| | Cond. 1.5 | The variation of the density of calcifications is moderate or marked. |
| | Cond. 1.6 | There is no ductal orientation. |
| | Cond. 1.7 | The number of the calcifications per $cm^3$ is less than 20. |
| | Cond. 1.8 | A comparison with previous exams shows a change in the number or character of calcifications or it is newly developed. |
| **#2** | Cond. 2.1 | The volume of the calcifications is more than 6.00 $cm^3$. |
| | Cond. 2.2 | The density of the calcifications is moderate. |
| | Cond. 2.3 | The variation in *shape* of individual calcifications is from mild to moderate. |
| | Cond. 2.4 | The variation in *size* of the calcifications is from mild to moderate. |
| | Cond. 2.5 | The *Le Gal* type is neither #1 nor #2. |
| **#3** | Cond. 3.1 | The volume of the calcifications is between 0.03 $cm^3$ and 0.18 $cm^3$. |
| | Cond. 3.2 | The number of calcifications per $cm^3$ is between 10 and 20. |
| | Cond. 3.3 | The *Le Gal* type is #4. |
| | Cond. 3.4 | The density of the calcifications is from low to moderate. |
| **#4** | Cond. 4.1 | The volume of the calcifications is between 2.4 $cm^3$ and 6.84 $cm^3$. |
| | Cond. 4.2 | There is no ductal orientation. |
| | Cond. 4.3 | The density of the parenchyma is from moderate to high. |
| | Cond. 4.4 | The *Le Gal* type is neither #1 nor #5. |
| **#5** | Cond. 5.1 | The volume of the calcifications is between 0.072 $cm^3$ and 0.288 $cm^3$. |
| | Cond. 5.2 | The variation in the size of the calcifications is from moderate to marked. |
| | Cond. 5.3 | The *Le Gal* type is #2. |
| | Cond. 5.4 | There is no ductal orientation. |
| | Cond. 5.5 | The total number of the calcifications is less than 30. |
| | Cond. 5.6 | The variation in the shape of individual calcifications is from mild to moderate. |
| | Cond. 5.7 | The density of the calcifications is from low to moderate. |
| | Cond. 5.8 | The comparison with previous exam(s) is: change in the number, or character, or there are newly developed calcifications. |

**Table 15.4b.** Sets of Conditions for the Inferred Rules for the "Intraductal Carcinoma" Diagnostic Class (Part (b); Rules #6 to #9).

| Rule Number | Condition Numbers | Interpretation |
|---|---|---|
| **#6** | Cond. 6.1 | The volume of the calcifications is greater than $0.180\,cm^3$. |
| | Cond. 6.2 | The variation in the shape of individual calcifications is from moderate to marked. |
| | Cond. 6.3 | The *Le Gal* type is #3 or #4. |
| | Cond. 6.4 | The variation in the density of the calcifications is from mild to moderate. |
| | Cond. 6.5 | The density of the calcifications is low. |
| | Cond. 6.6 | The density of the parenchyma is from low to moderate. |
| **#7** | Cond. 7.1 | The volume of the calcifications is greater than $0.030\,cm^3$. |
| | Cond. 7.2 | The density of the parenchyma is from moderate to high. |
| | Cond. 7.3 | The comparison with previous exam(s) is: newly developed calcifications or no previous exam is available. |
| | Cond. 7.4 | The associated finding is mass or none. |
| | Cond. 7.5 | The number of the calcifications per $cm^3$ is less than 10. |
| | Cond. 7.6 | The variation in the shape of individual calcifications is from mild to moderate. |
| | Cond. 7.7 | The variation in the size of the calcifications is mild. |
| | Cond. 7.8 | The variation in the density of the calcifications is mild. |
| | Cond. 7.9 | The *Le Gal* type is neither #1 nor #5. |
| **#8** | Cond. 8.1 | The volume of the calcifications is greater than $0.960\,cm^3$. |
| | Cond. 8.2 | The variation in the shape of individual calcifications is from moderate to marked. |
| | Cond. 8.3 | The variation in the density of the calcifications is from moderate to high. |
| | Cond. 8.4 | The *Le Gal* type is neither #1 nor #5. |
| | Cond. 8.5 | The comparison with previous exam(s) is: change in the number or character or newly developed calcifications. |
| **#9** | Cond. 9.1 | The volume of the calcifications is greater than $0.168\,cm^3$. |
| | Cond. 9.2 | The total number of the calcifications is greater than 10. |
| | Cond. 9.3 | The variation in the shape of individual calcifications is from moderate to marked. |
| | Cond. 9.4 | The variation in the size of the calcifications is from moderate to marked. |
| | Cond. 9.5 | The variation in the density of the calcifications is moderate. |
| | Cond. 9.6 | The *Le Gal* type is #3. |
| | Cond. 9.7 | The comparison with previous exam(s) reveals change in the number, or character, or there are newly developed calcifications. |

**Table 15.4c.** Sets of Conditions for the Inferred Rules for the "Intraductal Carcinoma" Diagnostic Class (Part (c); Rules #10 to #12).

| Rule Number | Condition Numbers | Interpretation |
|---|---|---|
| #10 | Cond. 10.1 | The volume of the calcifications is between $0.054\,cm^3$ and $0.072\,cm^3$. |
| | Cond. 10.2 | There are no associated findings. |
| | Cond. 10.3 | The number of the calcifications per $cm^3$ is less than 20. |
| | Cond. 10.4 | The *Le Gal* type is not #1. |
| | Cond. 10.5 | The density of the parenchyma is from low to moderate. |
| | Cond. 10.6 | The comparison with previous exam(s) is: no previous exam. |
| #11 | Cond. 11.1 | The volume of the calcifications is between $0.448\,cm^3$ and $0.540\,cm^3$. |
| | Cond. 11.2 | The number of the calcifications per $cm^3$ is less than 20. |
| | Cond. 11.3 | The density of the parenchyma is from low to moderate. |
| | Cond. 11.4 | There are no associated findings. |
| | Cond. 11.5 | The *Le Gal* type is not #1. |
| | Cond. 11.6 | The comparison with previous exam(s) reveals change in the number and character. |
| #12 | Cond. 12.1 | The volume of the calcifications is between $0.030\,cm^3$ and $0.032\,cm^3$. |

**RULE #2:**

**IF all of the following conditions hold true at the same time:**

(the volume of the calcifications is more than $6.00\,cm^3$);
**and** (the density of the calcifications is moderate);
**and** (the variation in shape of individual calcifications is from mild to moderate);
**and** (the variation in size of the calcifications is from mild to moderate);
**and** (the *Le Gal* type is neither #1 nor #2),

**THEN the case is:** *"intraductal carcinoma."*

**Figure 15.1.** A Diagnostic Rule (Rule #2) Inferred from the Breast Cancer Data.

of conditions to be satisfied for that rule to be applicable. For instance, the second set of conditions (comprised of 5 individual conditions) in Table 15.4 indicates the rule which is depicted in Figure 15.1.

A similar interpretation holds for the rest of the sets of conditions shown in Table 15.4. One may recall that OCAT first generates well generalizing classification rules and as it iterates, it generates less generalizing ones. That is, later rules may suffer from overfitting of very few data points. This phenomenon takes place due to the algorithm's greedy strategy at each iteration.

An expert radiologist (Dr. James F. Ruiz from Woman's Hospital in Baton Rouge, Louisiana) indicated that rules #10, #12, and possibly #11 (in this order) may not work. These rules were generated during the last iterations of the OCAT approach. However, it is interesting that the first nine rules generated at the first iterations (i.e., at iterations 1 to 9) are in agreement with the domain knowledge of the same expert.

## 15.4 Concluding Remarks

The above rules were extracted and are exhibited here for illustrative purposes only. They illustrate how a real-life data set can be transformed with the binarization process (as described in Section 2.2) into an equivalent binary data set and then be analyzed. We used the OCAT approach as part of the RA1 heuristic (Chapter 4) on the binary data set. The results were converted back into the original attributes described earlier and the inferred rules were extracted.

This experiment indicates the potential of the described logic-based data mining approaches for knowledge discovery. Finally, it should be stated here that this study complements the other studies described in various chapters of this book on the same medical problem (including the one in the next chapter).

# Chapter 16

# A Fuzzy Logic Approach to Attribute Formalization: Analysis of Lobulation for Breast Cancer Diagnosis

## 16.1 Introduction

In many data mining and knowledge discovery applications a critical task is how to define the values of the various attributes that the analyst believes may be of significance. For easily quantifiable attributes (such as, age, weight, cost, etc.) this task is a rather straightforward one as it involves simple measurements and expressing the results in terms of some units. For other attributes, however, this task may not be a simple one. This is the case when some of the data are *fuzzy*. For instance, although in common language one often uses terms such as "small," "large," "round," "tall," and so on, these terms may mean different concepts to different people or to the same person at different times.

This is particularly the case in the medical domain, where lots of data are defined in such fuzzy terms. Thus, an important question is how to, in an objective and consistent manner, quantify such fuzzy terms. This challenge is application specific. A reasonable avenue in dealing with this kind of challenges is provided by *fuzzy logic*.

This chapter describes an approach for quantifying some of the attributes involved in diagnosing breast cancer. This medical problem has attracted the interest of many researchers due to its societal importance, and the complexities of the problems it is associated with. Thus, the following sections describe some fundamental issues associated with the diagnosis of breast cancer and how one may proceed in formalizing some of the key attributes involved with the diagnostic process. This type of analysis can be extended into other application domains too. This chapter is based on results first published in [Kovalerchuk, Triantaphyllou, *et al.*, 1997] and [Kovalerchuk, Triantaphyllou, and Ruiz, 1996].

## 16.2 Some Background Information on Digital Mammography

Many of the current diagnostic methods in digital mammography [Doi, *et al.*, 1993], [Wu, *et al.*, 1993] are based primarily on neural networks without incorporating

E. Triantaphyllou, *Data Mining and Knowledge Discovery via Logic-Based Methods*, Springer Optimization and Its Applications 43, DOI 10.1007/978-1-4419-1630-3_16,

fuzzy logic. Nevertheless, it should be mentioned that these methods use degrees of irregularity and circularity which are similar to key concepts in fuzzy logic. These degrees are used as inputs to neural networks [Wu, *et al.*, 1994]. In this chapter we apply a fuzzy logic approach for classifying masses (lesions) found in mammograms as lobulated or microlobulated. The lobulated and microlobulated features of a mass are profoundly important in breast cancer diagnosis [Tabar and Dean, 1986].

The proposed analysis is based on the medical definitions of the previous two terms, as given by the American College of Radiology (ACR) Breast Imaging Lexicon. According to this Lexicon, a mass has "lobular" shape if "it has contours with undulations." Note that the Lexicon defines the notion "lobular" without any indication of the size or number of undulations and without defining the concept of "undulation." The descriptive words in each category describe a continuum from benign to malignant.

A lobular mass is most often benign, although a few malignancies will be lobular. Lobular malignancies are usually well differentiated pathologically. Furthermore, a mass with microlobulated margins has a lower chance for malignancy than one with indistinct or spiculated margins. A microlobulated mass would fit into the low-intermediate suspicion category #4 of the BI-RADS (for *Breast Imaging, Reporting, And Data System* of the American College of Radiology), and would have a 10–20% chance of malignancy.

In this chapter the concept of undulation is defined as the contour between the minima of adjacent concavities. The depth of such concavities may vary from small to very large (as explained in detail in Section 16.3). Therefore, for a formal computer algorithmic analysis, the above means that if a mass has any one of "small/medium/large undulation," then the algorithm should classify it as lobular. But this is not necessarily what occurs in a real-life situation because a radiologist may take into account the *size*, the *number of undulations*, and how *deep* they are.

However, the ACR Lexicon does not mention these attributes in the formal definition of lobulation. Therefore, it is likely that different radiologists may have different perceptions about the size and number of undulations sufficient to classify the shape of a mass as lobular.

The term *microlobulated margins* means (according to the ACR Lexicon) that "the margins undulate with **short cycles** producing **small undulations**." Again, different radiologists may have different perceptions of what "short cycles" and "small undulations" mean. The ACR Lexicon does not provide a unified framework for defining these terms in a consistent and objective manner and again radiologists are left making subjective and individual decisions regarding these characteristics. The following two hypothetical examples highlight the need for a unified framework for defining terms related to the shape of masses in mammograms.

*Example 1.* Suppose that a radiologist has found one "big" and two "small" undulations in a given mass. Does this mean that the mass is lobular or microlobular or do both features coexist? Also suppose that for the same mass a second radiologist has decided that there are two "big" and one "small" undulations. Again, we have the same question: "Is this mass lobular or microlobular or do both features coexist?"

*Example 2.* Suppose that in some study, five out of ten radiologists concluded that a particular mass is lobular, but the other five came to the opposite conclusion. How should we train a computerized system to detect a lobular mass by utilizing this contradictory experience? Should we exclude these cases from the training set? However, similar cases may appear again in a real-life situation. If we exclude these cases, any trained detection system will diagnose them arbitrarily, although most properly it should not identify lobular features.

The last example illustrates a typical source of intra- and extra-observer variability in mammography and some of its consequences. How can one minimize these problems? This chapter proposes a lobular/microlobular mass identification approach which addresses this methodological and practical problem. This approach can also become the basis for analyzing and formalizing other ACR Lexicon terms or any subjective attributes in a wide spectrum of data mining applications.

The proposed approach is designed in a manner which follows the way human experts make decisions regarding this particular medical problem. Therefore, this chapter will concentrate only on the development of an approach for formalizing lobularity and microlobularity in masses found in mammograms.

This chapter is organized as follows. The next section presents some basic information on fuzzy sets and related issues. Section 16.4 discusses the development of some concepts which can be used to characterize lobularity and microlobularity of breast masses and the formalization of these features in terms of a fuzzy logic approach. Section 16.5 develops the notions of degrees of lobularity and microlobularity based on the formalized features. Finally, the chapter ends with some concluding remarks.

## 16.3 Some Background Information on Fuzzy Sets

For a long time it has been recognized that an exact linguistic description of many real-life physical situations may be practically impossible. This is due to the high degree of imprecision involved in real-world situations. Zadeh, in his seminal papers [Zadeh, 1965; 1968], proposed fuzzy set theory as the means for quantifying the inherent fuzziness that is present in ill-posed problems (which by many accounts are the majority of the real-life problems in decision making and data mining). Fuzziness is a type of imprecision which may be associated with sets in which there is no sharp transition from membership to nonmembership [Bellman and Zadeh, 1970]. Examples of fuzzy sets are classes of objects (entities) characterized by such adjectives as “large,” “small,” “serious,” “simple,” “approximate,” and so on [Bellman and Zadeh, 1970].

As an indication of the importance of fuzzy set theory in engineering and scientific problems one could consider the more than 2,000 references given in [Chang, 1971], [Dubois and Prade, 1980], [Gupta, Ragade and Yager, 1979], [Xie and Berdosian, 1983], [Zadeh, Fu, Tanaka, and Shimura, 1975], [Zadeh, 1976; 1978; 1979], [Sanchez, 2006], [Ross, 2004], and [Cox, 2001]. This number is just a sample and the actual figure is much higher and it grows all the time.

Currently, an increasingly large number of researchers have been faced with the problem that either their data or their background knowledge is fuzzy. This is particularly critical to people who build intelligent systems and advanced decision support systems, for the knowledge they are dealing with is almost always riddled with vague concepts and judgmental rules (e.g., [Lee, 1971], [Lee, Grize, and Dehnad, 1987], [Prade and Negoita, 1986], [Ramsay, 1988], [Zadeh, 1983], [Zimmermann, 1985; 1996], [Klir and Yuan, 1995], [Nguyen and Walker, 2005], and [Szczepaniak, Lisboa, and Kacprzyk, 2000]. An overview of some applications of fuzzy sets in multicriteria decision making can be found in [Triantaphyllou, 2000].

The most critical step in any application of fuzzy set theory is to effectively estimate the pertinent data (i.e., the membership values). Although this is a fundamental problem, there is not a unique way of determining membership values in a fuzzy set. This is mainly due to the way different researchers perceive this problem.

For fuzzy numbers many people use *triangular fuzzy numbers* (that is, fuzzy numbers with lower, modal, and upper values, see also the next definition) because they are simpler when they are compared to the more flexible *trapezoid fuzzy numbers*. A triangular fuzzy number is formally defined as follows:

**Definition 16.1 (Dubois and Prade, 1980).** *A fuzzy number $M$ on $R \in (-\infty, +\infty)$ is defined to be a* **triangular fuzzy number** *if its membership function $\mu_M : R \rightarrow [0, 1]$ is equal to*

$$\mu_M(x) = \begin{cases} \frac{1}{m-l}x - \frac{l}{m-l}, & \textit{if } x \in [l, m] \\ \frac{1}{m-u}x - \frac{l}{m-u}, & \textit{if } x \in [m, u] \\ 0, & \textit{otherwise.} \end{cases}$$

In the previous formula the following holds true: $l \leq m \leq u$, where $l$ and $u$ stand for the lower and upper value of the support of fuzzy number $M$, respectively, and $m$ for the modal ("middle") value. A triangular fuzzy number, as expressed above, will be denoted as $(l, m, u)$. The graphical representation of the above concept is given in Figure 16.1.

However, in this chapter we will use trapezoid fuzzy numbers. Such a representation is a straightforward extension of the simpler triangular fuzzy numbers. Figure 16.2 illustrates a typical trapezoid fuzzy number. Observe that now we have the two modal ("middle") points $m_1$ and $m_2$. The definition of trapezoid fuzzy numbers is analogous to the previous one (i.e., now there are four cases to consider).

## 16.4 Formalization with Fuzzy Logic

In this section we slightly change the previous two definitions of a lobular mass and a microlobulated mass. We define a mass to be *lobular* if it has a contour with *some big* and *deep* undulations. The margins of a mass are *microlobulated* if they have *several*

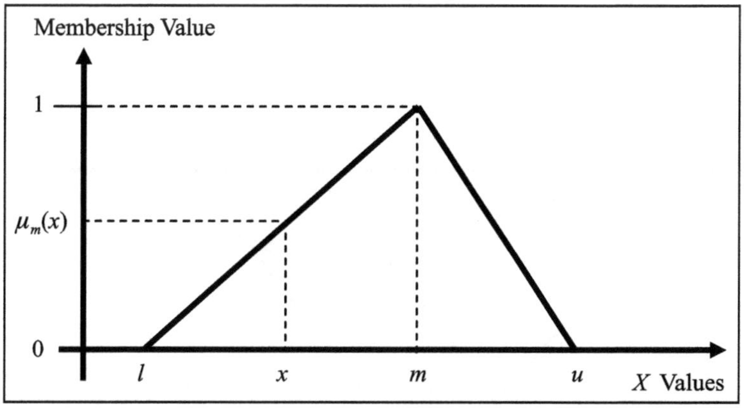


**Figure 16.1.** A Typical Triangular Fuzzy Number.

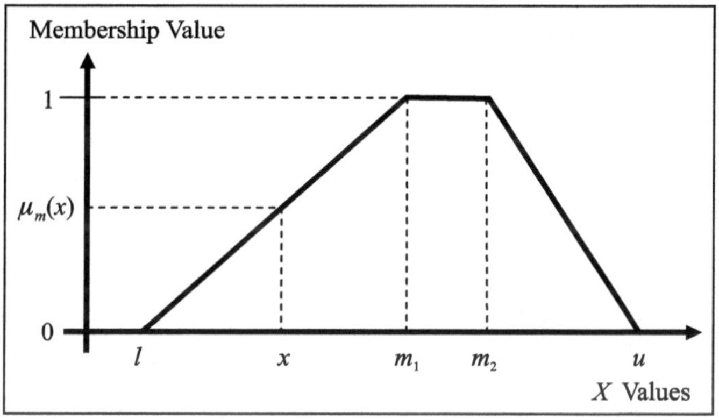


**Figure 16.2.** A Typical Trapezoid Fuzzy Number.

*small* concavities (cycles) producing *several small* and *shallow* undulations. At a first glance it may appear that we did not improve the precision of the definitions. However, these reformulations are of critical importance. They allow one to apply fuzzy logic and express the original two principal ACR definitions as functions of secondary and easily fuzzifiable terms.

The above considerations involve two important fuzzy terms, namely the terms *some* and *several*. These terms have a rather clear meaning when they are used in context with other terms of natural language [Kovalerchuk and Klir, 1995], [Kovalerchuk, 1996]. One can then define a fuzzy set with the fuzzy terms {*few, some, several, many*} for the *number of undulations*. Note that the number of undulations can be equal to 0, 1, 2, 3, . . . , etc.

For instance, for the fuzzy term *few* the number of undulations can be set equal to 0. That is, the corresponding family of the four (trapezoid) fuzzy membership functions are $\mu_{\text{few}}(x)$, $\mu_{\text{some}}(x)$, $\mu_{\text{several}}(x)$, and $\mu_{\text{many}}(x)$ (see also Figure 16.3).

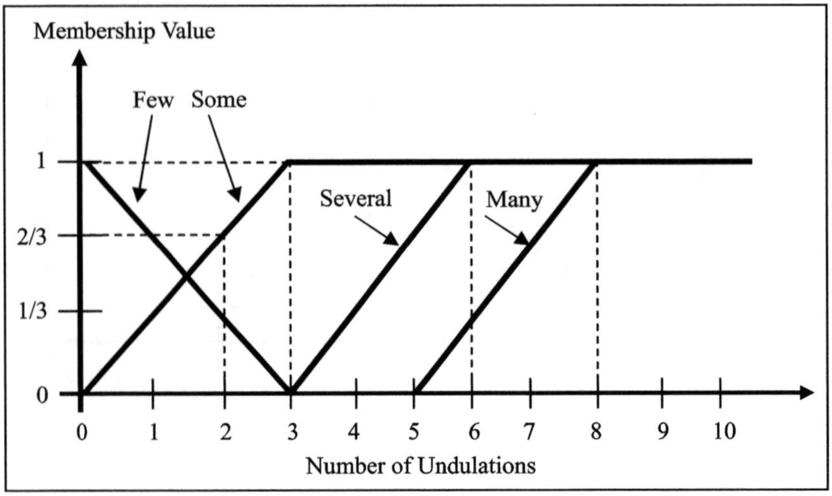


**Figure 16.3.** Membership Functions Related to the Number of Undulations.

Some possible sampled values of these membership functions could be $\mu_{\text{few}}(2) = 1/3$, $\mu_{\text{some}}(2) = 2/3$, $\mu_{\text{some}}(3) = 1$, $\mu_{\text{many}}(2) = 0$, etc. Some interviewed radiologists felt comfortable with this formalization.

Although one may argue with the specific numerical values of the above membership functions, the main issue here is that it is possible to effectively quantify fuzzy concepts which are critical for a consistent and objective classification of masses as lobular or microlobular.

Next we define the meaning of the terms of the fuzzy set {small, big}. This set is crucial in defining the *size of undulations*. First we need an adequate scale to measure the length of a given undulation. We consider the length of an undulation in *relative* terms since different masses may have different sizes. For instance, an undulation of 3 mm in length could be considered "microlobular" in a large mass while a small mass with the same undulation could be considered "lobular."

Therefore, we first need to compute $L$; the maximum length of a mass. This approach can allow one to estimate the undulation length as a fraction of $L$. In Figure 16.4a we present a mass with some undulations. Specifically, the curve between the points $A$ and $B$ is an undulation. We can formalize the fuzzy terms *small* and *big* by characterizing undulations on a scale determined by the relative undulation length (see also Figure 16.4b).

According to the membership functions in Figure 16.4b, a relative length of more than $L/4$ can be defined as a big undulation, while an undulation of relative length of less than $L/12$ could be considered as a small undulation. Undulations of intermediate length can be assigned intermediate membership values.

Since masses may have varying degree of depth of lobularity one can also define the fuzzy membership functions regarding the "shallow" or "deep" aspect of the undulations. Thus, we next introduce a relative measure of the *depth of lobularity*,

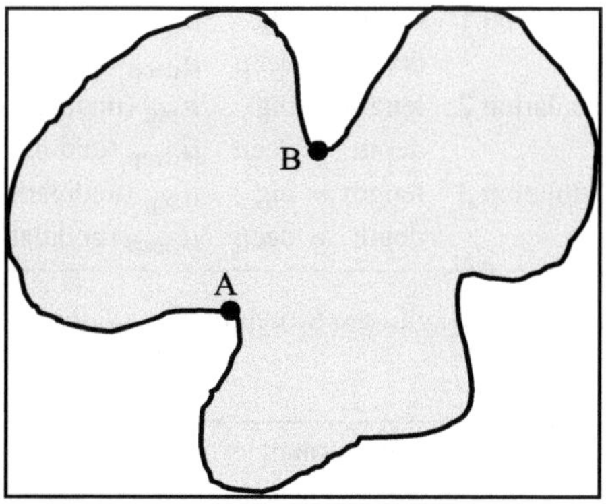


**Figure 16.4a.** A Diagrammatic Representation of a Mass with Undulations.

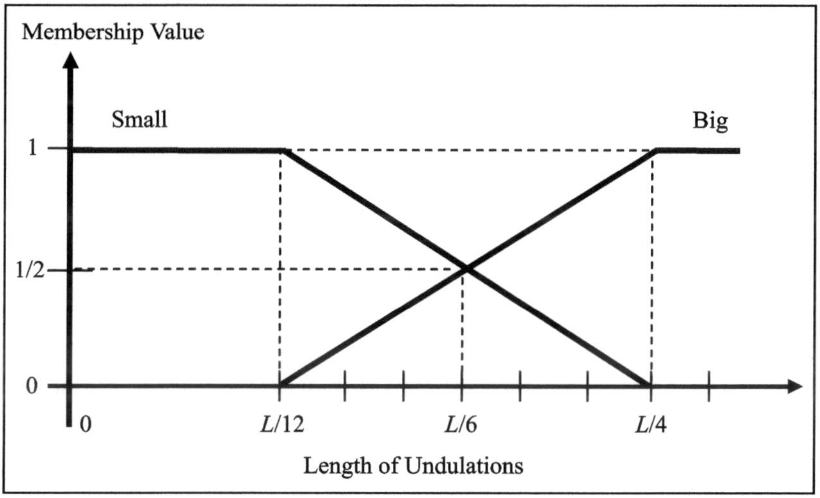


**Figure 16.4b.** Membership Functions Related to the Length of Undulations.

which is defined as a fraction of the maximum length (denoted as $L$) of the mass. This step is similar to those described in the previous fuzzy sets.

The concept of a lobular mass can now be formulated as follows: *A mass is lobular if it has at least three undulations with length and depth of not less than* $L/4$. We can also formulate the concept of microlobulated mass margins. *The mass margins are microlobulated if there are at least six undulations with length and depth of not more than* $L/12$. These definitions are based on the interdependence of the concepts of size, depth, and number of undulations and can be used to quantify the concepts of lobular and microlobular masses objectively and consistently. The fuzzy logic structures for the lobular and microlobular concepts are presented in Figures 16.5 and 16.6, respectively.

| | | | |
|---|---|---|---|
| | → undulation 1 | length = big | $\mu_{\text{big}}$ (undulation 1) = 1.00 |
| | | depth = deep | $\mu_{\text{deep}}$ (undulation 1) = 1.00 |
| **MASS** | → undulation 2 | length = big | $\mu_{\text{big}}$ (undulation 2) = 1.00 |
| | | depth = deep | $\mu_{\text{deep}}$ (undulation 2) = 1.00 |
| | → undulation 3 | length = big | $\mu_{\text{big}}$ (undulation 3) = 1.00 |
| | | depth = deep | $\mu_{\text{deep}}$ (undulation 3) = 1.00 |

**Figure 16.5.** Fuzzy Logic Structures for a *Lobular* Mass.

| | | | |
|---|---|---|---|
| | → undulation 1 | length = small | $\mu_{\text{small}}$ (undulation 1) = 1.00 |
| | | depth = shallow | $\mu_{\text{shallow}}$ (undulation 1) = 1.00 |
| | → undulation 2 | length = small | $\mu_{\text{small}}$ (undulation 2) = 1.00 |
| | | depth = shallow | $\mu_{\text{shallow}}$ (undulation 2) = 1.00 |
| **MASS** | → undulation 3 | length = small | $\mu_{\text{small}}$ (undulation 3) = 1.00 |
| | | depth = shallow | $\mu_{\text{shallow}}$ (undulation 3) = 1.00 |
| | → undulation 4 | length = small | $\mu_{\text{small}}$ (undulation 4) = 1.00 |
| | | depth = shallow | $\mu_{\text{shallow}}$ (undulation 4) = 1.00 |
| | → undulation 5 | length = small | $\mu_{\text{small}}$ (undulation 5) = 1.00 |
| | | depth = shallow | $\mu_{\text{shallow}}$ (undulation 5) = 1.00 |
| | → undulation 6 | length = small | $\mu_{\text{small}}$ (undulation 6) = 1.00 |
| | | depth = shallow | $\mu_{\text{shallow}}$ (undulation 6) = 1.00 |

**Figure 16.6.** Fuzzy Logic Structures for a *Microlobulated* Mass.

Figure 16.5 shows the fuzzy logic structures of a hypothetical mass with three undulations. Each undulation is presented with its length and depth. All these undulations are big and deep. Hence, all membership functions are equal to 1.00 and according to our formalization such a mass is lobular. Similarly, Figure 16.6 shows a hypothetical microlobulated mass with six undulations and all of them are small and shallow.

The previous definitions allow some masses to be classified as both *lobular and microlobulated* without any contradiction if the mass has at least nine undulations (of which three are lobular and six are microlobular). That is, one just needs to join the structures given in Figures 16.5 and 16.6. Cases of an intermediate nature can also be formalized. An example of such a case is depicted in Figure 16.7.

We take the three biggest and deepest undulations and compute the minimum of their membership function values for the terms *big* and *deep*. We define this value as the *degree of lobularity* (or *DL*). For instance, for the mass described in Figure 16.7 the minimum for the first three undulations is 0.70, that is, for this case $DL = 0.70$. Similarly, it can be easily verified that the *degree of microlobularity* (or *DM*) computed with the remaining 6 undulations is 0.60. Such estimates can be used as inputs for a breast cancer computer-aided diagnostic (CAD) system.

| | | | | |
|---|---|---|---|---|
| | → undulation 1 | length = big | $\mu_{big}$ (undulation 1) | = 0.80 |
| | | depth = deep | $\mu_{deep}$ (undulation 1) | = 0.70 |
| | → undulation 2 | length = big | $\mu_{big}$ (undulation 2) | = 0.73 |
| | | depth = deep | $\mu_{deep}$ (undulation 2) | = 0.71 |
| | → undulation 3 | length = big | $\mu_{big}$ (undulation 3) | = 0.90 |
| | | depth = deep | $\mu_{deep}$ (undulation 3) | = 0.80 |
| **MASS** | → undulation 4 | length = small | $\mu_{small}$ (undulation 4) | = 0.90 |
| | | depth = shallow | $\mu_{shallow}$ (undulation 4) | = 0.80 |
| | → undulation 5 | length = small | $\mu_{small}$ (undulation 5) | = 0.90 |
| | | depth = shallow | $\mu_{shallow}$ (undulation 5) | = 0.70 |
| | → undulation 6 | length = small | $\mu_{small}$ (undulation 6) | = 0.60 |
| | | depth = shallow | $\mu_{shallow}$ (undulation 6) | = 0.70 |
| | → undulation 7 | length = small | $\mu_{small}$ (undulation 7) | = 0.67 |
| | | depth = shallow | $\mu_{shallow}$ (undulation 7) | = 0.97 |
| | → undulation 8 | length = small | $\mu_{small}$ (undulation 8) | = 0.80 |
| | | depth = shallow | $\mu_{shallow}$ (undulation 8) | = 1.00 |
| | → undulation 9 | length = small | $\mu_{small}$ (undulation 9) | = 0.84 |
| | | depth = shallow | $\mu_{shallow}$ (undulation 9) | = 0.79 |

**Figure 16.7.** Structural Descriptions for a Fuzzy Lobular and Microlobulated Mass.

| | | | | |
|---|---|---|---|---|
| | → undulation 1 | length = big | $\mu_{big}$ (undulation 1) | = 0.80 |
| | | depth = deep | $\mu_{deep}$ (undulation 1) | = 0.70 |
| **MASS** | → undulation 2 | length = big | $\mu_{big}$ (undulation 2) | = 0.60 |
| | | depth = deep | $\mu_{deep}$ (undulation 2) | = 0.60 |

**Figure 16.8.** Fuzzy Logic Structures for a Mass with Less Than Three Undulations.

If the number of undulations is less than three, one can combine the membership functions for the length and depth with a *membership function for the number of undulations* (as defined in Figure 16.3). In this combination, we compute the minimum of these three values in accordance with standard fuzzy logic practice. We analyze the arguments for the use of the *min* (minimum) operator in the next section. Now let us consider, for instance, the mass with the two undulations described in Figure 16.8.

Figure 16.6 provides the means to compute 0.60 as the corresponding degree of lobularity (*DL*), while Figure 16.3 shows that $\mu_{some}(2) = 0.66$ for a case with two undulations. Thus, their minimum of 0.60 characterizes the lobularity of this mass. It is important to state here that the proposed fuzzy membership functions are *only indicative*. Their exact numerical forms can be determined from a consensus approach among radiologists and/or by using historic data. In the next section we present the *DL* and *DM* ideas formally.

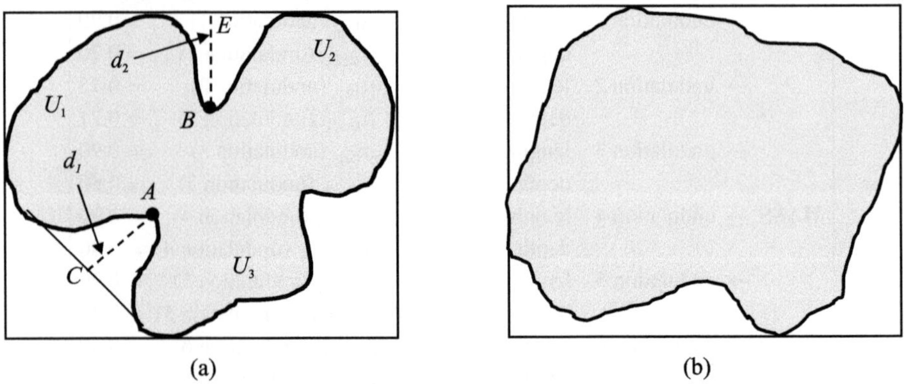


**Figure 16.9.** Diagrammatic Representation of Masses with (a) Deep and (b) Shallow Undulations.

## 16.5 Degrees of Lobularity and Microlobularity

Radiologists use an informal approach in determining the lobularity and microlobularity of a mass. To maintain consistency in these evaluations and increase objectivity, we need to formalize these concepts. Let us first consider the two masses depicted in Figure 16.9.

Intuitively, the first mass has deep undulations, while the second mass has shallow undulations. Different measures can be created to formalize this distinction. Figure 16.9(a) shows two distances $d_1$ and $d_2$, defined between the points $A$ and $C$ and between the points $B$ and $E$, respectively, for undulation 1 (i.e., $U_1$). If each of them is no less than $L/4$, then this undulation is deep (see also Figure 16.9(b)). If these distances are no more than $L/12$, then undulation 1 is shallow (see also Figure 16.9(a)).

This situation indicates that formally the *depth D of the undulation* closely depends on the pair of values for $d_1$ and $d_2$ (this concept is not to be confused with the one of depth of lobularity which was defined in Section 16.2). The method used to compute these values was considered in [Kovalerchuk, Triantaphyllou, and Ruiz, 1996].

The values of $\mu_{\text{deep}}(d_1)$ and $\mu_{\text{deep}}(d_2)$ are computed by using the corresponding membership functions in Figure 16.4. In this way the previous two measures can be transformed into a single degree of lobularity for a given undulation. Recall that we use the same membership functions for the length and depth of undulations. This is done by substituting the terms *big* for *deep* and *small* for *shallow*. Next, we compute $\min\{\mu_{\text{deep}}(d_1), \mu_{\text{deep}}(d_2)\}$, which could be considered as the *degree of depth of the undulation*. That is:

$$\mu_{\text{deep}}(\text{undulation}) = \min\{\mu_{\text{deep}}(d_1), \mu_{\text{deep}}(d_2)\}.$$

Similarly, we define the *degree of shallowness of an undulation* as

$$\mu_{\text{shallow}}(\text{undulation}) = \min\{\mu_{\text{shallow}}(d_1), \mu_{\text{shallow}}(d_2)\}.$$

Observe that the length of undulation 1 (denoted as $U_1$) is measured as the length of the mass margin between points $A$ and $B$ (see also Figure 16.9(a)).

Now one can define the *Degree of Lobularity (DL) of a mass* as follows:

$$DL(\text{mass}) = \min\{\mu_{\text{some}}(k), \min_{k \geq i \geq 1} \{\mu_{\text{big}}(U_i), \mu_{\text{deep}}(U_i)\}\}, \tag{16.1}$$

where $U_1, U_2, \ldots, U_k$ are undulations such that

$$\min_{k \geq i \geq 1} \{\mu_{\text{big}}(U_i), \mu_{\text{deep}}(U_i)\} \geq 0.50.$$

Similarly, one can define the *Degree of Microlobularity (DM) of a mass* with $k$ undulations:

$$DM(\text{mass}) = \min\{\mu_{\text{several}}(k), \min_{k \geq i \geq 1} \{\mu_{\text{small}}(U_i), \mu_{\text{shallow}}(U_i)\}\}, \tag{16.2}$$

where $U_1, U_2, \ldots, U_k$ are undulations such that

$$\min_{k \geq i \geq 1} \{\mu_{\text{small}}(U_i), \mu_{\text{shallow}}(U_i)\} \geq 0.5.$$

For the extreme case of $k = 0$, we have $\mu_{\text{some}}(k) = 0$ and $\mu_{\text{several}}(k) = 0$ (see also Figure 16.3). Thus, both degrees of lobularity and microlobularity are equal to 0, i.e., the outcome corresponds to what is expected with common sense.

There are some theoretical and experimental arguments for the general case (e.g., [Kovalerchuk and Klir, 1995], [Kovalerchuk and Dalabaev, 1993], and [Kovalerchuk and Taliansky, 1992]) justifying formulas (16.1) and (16.2). However, we can also use some additional arguments derived from this mammographic problem. A consistent computer-based breast cancer diagnostic system should refuse to diagnose a mammogram with a significant number of doubtful features. We can express how doubtful a given feature is by some degree between 0 and 1, with the highest degree of doubt given at 0.50. The values of *DL* and *DM* are examples of such degrees. For these uncertain (doubtful) features, a CAD system can suggest the presence of a particular feature, but only with some *degree of confidence*. This confidence can be very low. Also, this degree of confidence depends on the particular values of the *DL* and *DM* quantities. Therefore, the formulas used to define *DL* and *DM* become even more critical.

This situation can be explained with a modified example from Figure 16.6. Assume that the first five membership functions for undulations are equal to 1.00 and the sixth function is equal to 0.60 (i.e., $\mu_{\text{deep}}(\text{undulation } 3) = 0.60$). Then formula (16.1) gives us a "pessimistic" assessment, i.e., a low degree of certainty for the presence of lobularity. This is expressed as $DL = 0.60$. Substituting in (16.1) the minimum (min) operation with the maximum will give us an "optimistic" assessment, i.e.., high degree of lobularity, $DL = 1.00$ for this case. In the last "optimistic" assessment we *ignore and lose the warning* information (i.e., the fact that $\mu_{\text{deep}}(\text{undulation } 3) = 0.60$). The value 0.60 suggests that one should be cautious and study the case in great detail. However, no warning information is lost if we use the "pessimistic" minimum (min) operation in formulas (16.1) and (16.2).

Therefore, for critical questions regarding a very critical situation such as cancer diagnosis, we see that the "pessimistic" strategy is the safest as it is the most conservative one. We also consider statements with a low degree of confidence as a preliminary suggestion indicating that we need to *switch the set of features to a higher level of detail* in order to fully evaluate the complexity of a given case. Some experiments in [Doi, *et al.*, 1993] and [Wu, *et al.*, 1993] have shown that relatively simple cases can be diagnosed within a small feature space. However, for more complicated cases we need a pathologically confirmed training sample with more features and a specifically designed diagnostic method. A CAD system designed as above can have switching capabilities based on the described approach.

## 16.6 Concluding Remarks

Radiologists often make relatively subjective determinations for many features related to breast cancer diagnosis. We have formalized some important features from the ACR Breast Imaging Lexicon, i.e., lobulation and microlobulation of masses (nodules, lesions). This formalization is the basis of the following three steps: (i) extensive radiological validation; (ii) automatic detection of lobulation/microlobulation in a mammographic image; and (iii) similar formalizations of other terms from the ACR Breast Imaging Lexicon. This study suggests that fuzzy logic can be an effective tool in dealing with this kind of medical problems.

It should also be stated here that the ACR Breast Imaging Lexicon involves many concepts which could be defined in a fuzzy logic approach similar to the proposed lobulation and microlobulation analysis. However, as has been shown in our previous work with breast calcifications [Kovalerchuk, Triantaphyllou, and Ruiz, 1996], the various features presented in the Lexicon pose a broad range of problems requiring tailored solutions. Analyzing all the concepts covered in the ACR Breast Imaging Lexicon (which are approximately thirty) is outside the scope of this chapter and would require a sequence of similar developments.

However, such a goal would be of great importance in breast cancer diagnosis, as it has been demonstrated that traditional artificial intelligence and statistical methods of pattern recognition and diagnosis may be dramatically unreliable. Related to this subject is also Chapter 9 which elaborates on the reliability issue of inferred data mining and knowledge discovery models.

The proposed fuzzy logic approach is both feasible and effective because this type of approach (but not on this geometric/medical context) has been applied successfully in other areas. Its complete application in breast cancer diagnosis is only a matter of time. The proposed fuzzy logic approach has the potential to open a new and very exciting direction for effective and early breast cancer diagnosis and, in general, in data mining and knowledge discovery research and applications.

# Chapter 17

# Conclusions

## 17.1 General Concluding Remarks

Each of the previous chapters ends with a section with some concluding remarks tailored to the contents of the particular chapter. This section provides some comprehensive concluding remarks. As was mentioned earlier, there are many approaches to data mining and knowledge discovery from data sets. Such approaches include neural networks, closest neighbor methods, and various statistical methods. However, such approaches may have some severe limitations for a number of reasons.

First of all, for discovering new knowledge from data sets, methods based on logic offer a direct and often intuitive approach for extracting easily interpretable patterns (i.e., the new knowledge). For instance, translating a complex statistical model into a set of rules, which can easily be interpreted by domain experts, may be a challenge in many applications. The same is true when dealing with neural networks. It is not a coincidence that now there is a new trend in developing *hybrid* methods that combine such traditional data mining and knowledge discovery approaches with logic-based methods. Furthermore, most statistical methods rely on rather strict assumptions which may or may not hold. For the same reason, the available data sets should be sufficiently large to verify the justification for these assumptions.

Regarding the data mining aspect of such methods, one may be simply interested in extracting some type of model which could predict the class membership of new observations with high accuracy. The interest may not be in interpreting the structure of these extracted patterns, but only on having a model which is highly accurate. Even so, mathematical logic-based methods may be the way to proceed, as they can be very accurate.

As was shown in Chapters 10 and 11, which discussed the monotonicity property, logic-based methods may offer more accurate approaches in defining the borders between the different classes of observations. By focusing on the border determination issue, one may be able to develop a model that is more accurate in general, but perhaps more importantly, more accurate when the new cases are complicated and ambiguous ones and not of an obvious class. After all, one needs a good computer-aided system to deal with the more complicated cases which require specialized

E. Triantaphyllou, *Data Mining and Knowledge Discovery via Logic-Based Methods*, Springer Optimization and Its Applications 43, DOI 10.1007/978-1-4419-1630-3_17,

expertise to analyze. Finally, it should be noted here that relevant to this fundamental issue is the discussion on the "three major illusions on accuracy" as described in Chapter 11.

Another reason that supports the argument that logic-based methods have a bright future comes from the huge literature currently available on such methods for other applications. The immense advance in digital technology, with circuit design and minimization, is actually based on advances on logic methods (digital methods). When one aims at circuit minimization, the objective is to replace a digital system by another one of smaller size but with the same or even more advanced functionality. This objective is very similar to the objective of extracting a Boolean function (i.e., a pattern) of small size but still capable of satisfying the requirements of the two classes of observations and also accurately predicting the class membership of new observations.

## 17.2 Twelve Key Areas of Potential Future Research on Data Mining and Knowledge Discovery from Databases

The areas of research and application discussed in the previous chapters highlight some of the areas which are most likely to witness new and significant developments in the future. In this section we will summarize some of the areas most likely to lead to future developments. The following paragraphs describe some of these potential areas by focusing on twelve key research problems.

### 17.2.1 Overfitting and Overgeneralization

Many data mining and knowledge discovery methods which are based on neural network models have done a good job in finding a balance between overfitting and overgeneralization. Overfitting becomes a limitation when the inferred model can accurately deal with cases that are very close to the available training data, but does poorly with cases that are very different from such data. The opposite problem occurs when the problem is overgeneralization. In an overgeneralization situation the model claims to be accurate in cases which are very different than the data used for training, but in reality the model is not accurate.

Such problems may occur with any system inferred from training data, including systems derived by logic-based methods. An approach to deal with these problems is to first gain a deep understanding of them in the context of neural networks and then transfer this understanding into the domain of logic-based methods. An alternative way is to approach them from a geometric point of view, as is the case in the studies described in [Pham and Triantaphyllou, 2007; 2008; 2009a; 2009b].

### 17.2.2 Guided Learning

The approach proposed in Chapter 5 for guided learning is based on a technique that guarantees the modification of one and only one of a pair of systems (the so-called "positive" and "negative" Boolean functions). An interesting idea might be to try to locate as new examples ones that are very deeply inside the "gray" regions (i.e., the undecidable region left out by the two systems or the area in which the two systems conflict with each other). A key issue here is how to define "depth" in the context of the space of the examples (observations). This may not be a trivial task as some attributes may not have an easy geometric/Euclidian interpretation.

Another alternative approach is to seek new observations from nonconflicting regions in which case a contradiction with one of the two systems now leads to the modification of *both* systems. For instance, if a data point that is classified as positive by both systems turns out to be negative in reality, then both systems need to be modified. If such strategy is coupled with the previous geometric direction of seeking observations that are deeply inside a region, then the potential benefits might be very significant. Another approach might be to combine concepts from monotone Boolean functions and attempt to define the borders of the classes more accurately and by seeking as few new queries as possible.

### 17.2.3 Stochasticity

The proposed logic-based methods are mostly deterministic ones and do not handle stochastic data in a direct manner. An alternative approach might be to define observations that are, say, 10% positive and 90% negative, as a different class. Such classes are nested and not overlapping. Thus, the nested monotone Boolean functions discussed in Chapter 10 might be one of the ways to deal with stochastic data.

### 17.2.4 More on Monotonicity

Some of the previous chapters strongly suggest that monotonicity in the data may lead to attractive algorithms in the field of data mining and knowledge discovery. An intriguing idea comes from the observation that *any* Boolean function can be decomposed into a number of increasing and decreasing monotone Boolean functions. Furthermore, any Boolean function may be "sandwiched" by a pair of monotone Boolean functions. These observations may lead to new ways of extending the mathematically attractive algorithms on monotone Boolean functions to the realm of general Boolean functions. This, in turn, may lead to a better interpretation of the derived results and also to more accurate and efficient algorithms.

### 17.2.5 Visualization

A new trend in data mining and knowledge discovery methods is to use visualization approaches to better understand the nature of the inferred models. Monotonicity may offer some intriguing possibilities in developing new ways for visualizing such models. The poset idea discussed in Chapter 10 may offer a new way for visualization. Most visualization systems suffer from degradation when the number of attributes increases. This is not the case with posets. One has to organize the observations along the various layers of posets. Recall that there are $n + 1$ such layers and up to $2^n$ possible observations when the number of (binary) attributes is equal to $n$.

### 17.2.6 Systems for Distributed Computing Environments

Many databases are today distributed. The same is true with computing resources. The advent of the Internet and the Web along with the spread of Globalization make the idea of developing new data mining and knowledge discovery methods that analyze distributed databases and also use distributed computing resources very appealing. The use of mobile agents may be a way to go in this direction. Logic-based methods offer easy ways for checking for consistency among different results derived locally. Another related idea is to use the so-called "Grid computing" methods for the same goal.

### 17.2.7 Developing Better Exact Algorithms and Heuristics

A never-ending goal with any computational area is the development of more effective and efficient exact algorithms and heuristics. The claim that faster computers will be just sufficient to handle the needs of future applications is naïve. More gains can be achieved by making algorithmic developments that tackle the issues squarely. The only certain issue is that the sizes of future databases will increase more and more as data become readily available and storing media become more cost effective. Ways for partitioning large databases and achieving scalability are critical. Perhaps, the rejectability graph discussed in Chapter 8 and other graph-theoretic methods may offer some plausible opportunities in this direction.

### 17.2.8 Hybridization and Other Algorithmic Issues

The mere presence of many and diverse data mining and knowledge discovery algorithms signifies that different approaches may be more effective under different conditions and domains. An existing trend here is to combine different approaches in a way that the new, or hybrid, system will have the advantages of the individual approaches and none of their disadvantages. When such hybrid methods are built around a logic-based framework, the hybrid system may be able to better explain its decision-making process.

Another key problem is the identification of errors and outliers (which may not be the result of errors) in the data; also, to be able to handle problems with missing data and not just to ignore the observations which happen to have some of their fields missing. Surprisingly enough, an issue that has received relatively little attention despite its apparent importance is that of dealing with problems which involve multiple classes and not just two (i.e., the "positive" and "negative" class). If methods are developed that can solve such multiclass problems in a different way, then it is possible that the accuracy of multiclass models can be improved dramatically.

Finally, another critical issue is to develop inference methods that try to determine the border of the various classes of observations more accurately. In this way, the methodological problems described as the "three major illusions on accuracy" in Chapter 11, may be tackled more effectively.

### 17.2.9 Systems with Self-Explanatory Capabilities

In the past such systems were primarily built for advanced expert systems. Being able to better explain its behavior, a system becomes easier to validate. It also becomes easier to be trusted by the domain experts. Logic-based systems provide an intuitive framework for embedding self-explanatory capabilities in intelligent systems built by data mining and knowledge discovery methods.

### 17.2.10 New Systems for Image Analysis

The use of digital cameras for video and still images has become ubiquitous in recent years. Now it is almost a common occurrence for amateurs to possess high-definition (HD) video cameras, something that most professionals could not even dream of just a few years ago. Web pages which organize, promote, and share public and private photos and videos are everywhere. Services like those offered by *Flickr*, *Picasa*, and *YouTube* are the current main examples of such capabilities.

The above situation has led to a proliferation of digital images and videos on the Web. At the same time, methods for interpreting the context of digital images and videos are still in their infancy. Some approaches resort to attaching some textual descriptions (as generated by human operators) to images and videos, so they can be easier catalogued and become easier retrievable. Logic-based methods, combined with data mining and knowledge discovery methods, may offer ways for breaking this stumbling block and opening a totally new horizon of possibilities.

### 17.2.11 Systems for Web Applications

Since the Web has literarily invaded every corner of the planet and modern society, a new breed of data mining and knowledge discovery methods may provide new opportunities for harvesting the contextual and computational richness of the Web plus better communications. Some prime examples of this trend are the new generation of iPhone and Blackberry devices and other related Web/communication gadgets. Furthermore, the combination of the GPS (Global Positioning System) enables such applications to reach new highs. Again, logic-based methods may offer some unique opportunities in achieving these and also future goals.

### 17.2.12 Developing More Applications

New applications, not only on numbers but in new domains as well, may offer the background for defining new computational challenges and lead to the development of new algorithms. At the same time, new algorithms may make data mining and knowledge discovery methods appealing to new types of applications. This is a two-way relationship in a closed loop which can lead to new developments in new theories and applications alike. Logic-based methods can be the driving force in this bidirectional model of development.

The previous twelve areas of possible future developments are only indicative of the great potential that logic-based methods have in the data mining and knowledge discovery field. It should also be noted here that this field, in the way we know it now, is rather new and still developing. As is the case with many new ideas, there is a great deal of hype at the beginning. Next, there is a stage of excessive skepticism about the potential of the new field. After that there is the stage of maturity. It is hard to tell at which stage we are now. Perhaps there are signs of being in any one of these three stages! However, one issue is rather certain; the future will reveal many more opportunities and challenges for logic-based approaches to data mining and knowledge discovery from databases/data sets.

## 17.3 Epilogue

It seems like there is a very powerful relationship developing in modern society when one considers the areas of technology, science, and modern life. First of all, the generation, collection, and storage of data keep increasing at a torrid pace. At the same time, computing power becomes more effective and efficient. Although some scholars raise doubts on how much longer Moore's law may still be applicable, there is no end in sight on the growth of computing power. When these two driving factors are examined together, the only certain conclusion is that needs and methods for analyzing large and complex data sets will be even more important in the future. This is represented diagrammatically in Figure 17.1. For apparent reasons, we will call this the "Data Mining and Knowledge Discovery Equation for the Future."

The advent of the Web, the seemingly pervasive use of cameras and scanning systems which generate endless numbers of still images and videos all the time everywhere, are just some of the many new ways for collecting and storing data about nearly everything on a nearly continuous basis. Mobile phones and personal gadgets, such as music players, digital cameras, just to name a few, are means for

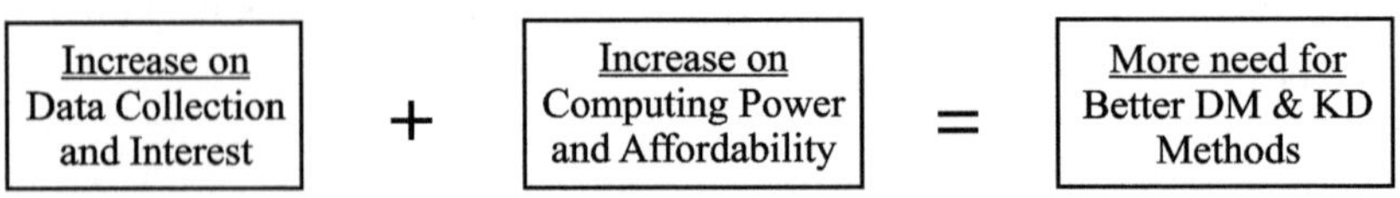


**Figure 17.1.** The "Data Mining and Knowledge Discovery Equation for the Future."

collecting and storing more and more data. Today, besides the traditional keys in the key chain that people carry everywhere with them, there are attached small cards with bar codes and a memory stick with storage capacity unbelievable just a few years ago even for desktop computers. There are even "wearable computers" for continuously monitoring or enhancing somebody's physical activities. Predicting the future in this area, even for the relatively short run, has already proven to be an elusive task. It is only natural to expect that the future is impossible to predict in terms of ways for collecting and storing data.

At the same time, computing power becomes more powerful and more affordable. It has reached the point that now many regular appliances do have complex CPUs embedded in them with advanced computing capabilities and their owners are not even aware of that. As the "Data Mining Equation for the Future" in Figure 17.1 illustrates, the synergy of these two factors will only force more interest and need for new data mining and knowledge discovery methods.

It should be clearly stated at this point that the subject of data mining and knowledge discovery methods is not a purely computer science/operations research issue. It is way too important to be confined to a single or a very closely related family of disciplines. It is also a social issue, as the analysis of data may have lots of social and/or ethical implications. For that, it is also a political issue and some governments have already been involved in heated debates regarding the use of such methods on some types of data sets. Such trends will naturally intensify even more in the future. All these developments, both algorithmic (as the ones discussed in this book) and also social, ethical, and legal, are crucial for the effective development and appropriate use of this very fast emerging and potentially ultrapowerful technology. What we have seen so far is just the beginning.

# References


1. Abe, S., (2005), *Support Vector Machines for Pattern Classification*, Series on Advances in Pattern Recognition, Springer, Heidelberg, Germany.
2. Ackoff, R.L., (1987), *The Art of Problem Solving*, John Wiley & Sons, New York, NY, U.S.A.
3. Adamo, J.-M., (2000), *Data Mining for Association Rules and Sequential Patterns: Sequential and Parallel Algorithms*, Springer, Heidelberg, Germany.
4. Agrawal, R., and R. Srikant, (1994), "*Fast Algorithms for Mining Association Rules*," Proceedings of the 20th VLDB Conference, Santiago, Chile.
5. Agrawal, R., T. Imielinski, and A. Swami, (1993), "*Mining Association Rules Between Sets of Items in Large Databases*," Proceedings of the 1993 ACM SIGMOD Conference, Washington, DC, U.S.A., May.
6. Aho, A.V., J.E. Hopcraft, and J.D. Ullman, (1974), "*The Design and Analysis of Computer Algorithms*," Addison-Wesley Publishing Company, Reading, MA, U.S.A.
7. Aldenderfer, M.S., (1984), *Cluster Analysis*, Sage Beverly Hills, CA, U.S.A.
8. Alekseev, V., (1988), "*Monotone Boolean Functions*," Encyclopedia of Mathematics, Kluwer Academic Publishers, Norwell, MA, U.S.A., Vol. 6, pp. 306–307.
9. Anderberg, M.R., (1973), *Cluster Analysis for Applications*, Academic Publishers, New York, NY, U.S.A.
10. Angluin, D., (1987), "*Learning Propositional Horn Sentences With Hints*," Technical Report, YALE/DCS/RR-590, Department of Computer Science, Yale University, New Haven, CT 06511, U.S.A.
11. Angluin, D., (1988), "*Queries and Concept Learning*," Mach. Learn., Vol. 2, pp. 319–342.
12. Angluin, D., (1992), "*Computational Learning Theory: Survey and Selected Bibliography*," Proceedings of the 24th Annual ACM Symposium on the Theory of Computing, Victoria, BC, Canada, May 4–6, pp. 351–369.
13. Angluin, D., and C.H. Smith, (1983), "*Inductive Inference: Theory and Methods*," *Comput. Surv.*, Vol. 15, pp. 237–265.
14. Arbib, M.A., (Editor), (2002), *The Handbook of Brain Theory and Neural Networks*, MIT Press, Cambridge, MA, U.S.A.
15. Ayer, M., H.D. Brunk, G.M. Ewing, W.T. Reid, and E. Silverman, (1955), "*An Empirical Distribution Function for Sampling with Incomplete Information*," Ann. Math. Statistics, Vol. 26, pp. 641–647.


E. Triantaphyllou, *Data Mining and Knowledge Discovery via Logic-Based Methods*, Springer Optimization and Its Applications 43, DOI 10.1007/978-1-4419-1630-3,

16. Babel, L., (1991), "*Finding Maximum Cliques in Arbitrary and in Special Graphs*," Computing, Vol. 46, pp. 321–341.
17. Babel, L., (1994), "*A Fast Algorithm for the Maximum Weight Clique Problem*," Computing, Vol. 51, pp. 31–38.
18. Babel, L., and G. Tinhofer, (1990), "*A Branch and Bound Algorithm for the Maximum Clique Problem*," Methods Models Oper. Res., Vol. 34, pp. 207–217.
19. Balas, E., and W. Niehaus, (1994), "*Finding Large Cliques by Bipartite Matching*," Management Science Research Report #MSRR-597, Carnegie Mellon University, Pittsburgh, PA 15213, U.S.A., 11 pages.
20. Balas, E., and J. Xue, (1993), "*Weighted and Unweighted Maximum Clique Algorithms with Upper Bounds From Fractional Coloring*," Management Science Research Report #MSRR-590, Carnegie Mellon University, Pittsburgh, PA 15213, U.S.A., 19 pages.
21. Barnes, J.W., (1994), *Statistical Analysis for Engineers and Scientists, A Computer-Based Approach*, McGraw-Hill, New York, NY, U.S.A.
22. Bartnikowski, S., M. Grandberry, J. Mugan, and K. Truemper, (2006), "*Transformation of Rational Data and Set Data to Logic Data*," in *Data Mining and Knowledge Discovery Approaches Based on Rule Induction Techniques*, E. Triantaphyllou and G. Felici, (Editors), Massive Computing Series, Springer, Heidelberg, Germany, pp. 253–278.
23. Baum, E.B., and D. Haussler, (1989), "*What Size Net Gives Valid Generalizations?*" Neural Comput., No. 1, pp. 151–160.
24. Bayardo, Jr., R.J., R. Agrawal, and D. Gunopulos, (1999), "*Constraint-Based Rule Mining in Large, Dense Databases*," Proceedings of the 15th International Conference on Data Engineering.
25. Bellman, R.E., and L.A. Zadeh, (1970), "*Decision-Making in a Fuzzy Environment*," Manage. Sci., Vol. 17B, No. 4, pp. 141–164.
26. Ben-David, A., (1992), "*Automatic Generation of Symbolic Multiattribute Ordinal Knowledge-Based DSSs: Methodology and Applications*," Decision Sci., Vol. 23, No. 6, pp. 1357–1372.
27. Ben-David, A., (1995), "*Monotonicity Maintenance in Information-Theoretic Machine Learning Algorithms*," Mach. Learn., Vol. 19, No. 1, pp. 29–43.
28. Berry, M.W., C. Kamath, and D. Skillicorn, (Editors), (2004), Proceedings of the Fourth SIAM International Conference on Data Mining, Lake Buena Vista, FL, U.S.A., April, published by the Society for Industrial and Applied Mathematics (SIAM).
29. Bezdek, J., (1981), *Pattern Recognition with Fuzzy Objective Function Algorithms*, Plenum Press, New York, NY, U.S.A.
30. Bioch, J.C., and T. Ibaraki, (1995), "*Complexity of Identification and Dualization of Positive Boolean Functions*," Inf. Comput., Vol. 123, pp. 50–63.
31. Bird, R.E., T.W. Wallace, and B.C. Yankaskas, (1992), "*Analysis of Cancer Missed at Screening Mammography*," Radiology, Vol. 184, pp. 613–617.
32. Blair, C.E., R.G. Jeroslow, and J.K. Lowe, (1986), "Some Results and Experiments in Programming Techniques for Propositional Logic," Comput. Oper. Res., Vol. 13, No. 5, pp. 633–645.
33. Blaxton, T., and C. Westphal, (1998), *Data Mining Solutions: Methods and Tools for Solving Real-World Problems*, John Wiley & Sons, New York, NY, U.S.A., pp. 186–189.
34. Bloch, D.A., and B.W. Silverman, (1997), "*Monotone Discriminant Functions and Their Applications in Rheumatology*," J. Am. Stat. Assoc., Vol. 92, No. 437, pp. 144–153.
35. Block, H., S. Qian, and A. Sampson, (1994), "*Structure Algorithms for Partially Ordered Isotonic Regression*," J. Comput. Graph. Stat., Vol. 3, No. 3, pp. 285–300.

36. Blumer, A., A. Ehrenfeucht, D. Haussler, and M.K. Warmuth, (1989), "Learnability and the Vapnik-Chernovenkis Dimension," J. Assoc. Comput. Mach., Vol. 36, No. 4, pp. 929–965.
37. Bollobás, B., (1979), *Graph Theory, An Introductory Course*, Springer-Verlag, New York, NY, U.S.A.
38. Bongard, M., (1970), *Pattern Recognition*, Spartan Books, New York, NY, U.S.A.
39. Boone, J., (1994), "*Sidetracked at the Crossroads*," Radiology, Vol. 193, No. 1, pp. 28–30.
40. Borland, (1991), Turbo Pascal 1.5 for Windows, Borland International, Inc., Scotts Valley, CA, U.S.A.
41. Boros, E., (1994), "*Dualization of Aligned Boolean Functions*," RUTCOR Research Report RRR 9-94, Rutgers University, New Brunswick, NJ, U.S.A.
42. Boros, E., V. Gurvich, V., P.L. Hammer, T. Ibaraki, and A. Kogan, (1994), "*Structural Analysis and Decomposition of Partially Defined Boolean Functions*," RUTCOR Research Report RRR 13-94, Rutgers University, New Brunswick, NJ, U.S.A.
43. Boros, E., P.L. Hammer, and J.N. Hooker, (1994), "*Predicting Cause-Effect Relationships from Incomplete Discrete Observations*." SIAM J. Discrete Math., Vol. 7, No. 4, pp. 531–543.
44. Boros, E., P.L. Hammer, and J.N. Hooker, (1995), "*Boolean Regression*," Ann. Oper. Res., Vol. 58, pp. 201–226.
45. Boros, E., P.L. Hammer, T. Ibaraki, and K. Makino, (1997), "Polynomial-Time Recognition of 2-Monotonic Positive Boolean Functions Given by an Oracle," *SIAM J. Comput.*, Vol. 26, No. 1, pp. 93–109.
46. Boros, E., P.L. Hammer, T. Ibaraki, A. Kogan, E. Mayoraz, and I. Muchnik, (2000), "*An Implementation of Logical Analysis of Data*," IEEE Trans. Knowl. Data Eng., Vol. 12, No. 2, March–April, pp. 292–306.
47. Bradshaw, G., R. Fozzard, and L. Ceci, (1989), "*A Connectionist Expert System that Really Works*," in *Advances in Neural Information Processing*, Morgan Kaufman, Palo Alto, CA, U.S.A.
48. Brayton, R., G. Hachtel, C. McMullen, and A. Sangiovanni-Vincentelli, (1985), "*Logic Minimization Algorithms for VLSI Minimization*," Kluwer Academic Publishers, Norwell, MA, U.S.A.
49. Brown, D., (1981), "*A State-Machine Synthesizer-SMS*," in Proceedings of the 18th Design Automation Conference, pp. 443–458.
50. Bshouty, N.H., S.A. Goldman, T.R. Hancock, and S. Matar, (1993), "*Asking Questions to Minimize Errors*," Proceedings of the Sixth Annual ACM Conference on Computational Learning Theory, pp. 41–50.
51. Buckley, C., and G. Salton, (1995), "*Optimization of Relevance Feedback Weights*," Proceedings of the SIGIR 1995, pp. 351–357.
52. Burhenne, H.J., L.W. Burhenne, D. Goldberg, T.G. Hislop, A.J. Worth, P.M. Ribeck, and L. Kan, (1994), "*Interval Breast Cancer in the Screening Mammography Program of British Columbia: Analysis and Calcification*," *AJR*, No. 162, pp. 1067–1071.
53. Carbonell, J.G., R.S. Michalski, and T.M. Mitchell, (1983), "*An Overview of Machine Learning from Examples*," R.S. Michalski, J.G. Carbonell, and T.M. Mitchell (Editors), *Machine Learning: An Artificial Intelligence Approach*, R.S. Tioga Publishing Company, Palo Alto, CA, U.S.A., pp. 3–23.
54. Carraghan, R., and P.M. Pardalos, (1990), "*An Exact Algorithm for the Maximum Clique Problem*," Oper. Res. Lett., Vol. 9, No. 11, pp. 375–382.

55. Cavalier, T.M., P.M. Pardalos, and A.L. Soyster, (1990), "*Modeling and Integer Programming Techniques Applied to Propositional Calculus*," Comput. Oper. Res., Vol. 17, No. 6, pp. 561–570.
56. Chan, H.P., S.C.B. Lo, B. Sahiner, K.L. Lam, and M.A. Helvie, (1995), "*Computer-Aided Detection of Mammographic Micro-Calcifications: Pattern Recognition with an Artificial Neural Network*," Medi. Phys., Vol. 22, No. 10, pp. 1555–1567.
57. Chandru, V., and J.N. Hooker, (1999), *Optimization Methods for Logical Inference*, John Wiley & Sons, New York, NY, U.S.A.
58. Chang, I., R. Engel, D. Kandlur, D. Pendarakis, and D. Saha, (1999), "*Key Management for Secure Internet Multicast Using Boolean Function Minimization Techniques*," INFOCOM'99, Eighteenth Annual Joint Conference of the IEEE, Vol. 2, pp. 689–698.
59. Chang, S.K., (1971), "*Proceedings of the Brooklyn Polytechnical Institute Symposium on Computers and Automata*," Fuzzy Programs, Brooklyn, New York, NY, U.S.A., Vol. 21, pp. 124–135.
60. Chen, H., (1996), "*Machine Learning Approach to Document Retrieval: An Overview and an Experiment*," Technical Report, University of Arizona, MIS Department, Tucson, AZ, U.S.A.
61. Chen, H., R. Hsu, R. Orwing, L. Hoopes, and J.F. Numamaker, (1994), "*Automatic Concept Classification of Text From Electronic Meetings*," Commun. ACM, Vol. 30, No. 10, pp. 55–73.
62. Church, R., (1940), "*Numerical Analysis of Certain Free Distributive Structures*," Duke Math. J., Vol. 6, pp. 732–734.
63. Church, R., (1965), "*Enumeration by Rank of the Free Distributive Lattice with 7 Generators*," Not. Am. Math. Soc., Vol. 11, pp. 724–727.
64. Cleveland, D., and A.D. Cleveland, (1983), *Introduction to Indexing and Abstracting*, Libraries Unlimited, Littleton, CO, U.S.A.
65. Cohn, D., L. Atlas, and R. Lander, (1994), "*Improving Generalizing with Active Learning*," Machine Learning, Vol. 15, pp. 201–221.
66. Cohn, D.A., (1965), "*Minimizing Statistical Bias with Queries*," A.I. Memo No. 1552, Artificial Intelligence Laboratory, Massachusetts Institute of Technology (MIT), Cambridge, MA, U.S.A.
67. Cohn, D.A., (1996), "*Neural Network Exploration Using Optimal Experiment Design*," Neural Networks, Vol. 9, No. 6, pp. 1071–1083.
68. Cole, K.C., (1999), *The Universe and the Teacup: The Mathematics of Truth and Beauty*, Harvest Books, Fort Washington, PA, U.S.A.
69. "*Computational Learning Theory*," (1995a), in Proceedings of the 8th Annual Conference on Computational Learning Theory, ACM, Santa Cruz, CA, U.S.A.
70. "*Computational Learning Theory*," (1995b), Second European Conference, Euro COLT'95, Springer-Verlag, New York, NY, U.S.A.
71. Cormen, T.H., C.H. Leiserson, and R.L. Rivest, (1997), *Introduction to Algorithms*, MIT Press, Cambridge, MA, U.S.A.
72. Cox, E., (2001), *Fuzzy Modeling Tools for Data Mining and Knowledge Discovery*, The Morgan Kaufmann Series in Data Management Systems, Morgan Kaufmann, San Francisco, CA, U.S.A.
73. Cramma, Y., P.L. Hammer, and T. Ibaraki, (1988), "*Cause-Effect Relationships and Partially Defined Boolean Functions*," RUTCOR Research Report #39-88, RUTCOR, Rutgers University, New Brunswick, NJ, U.S.A.
74. Date, C.J., (1995), *An Introduction to Database Systems*, Addison-Wesley Publishing Company, Reading, MA, U.S.A.

75. Dayan, P., and L.F. Abbot, (2001), *Theoretical Neuroscience: Computational and Mathematical Modeling of Neural Systems*, MIT Press, Cambridge, MA, U.S.A.
76. Dedekind, R., (1897), "*Ueber Zerlegungen von Zahlen durch ihre grossten gemeinsamen Teiler*," Festschrift Hoch. Braunschweig (in German), Gesammelte Werke, II. pp. 103–148.
77. Deshpande, A.S., and E. Triantaphyllou, (1998), "*A Greedy Randomized Adaptive Search Procedure (GRASP) for Inferring Logical Clauses from Examples in Polynomial Time and some Extensions*," Math. Comput. Model., Vol. 27, No. 1, pp. 75–99.
78. Dietterich, T.C., and R.S. Michalski, (1981), "*Inductive Learning of Structural Descriptions*," Artif. Intell., Vol. 16, p. 100–111.
79. Dietterich, T.C., and R.S. Michalski, (1983), "*A Comparative Review of Selected Methods for Learning from Examples*," in *Machine Learning: An Artificial Intelligence Approach*, R.S. Michalski, J.G. Carbonell, and T.M. Mitchell (Editors), Tioga Publishing Company, Palo Alto, CA, U.S.A., pp. 41–81.
80. DOE, (1995), General Course on Classification/Declassification, Student Syllabus, Handouts, and Practical Exercises, U.S. Department of Energy (DOE), Germantown, MD, U.S.A.
81. Doi, K., M.L. Giger, R.M. Nishikawa, K.R. Hoffmann, H. Macmahon, R.A. Schmidt, and K.-G. Chua, (1993), "*Digital Radiography: A Useful Clinical Tool for Computer-Aided Diagnosis by Quantitative Analysis of Radiographic Images*," Acta Radiol., Vol. 34, No. 5, pp. 426–439.
82. D'Orsi, C., D. Getty, J. Swets, R. Picket, S. Seltzer, and B. McNeil, (1992), "*Reading and Decision Aids for Improved Accuracy and Standardization of Mammographic Diagnosis*," *Radiology*, Vol. 184, pp. 619–622.
83. Dubois, D., and H. Prade, (1980), *Fuzzy Sets and Systems: Theory and Applications*, Academic Press, New York, NY, U.S.A.
84. Duda, R.O., and P.E. Hart, (1973), *Pattern Classification and Scene Analysis*, John Wiley & Sons, New York, NY, U.S.A.
85. Dynmeridian, (1996), Declassification Productivity Initiative Study Report, DynCorp Company, Report Prepared for the U.S. Department of Energy (DOE), Germantown, MD, U.S.A.
86. Eiter, T., and G. Gottlob, (1995), "*Identifying the Minimal Transversals of a Hypergraph and Related Problems*," SIAM J. Comput., Vol. 24, No. 6, pp. 1278–1304.
87. Elmore, J., M. Wells, M. Carol, H. Lee, D. Howard, and A. Feinstein, (1994), "*Variability in Radiologists' Interpretation of Mammograms*," New Engl. J. Med., Vol. 331, No. 22, pp. 1493–1499, 1994.
88. Engel, K., (1997), *Encyclopedia of Mathematics and its Applications* 65: Sperner Theory, Cambridge University Press, Cambridge, MA, U.S.A.
89. Fauset, L., (1994), *Fundamentals of Neural Networks*, Prentice Hall, Upper Saddle River, NJ, U.S.A.
90. Fayyad, U. M., G. Piatetsky-Shapiro, P. Smyth, and R. Uthurusamy, (Editors), (1996), *Advances in Knowledge Discovery and Data Mining*, AAAI Press and MIT Press, Cambridge, MA, U.S.A.
91. Federov, V.V., (1972), *Theory of Optimal Experiments*, Academic Press, New York, NY, U.S.A.
92. Felici, G., and K. Truemper, (2002), "*A MINSAT Approach for Learning in Logic Domains*," INFORMS J. Comput., Vol. 14, No. 1, pp. 20–36.
93. Feo, T.A., and M.G.C. Resende, (1995), "*Greedy Randomized Adaptive Search Procedures*," J. Global Optim., Vol. 6, pp. 109–133.

94. Fisher, R.A., (1936), "*The Use of Multiple Measurements in Taxonomic Problems*," Ann. Eugenics, Vol. 7, pp. 179–188.
95. Fisher, R.A., (1938), "*The Statistical Utilization of Multiple Measurements*," Ann. Eugenics, Vol. 8, pp. 376–386.
96. Floyed Carey, Jr., E., J.Y. Lo, A.J. Yun, D.C. Sullivan, and P.J. Kornguth, (1994), "*Prediction of Breast Cancer Malignancy Using an Artificial Neural Network*," Cancer, Vol. 74, No. 11, pp. 2944–2948.
97. Fox, C.H., (1990), "*A Stop List for General Text*," Special Interest Group on Information Retrieval, Vol. 24, No. 1–2, pp. 19–35.
98. Fredman, M.L., and L. Khachiyan, (1996), "*On the Complexity of Dualization of Monotone Disjunctive Normal Forms*," J. Algorithms, Vol. 21, pp. 618–628.
99. Freitas, A.A., (2002), *Data Mining and Knowledge Discovery with Evolutionary Algorithms*, Springer, Heidelberg, Germany.
100. Fu, L.M., (1991), "*Rule Learning by Searching on Adapted Nets*," Proceedings of the Ninth National Conference on Artificial Intelligence, AAAI Press, Anaheim, CA, U.S.A., pp. 590–595.
101. Fu, L.M., (1993), "*Knowledge-Based Connectionism for Revising Domain Theories*," IEEE Trans. Syst. Man Cybern., Vol. 23, No. 1, pp. 173–182.
102. Gainanov, D.N., (1984), "*On One Criterion of the Optimality of an Algorithm for Evaluating Monotonic Boolean Functions*," U.S.S.R. Comput. Math. Math. Phys., Vol. 24, No. 4, pp. 176–181.
103. Galant, S., (1988), "*Connectionist Expert Systems*," Commun. ACM, Vol. 31, No. 2, pp. 152–169.
104. Gale, D., E. Roebuck, and E. Riley, (1987), "*Computer aids to mammographic diagnosis*," Br. J. Radiol., Vol. 60, pp. 887–891.
105. Garcia, F.E., S.M. Waly, S.S. Asfour, and T.M. Khalil, (1997), "*Prediction of Localized Muscle Fatigue: Part II: A Learning Vector Quantization Approach*," in *Advances in Occupational Ergonomics and Safety II*, B. Das and W. Karwowski (Editors), IOS Press and Ohmsha, pp. 343–346.
106. Getty, D., R. Pickett, C. D'Orsi, and J. Swets, (1988), "*Enhanced Interpretation of Diagnostic Images*," *Invest. Radiol.*, Vol. 23, pp. 240–252.
107. Gimpel, J., (1965), "*A Method of Producing a Boolean Function Having an Arbitrarily Prescribed Prime Implicant Table*," IEEE Trans. Comput., Vol. 14, pp. 485–488.
108. Goldman, S., and R.H. Sloan, (1994), "*The Power of Self-Directed Learning*," Mach. Learn., Vol. 14, pp. 271–294.
109. Goldman, S.A., (1990), "*Learning Binary Relations, Total Orders, and Read-Once Formulas*," Ph.D. thesis, Massachusetts Institute of Technology (MIT). Available as Technical Report MIT/LCS/TR-483, MIT Laboratory for Computer Science, Cambridge, MA, U.S.A.
110. Golumbic, M.C., (1980), *Algorithmic Graph Theory and Perfect Graphs*, Academic Press, New York, NY, U.S.A.
111. Gorbunov, Y., and B. Kovalerchuk, (1982), "*An Interactive Algorithm for Restoring of a Monotone Boolean Function*," Izvestia AN USSR (Proceedings of the Academy of Science of the USSR), STN, 2, Tashkent, former USSR, pp. 3–16 (in Russian).
112. Gupta, M.M., R.K. Ragade, and R.Y. Yager, (Editors), (1979), *Fuzzy Set Theory and Applications*, North-Holland, New York, NY, U.S.A.
113. Gurney, J., (1994), "*Neural Networks at the Crossroads: Caution Ahead*," Radiology, Vol. 193, No. 1, pp. 27–28.

114. Hall, F.M., J.M. Storella, D.Z. Silverstone, and G. Wyshak, (1988), *"Non Palpable Breast Lesions: Recommendations for Biopsy Based on Suspicion of Carcinoma at Mammography,"* *Radiology*, Vol. 167, pp. 353–358.
115. Hall, L., and A. Romaniuk, (1990), *"A Hybrid Connectionist, Symbolic Learning System,"* Proceedings of the AAAI '90, Boston, MA, pp. 783–788.
116. Hammer, P.L., and A. Kogan, (1992), *"Horn Function Minimization and Knowledge Compression in Production Rule Bases,"* RUTCOR Research Report #8-92, Rutgers University, New Brunswick, NJ, U.S.A.
117. Hansel, G., (1966), *"Sur le Nombre des Fonctions Booleennes Monotones de n Variables,"* C.R. Acad. Sci., Paris, France, Vol. 262, pp. 1088–1090.
118. Harman, D., (1995), *"Overview of the Second Text Retrieval Conference (TREC-2),"* Inf. Process. Manage., Vol. 31, No. 3, pp. 271–289.
119. Hattori, K., and Y. Torri, (1993), *"Effective Algorithms for the Nearest Neighbor Method in the Clustering Problem,"* Pattern Recognition, Vol. 26, No. 5, pp. 741–746.
120. Haussler, D., (1988), *"Quantifying Inductive Bias: AI Learning Algorithms and Valiant's Learning Framework,"* Artif. Intell., Vol. 36, pp. 177–221.
121. Haussler, D., (1989), *"Learning Conjunctive Concepts in Structural Domains,"* Mach. Learn., Vol. 4, pp. 7–40.
122. Haussler, D., and M. Warmuth, (1993), *"The Probably Approximately Correct (PAC) and Other Learning Models,"* in *Foundations of Knowledge Acquisition: Machine Learning*, A.L. Meyrowitz and S. Chipman, (Editors), Kluwer Academic Publishers, Norwell, MA, U.S.A., pp. 291–312.
123. Helft, N., (1988), *"Learning Systems of First-Order Rules,"* Proceedings of the Fifth International Conference on Machine Learning, John Laird, (Editor), University of Michigan, Ann Arbor, MI, U.S.A., pp. 395–401, June 12–14.
124. Hojjatoleslami, A., and J. Kittler, (1996), *"Detection of Clusters of Microcalcifications Using a K-Nearest Neighbor Rule with Locally Optimum Distance Metric,"* in Digital Mammography '96, Proceedings of the 3rd International Workshop on Digital Mammography, K. Doi, M.L. Giger, R.M. Nishikawa, and R.A. Schmidt, (Editors), Chicago, IL, U.S.A., June, 9–12, pp. 267–272.
125. Holland, J.H., K.J. Holyoak, R.E. Nisbett, and P.R. Thagard, (1986), *Induction: Processes of Inference, Learning, and Discovery*, MIT Press, Cambridge, MA, U.S.A.
126. Hong, S., R. Cain, and D. Ostapko, (1974), *"MINI: A Heuristic Approach for Logic Minimization,"* IBM J. Res. Dev., pp. 443–458.
127. Hooker, J.N., (1988a), *"Generalized Resolution and Cutting Planes,"* Ann. Oper. Res., Vol. 12, No. 1–4, pp. 217–239.
128. Hooker, J.N., (1988b),*"Resolution vs. Cutting Plane Solution of Inference Problems: Some Computational Experience,"* Oper. Res. Lett., Vol. 7, No. 1, pp. 1–7.
129. Hooker, J.N., (1988c), *"A Quantitative Approach to Logical Inference,"* Decis. Support Syst., Vol. 4, pp. 45–69.
130. Hooker, J.N., (2000), *Logic Based Methods for Optimization*, John Wiley & Sons, New York, NY, U.S.A.
131. Horvitz, D.G., and D.J. Thompson, (1952), *"A Generalization of Sampling without Replacement from a Finite Universe,"* J. Am. Stat. Assoc., Vol. 47, pp. 663–685.
132. Hosmer, Jr., D.W., and S. Lemeshow, (2000), *Applied Logistic Regression*, Second Edition, Probability and Statistics Series, Wiley-Interscience, New York, NY, U.S.A.
133. Houtsma, H., and A. Swami, (1993), *"Set Oriented Mining of Association Rules,"* Technical Report RJ 9567, IBM, October.
134. Hsiao, D., and F. Haray, (1970), *"A Formal System for Information Retrieval from a File,"* Commun. ACM, Vol. 13, No. 2, pp. 67–73.

135. Hunt, E., J. Martin, and P. Stone, (1966), *Experiments in Induction*, Academic Press, New York, NY, U.S.A.
136. Huo, Z., M.L. Giger, C.J. Vyborny, D.E. Wolverton, R.A. Schmidt, and K. Doi, (1996), "*Computer-Aided Diagnosis: Automated Classification of Mammographic Mass Lesions*," in Digital Mammography '96, Proceedings of the 3rd International Workshop on Digital Mammography, K. Doi, M.L. Giger, R.M. Nishikawa, and R.A. Schmidt, (Editors), Chicago, IL, U.S.A., June, 9–12, pp. 207–211.
137. Jeroslow, R.G., (1988), "*Computation Oriented Reductions of Predicate to Propositional Logic*," Decis. Support Syst., Vol. 4, pp. 183–197.
138. Jeroslow, R.G., (1989), *Logic-Based Decision Support*, North-Holland, Amsterdam, The Netherlands.
139. Jiang, Y., R.M. Nishikawa, C.E. Metz, D.E. Wolverton, R.A. Schmidt, J. Papaioannou, and K. Doi, (1996), "*A Computer-Aided Diagnostic Scheme for Classification of Malignant and Benign Clustered Microclassifications in Mammograms*," in Digital Mammography '96, Proceedings of the 3rd International Workshop on Digital Mammography, K. Doi, M.L. Giger, R.M. Nishikawa, and R.A. Schmidt, (Editors), Chicago, IL, U.S.A., June, 9–12, pp. 219–224.
140. Johnson, N., (1991), "*Everyday Diagnostics: A Critique of the Bayesian Model*," Med. Hypotheses, Vol. 34, No. 4, pp. 289–96.
141. Johnson, R.A., and D.W. Wichern, (1992), *Applied Multivariate Statistical Analysis*, Third Edition, Prentice Hall, Upper Saddle River, NJ, U.S.A.
142. Judson, D.H., (2001), "*A Partial Order Approach to Record Linkage*," Federal Committee on Statistical Methodology Conference, November 14–16, Arlington, VA, U.S.A.
143. Judson, D.H., (2006), "*Statistical Rule Induction in the Presence of Prior Information: The Bayesian Record Linkage Problem*," in *Data Mining and Knowledge Discovery Approaches Based on Rule Induction Techniques*, E. Triantaphyllou and G. Felici, (Editors), Massive Computing Series, Springer, Heidelberg, Germany, pp. 655–694.
144. Kamath, A.P., N.K. Karmakar, K.G. Ramakrishnan, and M.G.C. Resende, (1990), "*Computational Experience with an Interior Point Algorithm on the Satisfiability Problem*," Annals of Operations Research, P.M. Pardalos and J.B. Rosen, (Editors). Special issue on: Computational Methods in Global Optimization, Vol. 25, pp. 43–58.
145. Kamath, A.P., N.K. Karmakar, K.G. Ramakrishnan, and M.G.C. Resende, (1992), "*A Continuous Approach to Inductive Inference*," Math. Progr., Vol. 57, pp. 215–238.
146. Kamath, A.P., N.K. Karmakar, K.G. Ramakrishnan, and M.G.C. Resende, (1994), "*An Interior Point Approach to Boolean Vector Synthesis*," Proceedings of the 36th MSCAS, pp. 1–5.
147. Kamgar-Parsi, B., and L.N. Kanal, (1985), "*An Improved Branch-and-Bound Algorithm for Computing k-Nearest Neighbors*," Pattern Recognition Lett., Vol. 3, pp. 7–12.
148. Karmakar, N.K., M.G.C. Resende, and K.G. Ramakrishnan, (1991), "*An Interior Point Algorithm to Solve Computationally Difficult Set Covering Problems*," Math. Progr., Vol. 52, No. 3, pp. 597–618.
149. Karnaugh, M., (1953), "*The Map Method for Synthesis of Combinatorial Logic Circuits*," Trans. AIEE Commun. Electron., Vol. 72, pp. 593–599.
150. Karzanov, A.V., (1974), "*Determining the Maximal Flow in a Network by the Method of Preflows*," Sov. Math. Dokl., Vol. 15, pp. 434–437.
151. Kearns, M., M. Li, L. Pitt, and L.G. Valiant, (1987), "*On the Learnability of Boolean Formulae*," J. Assoc. Comput. Mach., No. 9, pp. 285–295.
152. Kearns, M., Ming Li, L. Pitt, and L.G. Valiant, (1987), "*On the Learnability of Boolean Formulae*," J. ACM, Vol. 34, No. 9, 285–295.

153. Kegelmeyer, W., J. Pruneda, P. Bourland, A. Hills, M. Riggs, and M. Nipper, (1994), *"Computer-Aided Mammographic Screening for Spiculated Lesions,"* *Radiology*, Vol. 191, No. 2, pp. 331–337.
154. Keller, J.M., M.R. Gray, and J.A. Gigens, Jr., (1985), *"A Fuzzy K-Nearest Neighbor Algorithm,"* IEEE Trans. Syst. Man Cybern., Vol. SMC-15, pp. 580–585.
155. Kleinbaum, D.G., M. Klein, and E.R. Pryor, (2005), *Logistic Regression*, Second Edition, Springer, Heidelberg, Germany.
156. Kleitman, D., (1969), *"On Dedekind's Problem: The Number of Monotone Boolean Functions,"* Proc. Amer. Math. Soc., Vol. 21, pp. 677–682.
157. Klir, G.J., and B. Yuan, (1995), *Fuzzy Sets and Fuzzy Logic: Theory and Applications*, Prentice Hall, Englewood Cliffs, NJ, U.S.A.
158. Kopans, D., (1994), *"The Accuracy of Mammographic Interpretation,"* (Editorial), New Engl. J. Med., Vol. 331, No. 22, pp. 1521–1522.
159. Korobkov, V., (1965), *"On Monotone Boolean Functions of Algebra Logic,"* Problemy Kibernetiki, Vol. 12, pp. 5–28, Nauka, Moscow, Russia, (in Russian).
160. Korshunov, A.D., (1981), *"On the Number of Monotone Boolean Functions,"* Probl. Kibern., Vol. 38, pp. 5–108 (in Russian).
161. Kovalerchuk, B., (1996), *"Linguistic Context Spaces: Necessary Frames for Correct Approximate Reasoning,"* Int. J. Gen. Syst., Vol. 24, No. 4, pp. 23–33.
162. Kovalerchuk, B., and B. Dalabaev, (1993), *"T-Norms as Scales,"* The First European Congress of Fuzzy and Intelligent Technologies, Aachen, Germany, pp. 1482–1487.
163. Kovalerchuk, B., and G. Klir, (1995), *"Linguistic Context Spaces and Modal Logic for Approximate Reasoning and Fuzzy Probability Comparison,"* Proceedings of the Third International Symposium on Uncertainty Modeling and NAFIPS'95, College Park, MD, IEEE Computer Society Press, pp. a23–a28.
164. Kovalerchuk, B., and V. Taliansky, (1992), *"Comparison of Empirical and Computed Values of Fuzzy Conjunction,"* Fuzzy Sets Syst., Vol. 46, No. 2, pp. 49–53.
165. Kovalerchuk, B., and E. Vityaev, (2000), *Data Mining in Finance*, Kluwer Academic Publishers, Norwell, MA, U.S.A.
166. Kovalerchuk, B., E. Triantaphyllou, and E. Vityaev, (1995), *"Monotone Boolean Function Learning Techniques Integrated with User Interaction,"* Proceedings of the Workshop on Learning from Examples vs. Programming by Demonstrations 12th International Conference on Machine Learning, Lake Tahoe, CA, U.S.A., July 9–12, pp. 41–48.
167. Kovalerchuk, B., E. Triantaphyllou, A.S. Deshpande, and E. Vityaev, (1996), *"Interactive Learning of Monotone Boolean Functions,"* Inf. Sci., Vol. 94, Nos. 1–4, pp. 87–118.
168. Kovalerchuk, B., E. Triantaphyllou, and J.F. Ruiz, (1996a), *"Monotonicity and Logical Analysis of Data: A Mechanism for Evaluation of Mammographic and Clinical Data,"* Proceedings of the 13th Symposium for Computer Applications in Radiology (SCAR), Denver, CO, June 6–9, pp. 191–196.
169. Kovalerchuk, B., E. Triantaphyllou, J.F. Ruiz, and J. Clayton, (1996b), *"Fuzzy Logic in Digital Mammography: Analysis of Lobulation,"* Proceedings of the FUZZ-IEEE '96 International Conference, New Orleans, LA, September 8–11, Vol. 3, pp. 1726–1731.
170. Kovalerchuk, B., E. Triantaphyllou, J.F. Ruiz, V.I. Torvik, and E. Vityaev, (2000), *"The Reliability Issue of Computer-Aided Breast Cancer Diagnosis,"* Comput. Biomed. Res., Vol. 33, No. 4, August, pp. 296–313.

171. Kovalerchuk, B., E. Triantaphyllou, J.F. Ruiz, and J. Clayton, (1997), "*Fuzzy Logic in Computer-Aided Breast Cancer Diagnosis: Analysis of Lobulation*," Artif. Intell. Med., No. 11, pp. 75–85.
172. Kovalerchuk, B., E. Vityaev, and E. Triantaphyllou, (1996c), "*How Can AI Procedures Become More Effective for Manufacturing?*" Proceedings of the Artificial Intelligence and Manufacturing Research Planning Workshop, AAAI Press, Albuquerque, NM, U.S.A., June 24–26, pp. 103–111.
173. Kurita, T., (1991), "*An Efficient Agglomerative Clustering Algorithm Using a Heap*," Pattern Recognition, Vol. 24, No. 3, pp. 205–209.
174. Lee, C.I.C., (1983), "*The Min-Max Algorithm and Isotonic Regression*," Ann. Stat., Vol. 11, pp. 467–477.
175. Lee, C.T.R., (1971), "*Fuzzy Logic and the Resolution Principle*," Second International Joint Conference on Artificial Intelligence, London, U.K., pp. 560–567.
176. Lee, S.N., Y.L. Grize, and K. Dehnad, (1987), "*Quantitative Models for Reasoning Under Uncertainty in Knowledge Based Expert Systems*," Int. J. Intell. Syst., Vol. 2, pp. 15–38.
177. Luhn, H.P., (1957), "*A Statistical Approach to Mechanized Encoding and Searching of Literary Information*," IBM J. Res. Dev., Vol. 4, No. 4, pp. 600–605.
178. Luhn, H.P., (1958), "*The Automatic Creation of Literature Abstracts*," IBM J. Res. Dev., Vol. 5, No. 3, pp. 155–165.
179. "Machine Learning '95," (1995a), *The 12th International Conference on Machine Learning*, Tahoe City, CA, U.S.A., Morgan Kaufmann Publishers.
180. "Machine Learning: ECML_95," (1995b), *Eighth European Conference on Machine Learning*, N. Lavrac and S. Wrobel, (Editors), Springer, Berlin, Germany.
181. MacKay, D.J.C., (1992), "*Information-based Objective Functions for Active Data Selection*," Neural Comput., Vol. 4, No. 4, pp. 589–603.
182. Makino, K., and T. Ibaraki, (1995), "*A Fast and Simple Algorithm for Identifying 2-Monotonic Positive Boolean Functions*," Proceedings of ISAACS'95, Algorithms and Computation, Springer-Verlag, Berlin, Germany, pp. 291–300.
183. Makino, K., and T. Ibaraki, (1997), "*The Maximum Latency and Identification of Positive Boolean Functions*," SIAM J. Comput., Vol. 26, No. 5, pp. 1363–1383.
184. Makino, K., T. Suda, H. Ono, and T. Ibaraki, (1999), "*Data Analysis by Positive Decision Trees*," *IEICE Trans. Inf. Syst.*, Vol. E82-D, No. 1, pp. 76–88.
185. Mangasarian, O.L., (1993), "*Mathematical Programming in Neural Networks*," ORSA J. Comput., Vol. 5, No. 4, pp. 349–360.
186. Mangasarian, O.L., R. Setiono, and W.H. Woldberg, (1991), "*Pattern Recognition Via Linear Programming: Theory and Application to Medical Diagnosis*," in *Large-Scale Numerical Optimization*, T.F. Coleman and Y. Li, (Editors), SIAM, pp. 22–30.
187. Mangasarian, O.L., W.N. Street, and W.H. Woldberg, (1995), "*Breast Cancer Diagnosis and Prognosis Via Linear Programming*," Oper. Res., Vol. 43, No. 4, pp. 570–577.
188. Mansour, Y., (1992), "*Learning of DNF Formulas*," Proceedings of the Fifth Annual Workshop on Computational Learning Theory, pp. 53–59.
189. McCluskey, E., (1956), "*Minimization of Boolean Functions*," Bell Syst. Tech. J., Vol. 35, pp. 1417–1444.
190. Meadow, C.T., (1992), *Text Information Retrieval Systems*, Academic Press, San Diego, CA, U.S.A.
191. Michalski, R.S., (1973), "*Discovering Classification Rules Using Variable-Valued Logic System VL1*," Proceedings of the Third International Joint Conference on Artificial Intelligence, pp. 162–172, Stanford, CA, U.S.A.

192. Michalski, R.S., (1985), "*Knowledge Repair Mechanisms: Evolution vs. Revolution*," Proceedings of the Third International Workshop on Machine Learning, pp. 116–119, Skytop, PA, U.S.A.
193. Michalski, R.S., (1986), "*Machine Learning Research in the Artificial Intelligence Laboratory at Illinois*," in *Machine Learning: A Guide to Current Research*, T.M. Mitchell, J.G. Carbonell, and R.S. Michalski, (Editors), Kluwer Academic Publishers, Norwell, MA, U.S.A., pp. 193–198.
194. Michalski, R.S., and J.B. Larson, (1978), "*Selection of the Most Representative Training Examples and Incremental Generation of VL1 Hypotheses: The Underlying Methodology and the Description of the Programs ESEL and AQ11*," Technical Report No. UIUCDCS-R-78-867, University of Illinois at Urbana, Department of Computer Science, Urbana, IL, U.S.A.
195. Miller, A., B. Blott, and T. Hames, (1992), "*Review of Neural Network Applications in Medical Imaging and Signal Processing*," Med. Biol. Eng. Comput., Vol. 30, pp. 449–464.
196. Mitchell, T., (1980), "*The Need for Biases in Learning Generalizations*," Technical Report CBM-TR-117, Rutgers University, New Brunswick, NJ, U.S.A.
197. Motwani, R., and P. Raghavan, (1995), *Randomized Algorithms*, Cambridge University Press, New York, NY, U.S.A.
198. Murthy, P.M., and D.W. Aha, (1994), UCI Repository of Machine Learning Databases, *Machine Readable Data Repository*, University of California, Department of Information and Computer Science, Irvine, Calif., U.S.A.
199. Myers, R.H., (1990), *Classical and Modern Regression with Applications*, Second Edition, Duxbury Press, Belmont, CA, U.S.A.
200. Naidenova, X., (2006), "*An Incremental Learning Algorithm, for Inferring Logical Rules from Examples in the Framework of the Common Reasoning Process*," in *Data Mining and Knowledge Discovery Approaches Based on Rule Induction Techniques*, E. Triantaphyllou and G. Felici, (Editors), Massive Computing Series, Springer, Heidelberg, Germany, pp. 89–147.
201. Natarajan, B.K., (1989), "*On Learning Sets and Functions*," Mach. Learn., Vol. 4, No. 1, pp. 123–133.
202. Nguyen, H.T., and E.A. Walker, (2005), *A First Course in Fuzzy Logic*, Third Edition, Chapman & Hall/CRC, London, U.K.
203. Nieto Sanchez, S., E. Triantaphyllou, and D. Kraft, (2002a), "*A Feature Mining Approach for the Classification of Text Documents Into Disjoint Classes*," Inf. Process. Manage., Vol. 38, No. 4, pp. 583–604.
204. Nieto Sanchez, S., E. Triantaphyllou, J. Chen, and T.W. Liao, (2002b), "*An Incremental Learning Algorithm for Constructing Boolean Functions from Positive and Negative Examples*," Comput. Oper. Res., Vol. 29, No. 12, pp. 177–1700.
205. Pappas, N.L., (1994), *Digital Design*, West Publishing.
206. Pardalos, P.M., and C.S. Rentala, (1990), "Computational Aspects of a Parallel Algorithm to Find the Connected Components of a Graph," Technical Report, Dept. of Computer Science, Pennsylvania State University, PA, U.S.A.
207. Pardalos, P.M., and G.P. Rogers, (1992), "*A Branch and Bound Algorithm for the Maximum Clique Problem*," Comput. Oper. Res., Vol. 19, No. 5, pp. 363–375.
208. Pardalos, P.M., and J. Xue, (1994), "*The Maximum Clique Problem*," J. Global Optim., Vol. 4, pp. 301–328.
209. Perner, P., and A. Rosenfeld, (Editors), (2003), "*Machine Learning and Data Mining in Pattern Recognition*," Proceedings of the Third International Conference, MLDM 2003,

Leipzig, Germany, July 5–7, Lecture Notes in Artificial Intelligence, Springer, Heidelberg, Germany.

210. Peysakh, J., (1987), "*A Fast Algorithm to Convert Boolean Expressions into CNF*," IBM Computer Science RC 12913 (#57971), Watson, NY, U.S.A.
211. Pham, H.N.A., and E. Triantaphyllou, (2007), "*The Impact of Overfitting and Overgeneralization on the Classification Accuracy in Data Mining*," in *Soft Computing for Knowledge Discovery and Data Mining*, O. Maimon and L. Rokach, (Editors), Part 4, Chapter 5, Springer, New York, NY, U.S.A., pp. 391–431.
212. Pham, H.N.A., and E. Triantaphyllou, (2008), "*Prediction of Diabetes by Employing a New Data Mining Approach Which Balances Fitting and Generalization*," in *Studies in Computational Intelligence*, Roger Yin Lee, (Editor), Vol. 131, Chapter 2, Springer, Berlin, Germany, pp. 11–26.
213. Pham, H.N.A., and E. Triantaphyllou, (2009a), "*An Application of a New Meta-Heuristic for Optimizing the Classification Accuracy When Analyzing Some Medical Datasets*," Expert Syst. Appl., Vol. 36, No. 5, pp. 9240–9249.
214. Pham, H.N.A., and E. Triantaphyllou, (2009b), "*A Meta Heuristic Approach for Improving the Accuracy in Some Classification Algorithms*," Working Paper, Department of Computer Science, Louisiana State University, Baton Rouge, LA 70803, U.S.A.
215. Picard, J.C., (1976), "*Maximal Closure of a Graph and Applications to Combinatorial Problems*," Manage. Sci., Vol. 22, pp. 1268–1272.
216. Pitt, L., and L.G. Valiant, (1988), "*Computational Limitations on Learning from Examples*," J. ACM, Vol. 35, No. 4, pp. 965–984.
217. Prade, H., and C.V. Negoita, (Editors), (1986), *Fuzzy Logic in Knowledge Engineering*, Verlag TUV Rheinland, Berlin, Germany.
218. Quine, W., (1952), "*The Problem of Simplifying Truth Functions*," Am. Math. Monthly, Vol. 59, pp. 102–111.
219. Quine, W., (1955), "*A Way to Simplify Truth Functions*," Am. Math. Monthly, Vol. 62.
220. Quinlan, J.R., (1979), "*Discovering Rules by Induction from Large Numbers of Examples: A Case Study*," in *Expert Systems in the Micro-Electronic Age*, D. Michie, (Editor), Edinburgh University Press, Scotland.
221. Quinlan, J.R., (1983), "*Learning Efficient Classification Procedures and Their Application to Chess and Games*," in Machine Learning: An Artificial Intelligence Approach, R.S. Michalski, J.G. Carbonell, and T.M. Mitchell, (Editors), Tioga Publishing Company, Palo Alto, CA, U.S.A., pp. 3–23.
222. Quinlan, J.R., (1986), "*Induction of Decision Trees*," Mach. Learn., Vol. 1, No. 1, pp. 81–106.
223. Radeanu, S., (1974), *Boolean Functions and Equations*, North-Holland, New York, NY, U.S.A.
224. Ramsay, A., (1988), *Formal Methods in Artificial Intelligence*, Cambridge University Press, Cambridge, England.
225. Reine, R.E., and R.S. Michalski, (1986), *Incremental Learning of Concept Descriptions*, Machine Intelligence Series, No. 11, Oxford University Press, Oxford, U.K.
226. Resende, M.G.C., and T.A. Feo, (1995), "*Greedy Randomized Adaptive Search Procedures*," Global Optim., Vol. 6, No. 2, pp. 109–133.
227. Rivest, R.L., (1987), "*Learning Decision Trees*," Mach. Learn., Vol. 2, No. 3, pp. 229–246.
228. Robertson, T., F.T. Wright, and R.L. Dykstra, (1988), *Order Restricted Statistical Inference*, John Wiley & Sons, New York, NY, U.S.A.

229. Rosen, D., B. Martin, M. Monheit, G. Wolff, and M. Stanton, (1996), "*A Bayesian neural network to detect microclassifications in digitized mammograms*," in: Digital Mammography '96, Proceedings of the 3rd International Workshop on Digital Mammography, K. Doi, M.L. Giger, R.M. Nishikawa, and R.A. Schmidt, (Editors), Chicago, IL, U.S.A., June 9–12, pp. 277–282.
230. Ross, T.J., (2004), *Fuzzy Logic with Engineering Applications*, Second Edition, Wiley, New York, NY, U.S.A.
231. Sahiner, B., H.-P. Chan, N. Petrick, M.A. Helvie, M.M. Goodsitt, and D.D. Adler, (1996), "*Classification of Mass and Normal Tissue: Feature Selection Using a Genetic Algorithm*," in: Digital Mammography '96, Proceedings of the 3rd International Workshop on Digital Mammography, K. Doi, M.L. Giger, R.M. Nishikawa, and R.A. Schmidt, (Editors), Chicago, IL, U.S.A., June 9–12, pp. 379–384.
232. Salomon, D., (1998), *Data Compression: The Complete Reference*, Springer-Verlag, New York, NY, U.S.A.
233. Salton, G., (1968), *Automatic Information Organization and Retrieval*, McGraw-Hill, New York, NY, U.S.A.
234. Salton, G., (1989), *Automatic Text Processing. The Transformation, Analysis, and Retrieval of Information by Computer*. Addison-Wesley, Reading, MA, U.S.A.
235. Salton, G., and A. Wong, (1975), "*A Vector Space Model for Automatic Indexing*," Information Retrieval and Language Processing, Vol. 18, No. 11, pp. 613–620.
236. Sanchez, E., (Editor), (2006), *Fuzzy Logic and the Semantic Web*, Elsevier Science.
237. Savasere, A., E. Omiecinski, and S. Navathe, (1995), An Efficient Algorithm for Mining Association Rules in Large Databases, Data Mining Group, Tandem Computers, Inc., Austin, TX, U.S.A.
238. Savasere, A., E. Omiecinski, and S. Navathe, (1998), "*Mining for Strong Negative Associations in a Large Database of Customer Transactions*," Proceedings of the IEEE 14th International Conference on Data Engineering, Orlando, FL, U.S.A.
239. Schapire, R., (1992), *The Design and Analysis of Efficient Learning Algorithms*, MIT Press, Boston, MA, U.S.A.
240. Schlimmer, J.C., (1987), "*Incremental Adjustment of Representations for Learning*," Proceedings of the Fourth International Machine Learning Workshop, pp. 79–90. Irving, CA, U.S.A.
241. Schlimmer, J.C., and D. Fisher, (1986), "*A Case Study of Incremental Concept Learning*," Proceedings of the Fifth National Conference on Artificial Intelligence, Philadelphia, PA, U.S.A., pp. 496–501.
242. Schneeweiss, W., (1989), Boolean Functions with Engineering Applications and Computer Programs, Springer-Verlag, Berlin, Germany.
243. Scholtes, J.C., (1993), *Neural Networks in Natural Language Processing and Information Retrieval*, North-Holland, The Netherlands.
244. Shavlik, J.W., (1994), "*Combining Symbolic and Neural Learning*," Mach. Learn., Vol. 14, pp. 321–331.
245. Shaw, W.M., (1995), "*Term-Relevance Computations and Perfect Retrieval Performance*," Inf. Process. Manage., Vol. 31, No. 4, pp. 312–321.
246. Shawe-Taylor, J., M. Antony, and N. Biggs, (1989), "*Bounding Sample Size with the Vapnik-Chernovenkis Dimension*," Technical Report CSD-TR-618, University of London, Surrey, U.K.
247. Shmulevich, I., (1997),"*Properties and Applications of Monotone Boolean Functions and Stack Filters*," Ph.D. Dissertation, Department of Electrical Engineering, Purdue University, West Lafayette, IN, U.S.A.

248. Sokolov, N.A., (1982), "*On the Optimal Evaluation of Monotonic Boolean Functions*," U.S.S.R. Comput. Math. Math. Phys., Vol. 22, No. 2, pp. 207–220.
249. Solnon, C., and S. Fenet, (2006), "*A Study of ACO Capabilities for Solving the Maximum Clique Problem*," Heuristics, Vol. 12, No. 3, pp. 155–180.
250. Späth, H., (1985), *Cluster Dissection and Analysis: Theory, Fortran Programs, and Examples*, Ellis Harwood, Ltd., Chichester, U.K.
251. Sun, R., and F. Alexandre, (Editors), (1997), *Connectionist-Symbolic Integration: From Unified to Hybrid Approaches*, Lawrence Erilbaum Associates, Mahwah, NJ, U.S.A.
252. Swets, J., D. Getty, R. Pickett, C. D'orsi, S. Seltzer, and B. McNeil, (1991), "*Enhancing and Evaluating Diagnostic Accuracy*," Med. Decis. Mak., Vol. 11, pp. 9–18.
253. Szczepaniak, P.S., P.L.G. Lisboa, and J. Kacprzyk, (2000), *Fuzzy Systems in Medicine*, Studies in Fuzziness and Soft Computing, Physica-Verlag, Heidelberg, Germany.
254. Tabar, L., and P.B. Dean, (1986), *Teaching Atlas of Mammography*, Verlag, New York, NY, U.S.A.
255. Tan, P.-N., M. Steinbach, and V. Kumara, (2005), *Introduction to Data Mining*, Addison Wesley Reading, MA, U.S.A.
256. Tatsuoka, C., and T. Ferguson, (1999), "*Sequential Classification on Partially Ordered Sets*," Technical Report 99-05, Department of Statistics, The George Washington University, Washington, DC, U.S.A.
257. Toivonen, H., (1996), "*Sampling Large Databases for Association Rules*," Proceedings of the 22nd VLDB Conference, Bombay, India.
258. Torvik, V.I., and E. Triantaphyllou, (2002), "*Minimizing the Average Query Complexity of Learning Monotone Boolean Functions*," INFORMS J. Comput., Vol. 14, No. 2, pp. 142–172.
259. Torvik, V.I., and E. Triantaphyllou, (2003), "*Guided Inference of Nested Monotone Boolean Functions*," Inf. Sci., Vol. 151, pp. 171–200.
260. Torvik, V.I., and E. Triantaphyllou, (2004), "*Guided Inference of Stochastic Monotone Boolean Functions*," Technical Report, Louisiana State University, Dept. of Computer Science, Baton Rouge, LA, U.S.A.
261. Torvik, V.I., and E. Triantaphyllou, (2006), "*Discovering Rules that Govern Monotone Phenomena*," in *Data Mining and Knowledge Discovery Approaches Based on Rule Induction Techniques*, E. Triantaphyllou and G. Felici, (Editors), Massive Computing Series, Springer, Heidelberg, Germany, pp. 149–192.
262. Torvik, V.I., E. Triantaphyllou, T.W. Liao, and S.W. Waly, (1999), "*Predicting Muscle Fatigue via Electromyography: A Comparative Study*," Proceedings of the 25th International Conference of Computers and Industrial Engineering, New Orleans, LA, U.S.A., pp. 277–280.
263. Torvik, V.I., M. Weeber, D.R. Swanson, and N.R. Smalheiser, (2005), "*A Probabilistic Similarity Metric for Medline Records: A Model for Author Name Disambiguation*," JASIST, Vol. 52, No. 2, pp. 140–158.
264. Towell, G.G., and J.W. Shavlik, (1993), "*Extracting Refined Rules from Knowledge-Based Neural Networks*," Mach. Learn., Vol. 13, pp. 71–101.
265. Towell, G.G., J. Havlic, and M. Noordewier, (1990), "*Refinement Approximate Domain Theories by Knowledge-Based Neural Networks*," Proceedings of the AAAI '90 Conference, Boston, MA, U.S.A., pp. 861–866.
266. Triantaphyllou, E., (1994), "*Inference of a Minimum Size Boolean Function from Examples by Using a New Efficient Branch-and-Bound Approach*," Global Optim., Vol. 5, No. 1, pp. 69–94.

267. Triantaphyllou, E., (2000), *Multi-Criteria Decision Making Methods: A Comparative Study*, Kluwer Academic Publishers, Applied Optimization Series, Vol. 44, Norwell, MA, U.S.A., 320 pages.
268. Triantaphyllou, E., and G. Felici, (2006), (Editors), *Data Mining and Knowledge Discovery Approaches Based on Rule Induction Techniques*, Massive Computing Series, 750 pages, Springer, Heidelberg, Germany.
269. Triantaphyllou, E., and A.L. Soyster, (1995a), "*A Relationship Between CNF and DNF Systems Derivable from Examples*," ORSA J. Comput., Vol. 7, No. 3, pp. 283–285.
270. Triantaphyllou, E., and A.L. Soyster, (1995b), "*An Approach to Guided Learning of Boolean Functions*," Math. Comput. Model., Vol. 23, No. 3, pp. 69–86.
271. Triantaphyllou, E., and A.L. Soyster, (1996), "*On the Minimum Number of Logical Clauses Which Can be Inferred From Examples*," Comput. Oper. Res., Vol. 23, No. 8, pp. 783–799.
272. Triantaphyllou, E., A.L. Soyster, and S.R.T. Kumara, (1994), "*Generating Logical Expressions From Positive and Negative Examples Via a Branch-and-Bound Approach*," Comput. Oper. Res., Vol. 21, No. 2, pp. 185–197.
273. Triantaphyllou, E., B. Kovalerchuk, and A.S. Deshpande, (1996), "*Some Recent Developments in Logical Analysis*," *Interfaces in Computer Science and Operations Research*, R. Barr, R. Helgason, and J. Kennington, (Editors), Kluwer Academic Publishers, New York, NY, U.S.A., pp. 215–236.
274. Trick, M., and D. Johnson, (1995), "*Second DIAMACS Challenge on Cliques, Coloring and Satisfiability*," American Mathematical Society, (Summer), Rutgers University, New Brunswick, NJ, U.S.A.
275. Truemper, K., (2004), *Design of Logic-Based Intelligent Systems*, John Wiley & Sons, Hoboken, NJ, U.S.A.
276. Turksen, I.B., and H. Zhao, (1993), "*An Equivalence Between Inductive Learning and Pseudo-Boolean Logic Simplification: A Rule Generation and Reduction Scheme*," IEEE Trans. Syst. Man Cybern., Vol. 23, No. 3, pp. 907–917.
277. Utgoff, P.E., (1988), "*ID5: An Incremental ID3*," John Laird (Editor), Proceedings of the Fifth International Conference on Machine Learning, June 12–14, University of Michigan, Ann Arbor, MI, U.S.A., pp. 107–120.
278. Utgoff, P.E., (1989), "*Incremental Induction of Decision Trees*," Mach. Learn., Vol. 4, pp. 161–186.
279. Utgoff, P.E., (1998), "*Decision Tree Induction Based on Efficient Tree Reconstructing*," Mach. Learn. J., Vol. 29, pp. 5–40.
280. Valiant, L.G., (1984), "*A Theory of the Learnable*," Commun. ACM, Vol. 27, No. 11, pp. 1134–1142.
281. Valiant, L.G., (1985), "*Learning Disjunctions of Conjunctives*." Proceedings of the 9th IJCAI, pp. 560–566.
282. Van Rijsbergen, C.J., (1979), *Information Retrieval*, Second Edition, Butterworths, London, U.K.
283. Vapnik, V.N., (1982), *Estimating of Dependencies Based on Empirical Data*, Springer-Verlag, New York, NY, U.S.A.
284. Voorhees, E., (1998), "*Overview of the Sixth Text Retrieval Conference (TREC-6)*," Proceedings of the Sixth Text Retrieval Conference (TREC-6), pp. 1–27, Gaithersburg, MD, U.S.A.
285. Vyborny, C., (1994), "*Can Computers Help Radiologists Read Mammograms?*" Radiology, Vol. 191, pp. 315–317.
286. Vyborny, C., and M. Giger, (1994), "*Computer Vision and Artificial Intelligence in Mammography*," AJR, Vol. 162, pp. 699–708.

287. Waly, S.M., S.S. Asfour, T.M. Khalil, and F.E. Garcia, (1997), "*Prediction of Localized Muscle Fatigue: Part I: Discriminant Analysis Approach*," *Advances in Occupational Ergonomics and Safety II*, B. Das and W. Karwowski, (Editors), IOS Press and Ohmsha, pp. 339–342.
288. Wang, G., Q. Liu, Y. Yao, and A. Skowron, (Editors), (2003), "*Rough Sets, Fuzzy Sets, Data Mining, and Granular Computing*," Proceedings of the 9th International Conference, RSFDGrC, Chongqing, China, May 26–29, 2003, Lecture Notes in Artificial Intelligence, Springer, Heidelberg, Germany.
289. Wang, L., (Editor), (2005), *Support Vector Machines: Theory and Applications*, Series on Studies in Fuzziness and Soft Computing, Springer, Heidelberg, Germany.
290. Ward, M., (1946), "*A Note on the Order of the Free Distributive Lattice*," Bull. Am. Math. Soc., Vol. 52, No. 135, pp. 423.
291. Wiedemann, D., (1991), "*A Computation of the Eight Dedekind Number*," Order, Vol. 8, pp. 5–6.
292. Williams, H.P., (1986), "*Linear and Integer Programming Applied to Artificial Intelligence*," ReprintSeries, University of Southampton, Faculty of Mathematical Studies, Southampton, U.K., July Issue, pp. 1–33.
293. Williams, H.P., (1987), "*Linear and Integer Programming Applied to the Propositional Calculus*," Int. J. Syst. Res. Inf. Sci., Vol. 2, pp. 81–100.
294. Wingo, P.A., T. Tong, and S. Bolden, (1995), "*Cancer Statistics*," Ca-A Cancer Journal for Clinicians, Vol. 45, No. 1, pp. 8–30.
295. Witten, I.H., and F. Eibe, (2005), *Data Mining: Practical Machine Learning Tools and Techniques*, Second Edition, Series in Data Management Systems, Morgan Kaufmann.
296. Woldberg, W.W., and O.L. Mangasarian, (1990), "*A Multisurface Method of Pattern Separation for Medical Diagnosis Applied to Breast Cytology*," Proc. Nat. Acad. Sci. U.S.A., Vol. 87, No. 23, pp. 9193–9196.
297. Wu, Y., M. Giger, K. Doi, C. Vyborny, R. Schmidt, and C. Metz, (1993), "*Artificial Neural Networks in Mammography: Application to Decision Making in the Diagnosis of Breast Cancer*," Radiology, Vol. 187, No. 1, pp. 81–87.
298. Wu, Y., K. Doi, M. Giger, C. Metz, and W. Zhang, (1994), "*Reduction of False Positives in Computerized Detection of Lung Nodules in Chest Radiographs Using Artificial Neural Networks, Discriminant Analysis and a Rule-Based Scheme*," J. Digital Imaging, Vol. 17, No. 4, pp. 196–207.
299. Xie, L.A., and S.D. Berdosian, (1983), "*The Information in a Fuzzy Set and the Relation Between Shannon and Fuzzy Information*," Proceedings of the 7th Annual Conference on Information Sciences and Systems, Johns Hopkins University, Baltimore, MD, U.S.A., pp. 212–233.
300. Yablonskii, S., (1986), *Introduction to Discrete Mathematics*, Nauka Publishers, Moscow, Russia, (in Russian).
301. Yilmaz, E., E. Triantaphyllou, J. Chen, and T.W. Liao, (2003), "*A Heuristic for Mining Association Rules in Polynomial Time*," Mathematical and Computer Modelling, No. 37, pp. 219–233.
302. Zadeh, L.A., "*The Concept of a Linguistic Variable and Its Applications to Approximate Reasoning*," Parts 1–3, Inf. Sci., Vol. 8, pp. 199–249, 1975; Vol. 8, pp. 301–357, 1975; Vol. 9, pp. 43–80, 1976.
303. Zadeh, L.A., (1965), "*Fuzzy Sets*," Inf. Control, Vol. 8, pp. 338–353.
304. Zadeh, L.A., (1968), "*Fuzzy Algorithms*," Inf. Control, Vol. 12, pp. 94–102.
305. Zadeh, L.A., (1978), "*Fuzzy Sets as a Basis for a Theory of Possibility*," Fuzzy Sets Syst., Vol. 1, pp. 3–28.

306. Zadeh, L.A., (1979), "*A Theory of Approximate Reasoning*," *Machine Intelligence*, Edited by J. Hayes, D. Michie, and L.I. Mikulich, John Wiley & Sons, New York, NY, U.S.A., Vol. 9, pp. 149–194.
307. Zadeh, L.A., (1983), "*Can Expert Systems be Designed Without Using Fuzzy Logic?*" Proceedings of the 17th Annual Conference on Information Sciences and Systems, Johns Hopkins University, Baltimore, MD, U.S.A., pp. 23–31.
308. Zadeh, L.A., K.S. Fu, K. Tanaka, and M. Shimura, (Editors), (1975), *Fuzzy Sets and their Applications to Cognitive and Decision Processes*, Academic Press, New York, NY, U.S.A.
309. Zakrevskij, A.D., (1971), *Algorithms of Discrete Automata Synthesis*, Nauka Publishers, Moscow, Russia (in Russian).
310. Zakrevskij, A.D., (1981), *Logical Synthesis of Cascade Networks*, Nauka Publishers, Moscow, Russia (in Russian).
311. Zakrevskij, A.D., (1988), *Logic of Recognition*, Nauka Publishers, Minsk, Belarus (in Russian).
312. Zakrevskij, A.D., (1994), "*A Method of Logical Recognition in the Space of Multi-valued Attributes*," in Proceeding of the Third Electro-technical and Computer Science Conference ERK'94, Slovenia Section IEEE, Ljubljana, Slovenia, Vol. B, pp. 3–5, 1994.
313. Zakrevskij, A.D., (1999), "*Pattern Recognition as Solving Logical Equations*," Special Issue 1999 - SSIT'99 (AMSE), pp. 125–136.
314. Zakrevskij, A.D., (2001), "*A Logical Approach to the Pattern Recognition Problem*," in Proceedings of the International Conference KDS-2001 "Knowledge – Dialog – Solution," Saint Petersburg, Russia, Vol. 1, pp. 238–245.
315. Zakrevskij, A.D., (2006), "*A Common Logic Approach to Data Mining and Pattern Recognition*," in *Data Mining and Knowledge Discovery Approaches Based on Rule Induction Techniques*, E. Triantaphyllou and G. Felici, (Editors), Massive Computing Series, Springer, Heidelberg, Germany, pp. 1–43.
316. Zhang, Q., J. Sun, and E. Tsang, (2005), "*An Evolutionary Algorithm with Guided Mutation for the Maximum Clique Problem*," IEEE Trans. Evol. Comput., Vol. 9, No. 2, pp. 192–200.
317. Zhang, W., K. Doi, M. Giger, Y. Wu, R.M. Nishikawa, and C. Metz, (1994), "*Computerized Detection of Clustered Microclassifications in Digital Mammograms Using a Shift-Invariant Artificial Neural Network*," Med. Phys., Vol. 21, No. 4, pp. 517–524.
318. Zimmermann, H.-J., (1985), *Fuzzy Set Theory and Its Applications*, Kluwer Academic Publishers, Norwell, MA, U.S.A.
319. Zimmermann, H.-J., (1996), *Fuzzy Set Theory and Its Applications*, Kluwer Academic Publishers, Third Revised Edition, Norwell, MA, U.S.A.
320. Zipff, H.P., (1949), *Human Behavior and the Principle of Least Effort*, Addison-Wesley, Menlo Park, Calif., U.S.A.

# Subject Index

# Author Index

## About the Author

**Dr. Evangelos Triantaphyllou** did all his graduate studies at Penn State University from 1984 to 1990. While at Penn State, he earned a Dual M.S. in Environment and Operations Research (OR) (in 1985), an M.S. in Computer Science (in 1988), and a Dual Ph.D. in Industrial Engineering and Operations Research (in 1990). His Ph.D. dissertation was related to data mining by means of optimization approaches. Since the spring of 2005 he is a Professor in the Computer Science Department at the Louisiana State University (LSU) in Baton Rouge, LA. Before that, he had served as an Assistant, Associate, and Full Professor in the Industrial Engineering Department at the same university. Before coming to LSU, he had served for three years as an Assistant Professor of Industrial Engineering at Kansas State University. He had also served as an Interim Associate Dean for the College of Engineering at LSU.

His research is focused on decision-making theory and applications, data mining and knowledge discovery, and the interface of operations research and computer science. He has developed new methods for data mining and knowledge discovery and has also explored some of the most fundamental and intriguing subjects in decision-making. In 1999 he received the prestigious IIE (Institute of Industrial Engineers), Operations Research Division, Research Award for his research contributions in the above fields. In 2005 he received an LSU Distinguished Faculty Award as recognition of his research, teaching, and service accomplishments. His biggest source of pride are the numerous and great accomplishments of his graduate students.

Some of his graduate students have received awards and distinctions including the Best Dissertation Award at LSU for Science, Engineering and Technology for the year 2003. Former students of his hold top management positions at GE Capital, Motorola and held faculty positions at prestigious universities (such as the University of Illinois at Urbana).

In 2000 Dr. Triantaphyllou published a highly acclaimed book on multicriteria decision-making. Besides the previous monograph on decision-making and the present monograph on data mining, he has co-edited two books, one on data mining by means of rule induction (published in 2006 by Springer) and another one on the mining of enterprise data (published in 2008 by World Scientific). He has also published special issues of journals in the above areas and served or is serving on the editorial boards of some important research journals.

He always enjoys doing research with his students from whom he has learned and still is learning a lot. He has received teaching awards and distinctions. His research has been funded by federal and state agencies and the private sector. He has extensively published in some of the top refereed journals and made numerous presentations in national and international conferences.

Dr. Triantaphyllou has a strong interdisciplinary and also multidisciplinary background. He has always enjoyed organizing multidisciplinary teams of researchers and practitioners with complementary expertise. These groups try to comprehensively attack some of the most urgent problems in the sciences and engineering. He is firmly convinced that graduate and even undergraduate students, with their fresh and unbiased ideas, can play a critical role in such groups. He is a strong believer of the premise that the next round of major scientific and engineering discoveries will come from the work of such interdisciplinary groups. More details of his work, and those of his students can be found on his web site (*http://www.csc.lsu.edu/trianta*).

MIX
Papier aus verantwortungsvollen Quellen
Paper from responsible sources
FSC® C105338

If you have any concerns about our products,
you can contact us on
**ProductSafety@springernature.com**

In case Publisher is established outside the EU,
the EU authorized representative is:
**Springer Nature Customer Service Center GmbH**
**Europaplatz 3, 69115 Heidelberg, Germany**

Printed by Libri Plureos GmbH
in Hamburg, Germany